to be

Circadian Rhythms and the Human

D. S. Minors BSc, PhD
Lecturer in Physiology,
Department of Physiology,
University of Manchester

J. M. Waterhouse BA, DPhil
Senior Lecturer in Physiology,
Department of Physiology,
University of Manchester

With a foreword by
Professor R. T. W. L. Conroy

WRIGHT·PSG

Bristol   London   Boston
1981

Published by John Wright & Sons Ltd, 42–44 Triangle West, Bristol, BS8 1EX, England. John Wright PSG Inc., 545 Great Road, Littleton, Massachusetts 01460, USA.

British Library Cataloguing in Publication Data

Minors, D. S.
    Circadian rhythms and the human.
    1. Circadian rhythms
    I. Title   II. Waterhouse, J. M.
    612'.014   QP84.6

ISBN 0 7236 0592 0

Printed in Great Britain by
John Wright & Sons Ltd, at the Stonebridge Press, Bristol, BS4 5NU.

John Norton Mills

*Brackenbury Professor of Physiology
at the University of Manchester*

*was killed in a mountaineering accident
in North Wales, December 1977*

This book is dedicated to his memory

Preface

In recent years, the amount of research in the area of circadian rhythms has increased dramatically. As a result, considerable advances in our understanding of the processes responsible for these rhythms and the ways in which they affect the lives of some of us (shift workers, intercontinental travellers and those who provide services around-the-clock) have taken place.

In the decade since *Human Circadian Rhythms* by Conroy and Mills was published, a number of textbooks have appeared and these have been referred to, where appropriate, in the present text. Some of these have considered circadian rhythms in a broad context by covering the whole of the plant and animal kingdoms whereas others have concentrated in far more detail upon selected areas of circadian rhythms (for example, endocrinological rhythms), or upon special groups of experiments (for example, the work upon humans in isolation in Germany).

In the present text we have attempted to provide an introduction to human circadian rhythms that is suitable for the student, the person embarking upon research in the field and the intelligent layman who, by virtue of employment or otherwise, has an interest. Accordingly we have tried to expound and clarify the basic concepts involved as well as to attempt an assessment of some of the controversies that are known to exist. In particular we have tried to maintain a balance between data and speculation and to underline the importance of methodology in devising and interpreting experiments. So that the relevant data can be assessed by the reader himself, we have added considerable reference lists which include modern reviews where possible.

The book falls essentially into four main sections. In the first (Chapter 1) we outline the basic principles of the field together with their means of study. In the second (Chapters 2–8) these principles are illustrated by reference to the rhythmicity found in the different systems of the body (for example, temperature, endocrine). The application of these studies of circadian rhythm to the 'real world' is considered in Chapters 9–11. It is hoped that this section in particular will be of interest to a wide readership, not just those contemplating research. Finally, in Chapter 12, we consider some of the more fundamental problems in the study of circadian rhythms that have been introduced and dealt with incidentally throughout the book.

D.S.M.
J.M.W.

Acknowledgements

In an undertaking of this size, it is obvious that many people, too numerous to mention by name, will have offered advice and encouragement, for which we are most grateful. Among them, the following have been particularly helpful:

Professor Stanley Thomas who, especially in the months immediately after the death of John Mills, provided the opportunity for us to continue research into the field of human circadian rhythms.

Dr Roger Green (Reader in this Department) and Mr Neville Wright (medical student and BSc Hons graduate) who read the manuscript and offered constructive criticism.

Our wives, Colleen and Liz, who allowed us the time to bring our ideas to fruition and helped clarify our style of presentation.

Mrs Anne Mogie, the typist, who was mainly responsible for the many stages of our manuscript. Rarely did our illegibility confound her, but we did achieve this in one version of the legend to *Fig.* 5.6 in which the last word was typed as 'censors' rather than 'sensors' (q.v.)!

To all these people, both named and unnamed, we extend our thanks.

Contents

Foreword

by **Professor R. T. W. L. Conroy**
Department of Physiology
Royal College of Surgeons in Ireland

The term circadian is now so widely known that it is hard to realize that it is scarcely twenty years since Halberg introduced it to describe those biological rhythms that have a period of approximately 24 hours (*circa diem*).

Many advances in our knowledge of chronobiology in general and of circadian rhythms in particular have been made during the intervening years, both by Halberg and his colleagues in Minneapolis and by workers in an increasing number of centres throughout the world. One of the most outstanding of these is the Department of Physiology in the University of Manchester where the late John Mills pioneered an extensive series of investigations into human circadian rhythms. These studies are being most ably continued and extended by Drs Minors and Waterhouse, the authors of this work.

The study of circadian rhythms is, indeed, coming to be recognized as one of the most challenging and exciting areas in present day science and medicine. The scope of circadian research is extremely wide, varying from abstruse and difficult investigations into neurophysiological mechanisms to studies, often equally difficult, but of immediate practical relevance, on such everyday matters as 'jet lag' and shift working. Clinical studies, especially in chronopharmacology, are also being undertaken in increasing numbers.

An extensive literature on chronobiology has come into existence and many papers on circadian rhythms are published in a broad spectrum of scientific and medical journals. Also, an active International Society for Chronobiology exists, bringing together investigators from many disciplines.

Drs Minors and Waterhouse have written a clear and comprehensive account of the basic and applied aspects of human circadian rhythms. It is a most timely and welcome addition to the literature on a fundamental and rapidly developing area of human knowledge.

chapter 1 *Introduction and Methods of Study*

1. Rhythms in Living Organisms

Perhaps the most ubiquitous feature of nature is that of rhythmicity. Most men must be at least aware of the rhythmical nature of their environment, for in most communities social behaviour is organized to the 24-hour day and agricultural or horticultural activities require a knowledge of the alternation of the seasons. Rhythmical fluctuations occur not only at an environmental level but also at every level of nature's organization—even at the atomic level, a property which has been exploited recently with the advent of quartz watches and atomic clocks. It is not surprising, therefore, that rhythms are widespread in the whole arena of biology and are found throughout the whole evolutionary series, from simple unicellular organisms to complex multicellular organisms, in both the plant and animal kingdoms (Aschoff, 1963; Palmer, 1976). The study of these biological rhythms has itself become a science termed 'Chronobiology'.

That such rhythmicity occurs in man was, to a large extent, rejected in the mid-nineteenth century following the many exhaustive studies of Claude Bernard. At this time physiologists were urged to consider the body's internal environment (milieu intérieur) as constant and resisting any change in the external environment. This theme was further taken up by Walter B. Cannon who, in his book *The Wisdom of the Body*, introduced a new term, homeostasis, to describe the constancy of the internal environment, although he does qualify this new word: 'the word does not imply something set and immobile, a stagnation. It means a condition—a condition which may vary, but which is relatively constant' (Cannon, 1939).

Today, although the basic concepts of homeostasis hold true, we envisage the internal environment to be constantly changing with, in many cases, a regular oscillatory behaviour. In certain cases (*see,* for example, Chapter 2) this rhythmicity results from rhythmic variations in the set-point of homeostatic mechanisms, whereas for others, rhythmicity is superimposed upon the homeostatic mechanism (for example, the secretion of certain hormones, *see* Chapter 7).

The frequencies of rhythms in nature cover our every division of time. Thus, in man we find rhythms which oscillate once per fraction of a second (for example, the alpha rhythm of the electroencephalogram), those which oscillate once per several seconds (for example, the respiratory rhythm) and so on, through to those which oscillate once a year, the circannual rhythms (for review, *see* Reinberg, 1974). In most living organisms, however, those rhythms which seem to predominate are those which oscillate with a frequency corresponding with that of a major environmental periodicity. Thus many species which inhabit the shoreline exhibit rhythms which derive their frequencies from some aspect of the changing tides that result from the combined effects of the regular movement of the moon around the earth and the earth around the sun. Similarly, the annual alternation of summer and winter in temperate latitudes, which results from the movement of the earth around the sun, gives rise to the annual reproductive behaviour of plants and hence often also of the animals dependent upon them for their food. For most living organisms, however, and most certainly for man, the most evident environmental change is that which results from the regular spin of the earth about its central axis, namely the alternation between day and night—the solar day—which oscillates with a frequency of once per 24 hours. Since our habits of sleep, rest and activity, work and leisure, eating and drinking, largely follow a routine which is governed by the alternation of day and night and the social organization which results from this, it is hardly surprising that many human processes, physiological, psychological and biochemical, oscillate with a frequency similar to that of the solar day. It is these rhythms that have become known as circadian, from the latin *circa* (about) and *dies* (a day) (Halberg, 1959).

2. Definition of Terms

As with other fields of study, those who study circadian rhythms have developed specialist terms to facilitate concise communication and to avoid unwieldy descriptions. A glossary of terms used in this text is to be found at the end of the book (pp. 320—21), but for a fuller description of the entire vocabulary of terms used one should consult the *Glossary of Chronobiology* (Halberg and Katinas, 1973; Halberg et al., 1977). The following section, however, will briefly describe those terms which will be used frequently; other terms will be described when they arise in the text.

Before describing variables which exhibit circadian rhythmicity and the mechanisms which control these rhythms, it is of course necessary first to understand what is meant by a rhythm.

A rhythm (synonyms are oscillation, cycle, periodicity) has been defined by some (for example, Palmer, 1976) as a sequence of events that repeat themselves through time in the same order and at the same interval. Though this definition is true for a stationary time series in the long term, it should not be interpreted as meaning that a rhythm cannot change its characteristics; rather, a biological rhythm may, for example, change its cycle length (for example in the transition from the entrained state to a free-running state, *see* s. 6.2).

Some simple waveforms which may represent a rhythm are shown in *Fig.* 1.1. Such rhythms may be quantified as shown in *Fig.* 1.1 by:

1. The *period* of the rhythm. This is the reciprocal of the frequency and is the time to complete one cycle. Halberg (1959) has defined a rhythm as being circadian if its period lies in the range 20–28 hours. Rhythms with a

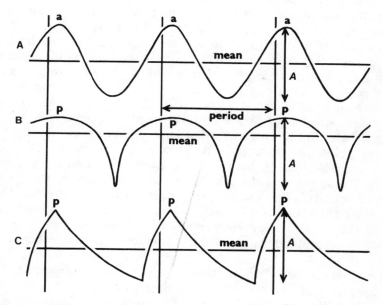

Fig. 1.1. Different forms of a rhythm. A, sinusoidal; B, symmetrical but non-sinusoidal; C, asymmetrical. a = acrophase; p = peak; A = amplitude. The vertical lines divide the traces up into two cycles. (*From* Conroy and Mills, 1970, Fig. 0.1.)

period less than this are then termed *ultradian* and those with a longer period, *infradian*.

2. The *mean* or *level* of the rhythm. This is the average value of a continuous variable over a single cycle. When the rhythm is described by the fitting of a cosine curve (*see* s. 4) the level of the best-fitting cosine curve is known as the *mesor*. Only when the data are measured equidistantly and over an integral number of cycles will the mesor equal the arithmetic mean.

3. The *amplitude* of the rhythm. In its strictest mathematical usage the amplitude refers to the magnitude of the variable between its mean value and the trough or peak. Such mathematical usage, however, is limited to rhythms which oscillate symmetrically about the mean level and would be inappropriate, for example, to the waveform shown in *Fig.* 1.1B. The term

'amplitude' has consequently been used by some to designate the range of oscillation from peak to trough. To avoid ambiguity, therefore, it is suggested that the term 'amplitude' should be used when the rhythm is represented by a symmetrical mathematical model and otherwise 'range of oscillation' be used.

4. The *phase* of the rhythm. This word has several meanings. In its strictest definition, the phase refers to the instantaneous state of the rhythm within a cycle, represented by the value of the variable—for example, the maximum or minimum. The phase of the rhythm thus tells us of the position of the rhythm in time. (In the special case where the data are represented by a fitted cosine curve the phase of the rhythm may be defined by the *acrophase* which is the time of maximum of the cosine function.) A *phase-shift* implies that whilst the rhythm retains a similar shape, it is displaced along the time axis. A phase-shift can be further described as a *phase-advance* or *phase-delay* to describe the direction of the displacement along the time axis. Thus a phase-delay means that every point in the rhythm occurs somewhat later, whilst a phase-advance means that every point occurs earlier. The term 'phase' is also used to describe the temporal relationship between two rhythms. When unqualified, for example, 'the rhythm of deep body temperature was in phase with the rhythm of activity', this means that similar aspects of the two rhythms (for example the times of maxima) occurred simultaneously. (When phase is used in this sense, the opposite, when the maximum of the one rhythm would occur simultaneously with the minimum of the other, is called *antiphase*.)

In describing a circadian rhythm it is often necessary to define the conditions under which the rhythm was observed. Since often rhythms are measured in individuals exposed to all the rhythmic 24-hour influences of the solar day, such as the alternation between day and night, day-to-night environmental temperature changes and rhythms of social behaviour, it is more convenient to use a single adjective to describe such conditions. The word which has come into general usage is *nychthemeral*. Thus, if a rhythm is measured in nychthemeral conditions, this implies that at least some of the normal periodicities of the solar day are present. Similarly, a nychthemeral rhythm refers to a bodily rhythm measured in nychthemeral conditions.

3. Detection of Rhythms

3.1. Timing and frequency of sampling

As described in the previous section, by definition, a rhythm must repeat itself in time. This implies that to define a rhythm in a variable, measurements of that variable must be made over more than one complete cycle. For this reason, the life cycle from birth to death cannot truly be regarded as a rhythm since it does not repeat itself. There are circumstances, however, where a variable can be sampled but once in any individual, and yet a rhythm described for that variable. For example, although birth or death occur only once in any individual, one can describe a circadian rhythm in the time of birth or the time of death for a population (*see Figs.* 11.6 and 11.7). The reason for this is that although any individual is sampled once only, different individuals are sampled on different days and therefore different cycles. Similarly, in determining whether the LD_{50} of a drug administered to animals is dependent upon the time of day of drug administration (termed 'chronotherapeutics', *see* Chapter 11), although each animal contributes only one datum point, the reproducibility of the oscillation can be demonstrated by administering the drug to different samples of the species on different days. It must be noted, however, that if no rhythm can be found when one is forced to rely upon a single datum point from each individual, this does not necessarily indicate that no rhythm exists in the individual. For example, the population would appear arrhythmic if the phase of the rhythm was very different in each individual used in the sample of the population.

Having thus described that, to detect a rhythm, individuals should ideally be sampled over more than one cycle, it is now pertinent to choose the sampling frequency, that is, the smallest number of samples that may represent the oscillation. Here one has often to compromise between what is ideal and what is practicable. The reliance, however, on few measurements per cycle is rarely sufficient. For instance, although a time-of-day variation can be shown by making but two measurements per cycle, it is necessary to make measurements over many cycles such that a statistically significant difference can be demonstrated between the two times of measurement. Making only two measurements per cycle, however, is beset with other problems (*see also* Chapter 8). For example, in the above case, if no statistical difference can be found between the measurements made at the two times of day, this does not necessarily imply that no rhythm is present; it may be that the variable was sampled at two times of day when it was at an identical value. It is obvious, therefore, that two measurements per cycle should never be relied upon. At the other extreme, a rhythm can be truly defined in all characteristics only if it is measured continuously. The impracticability of such continuous monitoring for many variables means that a compromise must be taken. The extent of this compromise depends upon the variable being measured. For example, as is the case with deep body temperature, if it is known that the variable cannot undergo rapid, large fluctuation (in the case of deep body temperature because of the thermal capacity of the body) then a rapid rate of sampling is not required to describe the circadian variation with some degree of accuracy. In other cases, if rapid, large fluctuations in the variable can occur, then the frequency of sampling must be high. This is particularly the case in describing the changes over the day in the plasma concentration of several hormones (particularly those with a short half-life) because the secretion of these occurs in short bursts or episodes (*see* Chapter 7).

In choosing the frequency of sampling, a factor which must always be borne in mind is the statistical analyses to be performed on the rhythm, for many statistical techniques (*see* Appendix) require data sampled at equidistant intervals and/or at a high frequency of sampling. Ofter frequent sampling over a single cycle may not be practical. This problem may be overcome by staggering samples over successive cycles. For example, if one requires hourly values but 2-hourly sampling only is possible, then on the first day measurements can be made on the even hours and on the second day on the odd hours. All data may then be represented over a single day. In doing this, however, it must be assumed that the rhythm is stable and reproducible from day to day.

In conclusion, therefore, it is not possible to generalize and allocate a specific frequency of sampling which must be used to demonstrate circadian fluctuations in a particular variable. However, the adage 'better too many than too few' must apply to the study of circadian rhythms also.

3.2. Experimental designs

The first step in the investigation of a possible circadian rhythm in the human is to establish that it exists in normal man living a conventional nychthemeral existence. Two methods of study thus become available. In the first, the longitudinal design, one or a few subjects are studied over many cycles, whereas in the other, the transverse design, many subjects are studied over one or a few cycles. Thus, the longitudinal design allows assessment of the reproducibility of a rhythm from day to day in a given individual whereas the transverse design allows assessment of the reproducibility of a rhythm between individuals. One must be careful, therefore, in the interpretation of results from the two types of experiment. For example, one cannot extrapolate from the longitudinal design that the observed rhythm might represent any individual in a population, and a lack of detectable rhythmicity from the transverse design may do nothing more than confirm that no single rhythmic parameter can be assigned to the population, for inferences about the rhythm in an individual cannot be made from this design. It is obvious then, that each design has its own strength and weakness and therefore, where possible, both designs should be employed. Despite this, there are many occasions when each individual can contribute one datum only, as when describing a rhythm in mortality, or when one measurement per day only can be made, as often occurs when investigating variations in the efficacy of a drug. In these cases, the transverse design must be used and therefore detection of a rhythm is dependent on the phase of the rhythm being reproducible from day to day or between different individuals. An example of the pitfalls in the interpretation of results from such transverse designs will be given in Chapter 11 (s. 5.1) in relation to the observed phase difference between the time of maximum onset of labour and that of birth (*Fig.* 11.6).

4. Analysis of Rhythms

The first step in any analysis to detect whether a rhythm is present in a variable measured over time, is to perform a 'macroscopic' inspection (Halberg, 1965). This involves plotting the data as a function of time on rectangular coordinates. If the period of the expected rhythm is known and data have been collected from several cycles, the average waveform can be derived from averaging corresponding points from each cycle and the averages (with standard errors) plotted over a single cycle. This has been referred to as 'waveform eduction' by Czeisler (1978) (*see* Chapter 7). Statistical analyses may then be performed to test whether a rhythm is truly present or the data simply represent random variation. The use of the Student's *t* test to compare values at different times of the day is inappropriate here. The reason for this is that, if, for example, 8 measurements have been made over a single cycle, there are 28 possible combinations of pairs of points; it can be calculated from the binomial distribution that on about 3 out of 4 occasions on which such measurements were made, one of the pairs would be significantly different at the $P<0.05$ level, even if there was no rhythm present. A more appropriate analysis is the analysis of variance. If data are collected from a single cycle a two-way analysis of variance is used to test whether the differences between values at various times are significant. If data are obtained for more than one cycle, it is also possible to determine whether there is any significant difference between cycles. This analysis, however, does not give any information about the shape of the rhythm or its phase, amplitude, mean level or period (if this is unknown); it merely indicates whether the data are different from random variation. It is obvious then that to quantify the rhythm more objectively, other more complex statistical techniques are required. Halberg (1965) has likened this to the use of a microscope, so that we can term this 'microscopic' inspection of the data.

Since the statistical analysis of rhythms is often complex and may be beyond the comprehension of the casual reader, we have included, for those with a modest mathematical background, an Appendix (pp. 311–19) which outlines some of the methods which may be used. However, since it will be referred to on many occasions in subsequent chapters, it must be noted at this stage that the analysis most popularly used is the cosinor analysis introduced by Halberg et al. (1967). This analysis is based upon the fitting of a cosine curve to the data such that the rhythm may be represented as in *Fig.* 1.1A. The rhythm may thus be characterized by the mesor, amplitude and acrophase, which are the mean level, distance from peak to mesor and the time of maximum, respectively, of the fitted cosine curve. Further details of this method will be given in the Appendix.

5. Exogenous and Endogenous Rhythms

Most organisms are usually forced in some way to interact with their environment. Thus, many mammals remain active diurnally since their eyes or other sensory apparatus are not adapted to nocturnal life, or their prey is active diurnally, and plants show photosynthetic activity during the hours of daylight only. In both these cases, therefore, the fluctuations between activity and inactivity or photosynthetic activity and a lack of it are a direct consequence of a rhythmic environment—in these particular cases, the alternation between light and dark. Such biological rhythms which are driven by external influences are termed exogenous. In man, although by the use of electric lighting we can become independent of the natural cycle of light and dark, our social behaviour normally adheres to the 24-hour solar day. It might be argued, therefore, that in man too, any internal rhythm exhibiting a period of 24 hours is exogenously timed; for example, since our social behaviour is geared to diurnal activity as a result of which we eat and drink diurnally and fast nocturnally, it might be argued that renal excretory rhythms (see Chapter 4) result from this rhythm in feeding times. Indeed, belief that biological rhythms may be timed externally predates the birth of Christ. Thus, Greek mythology pertains to the daily changes in leaf position, 'sleep' rhythm, of the plants now called heliotropes as a continuing expression of the unrequited love which Clytie had for the Greek god of the sun, Helios (Czeisler, 1978). This indicates some belief in the external timing of this rhythm. It was not until the eighteenth century, however, that this belief in the external timing of rhythms was challenged, when an astronomer, de Mairan (1729), studied these 'sleep' movements in plants maintained in constant darkness. Since this early discovery, biological rhythms which oscillate independent of environmental rhythms have been confirmed at most levels of the evolutionary series (for reviews, see Pittendrigh, 1960; Bunning, 1973; Palmer, 1976; Saunders, 1977; Cloudsley-Thompson, 1980). As a result, it has been proposed that the capacity for rhythmic change is an inherent characteristic of living organisms and that the control is endogenous and innate rather than exogenous. Such endogenous control presupposes the presence of a self-sustaining oscillator within the body which is capable of controlling such rhythms. We metaphorically refer to this structure as the 'clock' and those overt rhythms which result from the oscillations of this clock are termed endogenous.

A century after the discovery of the endogeneity of rhythms by de Mairan, a further property of the clock was revealed; it was found that the spontaneous period of the rhythm of leaf 'sleep' movement in constant darkness was not exactly 24 hours (De Candolle, 1832). It was for this reason that Halberg (1959) used the prefix *circa*, to denote that the circadian rhythms oscillated *about* once a day. Since this early observation of De Candolle, it has been confirmed for most animal species, including man, that the inherent period of the clock controlling endogenous rhythms is not exactly 24 hours. This will be discussed further below (s. 6.2).

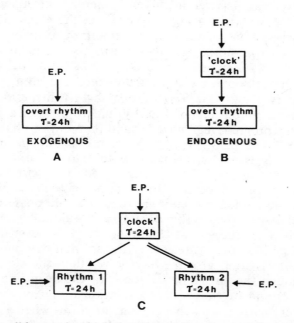

Fig. 1.2. Possible ways in which circadian rhythms may arise when observed in a nychthemeral environment. E.P. = external periodicities which oscillate with a period of 24 hours, τ = period. A, Rhythm exogenously timed: B, rhythm endogenously timed; C, rhythm resulting from an interaction of endogenous and exogenous influences. In C, $\Rightarrow$ indicates dominant influence and $\rightarrow$ indicates weak influence. Those external periodicities (E.P.) which entrain the 'clock' to an exact 24-hour periodicity are referred to as zeitgeber. (*See text* for further details.)

The way in which exogenous and endogenous rhythms result, when observed in a nychthemeral environment, is diagrammatically shown in *Figs.* 1.2A and B. It can be seen in *Fig.* 1.2B that in a nychthemeral environment an endogenous rhythm, although driven by an innate oscillator, is not completely independent of environmental periodicities. Rather, rhythmic environmental influences synchronize the biological clock from its inherent about-24-hour periodicity to an exact 24-hour periodicity. This process of synchronization of the clock by environmental influences we term *entrainment* and those external influences which are capable of entraining the clock are referred to as *zeitgeber* (Aschoff, 1954, 1958) —the German word meaning time-giver—or *synchronizers* (Halberg et al.,

1954). The way in which zeitgeber are thought to entrain the clock will be described in Chapter 12 (s. 4) and those zeitgeber which seem to be important to the synchronization of rhythms in man will be discussed in Chapter 12 (s. 5).

As is shown in *Fig.* 1.2B, it is thought that a 24-hour period results in endogenous rhythms observed in a nychthemeral environment because the zeitgeber exert their effect at, or very near to, the clock. The alternative hypothesis, namely that the clock is never entrained to an exact 24 hours and that twenty-four-hour periodicity can result in an overt rhythm through external 24-hour influences directly affecting the rhythm, must be rejected. The reason for this is that a single rhythm would then be influenced by two periodicities differing slightly in period; under such circumstances the amplitude of the rhythm would wax and wane as the two periodicities progressively moved in and then out of phase (an example of this will be shown below, s. 6.2). Since this is not a generally recorded phenomenon, this is evidence that it is the clock, rather than its final manifestation, which has been entrained to a precise 24-hour cycle. Further evidence for a direct synchronization of the clock is the observation that when nychthemeral cues are removed, the phase of the rhythm often shifts relative to the rhythm of sleep and wakefulness (*see* s. 6.2).

Although the hypothesis that the innate clock is never entrained to an exact 24-hour period must be rejected, the fact that external rhythms may exert effects on rhythms at some level lower than the clock cannot. For example, although the circadian variations of the deep body temperature (*see* Chapter 2) and of the urinary excretion of many electrolytes (*see* Chapter 4) will continue to show similar variations when a subject is kept awake continuously to those when the subject is observed in a customary environment (namely, lower values during the hours associated with sleep), our normal social behaviour of diurnal activity and feeding and nocturnal inactivity and abstinence will reinforce the basic rhythm resulting from the innate oscillator. In other words, an overt rhythm observed in a nychthemeral environment results from a simultaneous action of external periodic influences and those from the circadian clock. This is represented diagrammatically in *Fig.* 1.2C. Two possible rhythms are represented here; the block is assumed to be entrained to the solar day, although for rhythm 1 a large exogenous influence is also present, whilst rhythm 2 is highly endogenous in nature, being only weakly influenced by exogenous factors. In both cases a rhythm synchronized with the environment results. A more detailed example of the possible interaction of environmental periodic influences and the circadian clock which may be conducive to a rhythm in urine flow is shown by Mills (1966).

In *Fig.* 1.2C, the connection between each of the overt rhythms and the innate timing mechanism is shown as a direct one and the exogenous influences are represented as acting directly on the external manifestations of the clock activity. Most overt rhythms, however, are not directly driven by the clock. Rather they are the end of a chain of events where one rhythm results in the next, only the first rhythm in the chain originating directly from the activity of the clock. Thus, for example, the circadian rhythm of plasma cortisol concentrations results, at least in part, from an appropriately phased rhythm in adrenocorticotrophic hormone secretion, which in turn results from a similar rhythm in secretion of the hypothalamic corticotrophic releasing factor (*see* Chapter 7). In succeeding chapters (2–7) we will indicate those occasions where one internal rhythm seems to be, at least in part, responsible for another and in Chapter 12 (s. 1) we will show the possible causal nexus between rhythms and the ways in which this nexus is revealed. For the moment, however, it should be noted that exogenous influences may operate at any stage in the causal sequence of the transmission mechanisms between the clock and a final rhythmic manifestation. It follows that it is likely that the greater the number of stages in the transmission processes, the greater will be the exogenous component in the final overt rhythm.

6. Establishing the Endogeneity of Rhythms

To establish whether a rhythm is exogenous or endogenous is not just a theoretical exercise but is of practical importance. The reason for this is that if ever we are to manipulate these rhythms to our advantage we must have a knowledge of the mechanisms controlling the rhythms. Thus, if a rhythm is simply exogenous, it is an easy task to manipulate the phase or the period of the rhythm by changing the phase or the period of the external periodic influences which determine it. If, on the other hand, a rhythm arises endogenously, it is more difficult to influence since then we must manipulate the clock or the transmission processes between the clock and the overt rhythm.

Such manipulation of rhythms is becoming an essential requisite of modern society in two respects: first, with the ever-increasing emphasis on round-the-clock utilization of capital equipment and a shortening working week, there is an increased incidence of shift working; secondly, with ever-increasing frequency, those who hold high positions of responsibility in society are making long-distance transmeridian flights. In each case, if the endogenous clock resists a shift in phase of activity or environment (*see below,* s. 6.2) many endogenous rhythms will not maintain their usual phase relationship with the rhythm of rest and activity. The problems associated with such dissociation of circadian rhythms from environmental rhythms will be discussed in Chapters 9 and 10.

It is desirable, therefore, to establish the lines of evidence which lead one to a belief that a rhythm is endogenous. In addition, such evidence also often yields certain properties of the biological clock, such as its range of entrainment of both period and phase.

The main lines of evidence have been outlined elsewhere (Aschoff, 1960; Mills, 1966; Conroy and Mills, 1970). Briefly, an observed rhythm is established as endogenous if:

(i) it persists in the absence of all rhythmic influences likely to affect it;

(ii) when zeitgeber are sufficiently excluded, as in (i), its period is not an exact 24 hours;

(iii) it retains an about-24-hour period when a different period is adopted for the rhythm of habit and environment;

(iv) it does not immediately change its phase after an abrupt shift in the phase of the rhythm of habit and environment; and

(v) if it has become entrained to a new period or phase, as in (iii) or (iv) above, it does not immediately revert to its former period or phase when the former habits and environment are restored.

In the following sections, the types of experiment which must be performed to verify these lines of evidence will be expounded and possible pitfalls in the interpretation of results will be discussed. The way in which these experiments have then been used to establish the endogeneity of any particular rhythm will be given later in the relevant chapters (2—7).

6.1. Methods of temporal isolation of subjects

It is obvious from *Fig.* 1.2 that, no matter whether a rhythm is purely exogenous or endogenous or results from an interaction of both influences, when observed in a nychthemeral environment, a rhythm with a period of exactly 24 hours always results. It is thus impossible to distinguish the origin of an overt rhythm under such circumstances. To establish the endogeneity of a rhythm in a subject, therefore, it is first necessary to isolate the subject from all cues indicative of his habitual, nychthemeral environment. One can then observe the behaviour of his rhythms when all such zeitgeber have been excluded (*see below*, s. 6.2) or when artificial zeitgeber changes in phase and/or period (*see below*, s. 6.2) are introduced.

Basically three methods have been used to temporally isolate subjects.

Experiments performed in deep caves

One way of insulating a subject from all cues indicative of the alternation of day and night and the rhythm of the remainder of the community is to study the subject in a deep cave where no external light can penetrate and ambient temperature oscillations are minimal. The first study of a subject under such conditions was made in 1962 (Siffre, 1963, 1965; Halberg et al., 1965). Since this early study many others have been performed and have been enumerated elsewhere (Table 1/I in Wever, 1979). The results obtained from most of these experiments will be discussed in Chapter 5.

Experiments performed in polar regions

A second, naturally occurring environment in which there is little indication of the time of day is that in the polar regions, usually north of the Arctic circle. In these areas there is no alternation of day and night except at the equinoxes and experiments are therefore most conveniently performed in the summer months when there is continuous daylight. During these months the only indication of the alternation of day and night seems to be a movement of the sun round the horizon (though this is often obscured by cloud). The results from the experiments performed in such polar regions will be described in Chapters 2–7. It is to be noted at this stage, however, that whilst both the polar regions and deep caves present all the conditions necessary for temporal isolation, they are environments considered by most to be inhospitable and so considerably reduce the number of potential volunteers.

Experiments performed in isolation units

The lack of naturally occurring environments in which subjects might be isolated from habitual nychthemeral cues and which most subjects would consider as comfortable has led to the construction of units specifically designed for temporal isolation of subjects. There are but a few of such units in the world and the experiments which have been performed in them have been enumerated by Wever (1979, Table 1/II). A plan of our own unit is shown in *Fig.* 1.3. In the design of such a unit it must be borne in mind that all possible zeitgeber and indications of the rhythmic habits of the remainder of society outside the unit must be removed. Hence our own unit has the following features.

1. It is fully air-conditioned so that ambient temperature and humidity are maintained constant.

2. It is fully sound-proofed. The cavity walls of the isolation room and the external walls are filled with dense concrete. In addition, the ducting of the air-conditioning system contains sound attenuators such that no sound can penetrate the isolation room by this means.

3. The electrical supply is stabilized such that there are no fluctuations in the mains supply voltage which might be indicative of peak demand periods.

4. The entire inner isolation room is mounted on a series of cork and rubber pads so that ground-borne vibrations due to external road traffic are not conducted into the isolation room.

Not only do such isolation units allow us to temporally isolate people so as to establish the endogeneity of rhythms but also we can introduce artificial zeitgeber (for example, a changed light–dark cycle) to determine the effect of these on the human circadian system. In this way, information as to the zeitgeber important to the entrainment of the circadian system

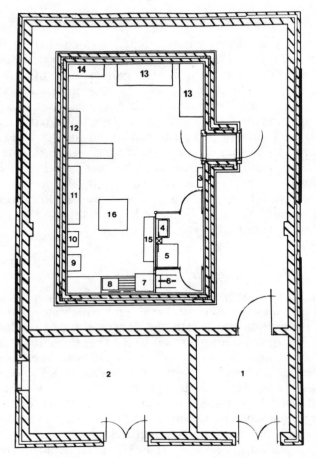

Fig. 1.3. Plan of the isolation unit at the University of Manchester. 1, Entrance hall. 2, Plant room. 3, Bookcase. 4, Wash hand basin with H & C. 5, Shower cubicle. 6, Toilets. 7, Cylinder cupboard. 8, Sink unit with H & C. 9, Refrigerator. 10, Cooker. 11, Cupboards. 12, Chairs. 13, Bunks. 14, Wardrobe. 15, Cupboards. 16, Table.

can be obtained as well as information on the effects of changed routines (for example, during shift work or following a time-zone transition after a transmeridian flight) on the circadian system in the absence of other confounding factors which may disturb rhythms. These points will be taken up again in Chapters 9, 10 and 12.

6.2. Experiments to establish the endogeneity of rhythms

Having described the ways in which subjects may be isolated from their customary nychthemeral environment, it is now pertinent to describe the experimental procedures which must be performed on the subjects in order to establish the endogeneity of rhythms.

'Free-running' experiments

If a rhythm is truly endogenous, then when all external rhythmic cues are removed the rhythm should continue oscillating since the clock which drives it is autonomous and self-sustaining. In the first type of experiment, therefore, the subject is studied in constant conditions; he is isolated without any timepiece or other indication of the passage of time, that is, all the more obvious zeitgeber are removed. As discussed by Aschoff (1960), three results of such an experiment are possible.

1. The rhythm may suddenly cease or dampen out over a few cycles. On the face of it, this would suggest that the rhythm was exogenous. However, this is not necessarily the case; it must also be considered that, although the rhythm is endogenous, it may have become unobservable because the environmental factors chosen for constancy are too unfavourable (for example, too hot or too cold).

2. The rhythm continues oscillating with a period of exactly 24 hours. In this case we cannot be sure that all external rhythmic influences have been excluded. A particularly striking case of this is reported by Wever (1979, *Fig.* 22) in which an exact 24-hour rhythm of rectal temperature and sleep—wakefulness was recorded in a female subject in temporal isolation in a unit. The reason for this became apparent later when it was discovered that the equipment for stabilizing the mains voltage feeding the lighting in the unit was not switched on, such that voltage fluctuations due to peak demand (hence giving temporal cues) were present during the experiment.

3. The rhythm continues oscillating with a period deviating from an exact 24 hours. If there is no other periodicity in the environment with a similar period, it is argued by most researchers (but *see below*) that the only way in which such a rhythm can result is from some innate, self-sustaining oscillator; indeed, this is the evidence on which the existence of the biological clock is based.

This last possibility then is the only result from which it may be concluded that a rhythm is endogenous.

When, under constant conditions, the innate clock takes up its natural frequency, any endogenous rhythm controlled by it will of course take up the same frequency so that the rhythm is no longer synchronized with the external environment—we term this *external desynchronization* and the rhythm is described as *free-running* (Bruce and Pittendrigh, 1957). We may thus describe those experiments designed to make the clock take up its free-running period as 'free-running' experiments.

By far the largest number of free-running experiments have been performed by Aschoff and Wever in an isolation unit and the results from these have recently been reviewed (Wever, 1979) and will be considered further in Chapter 5. The general finding has been that in the vast majority of subjects the free-running period is greater than 24 hours.

Similarly, the rhythms of subjects in caves have usually free-run with a period in excess of 24 hours. Examples for three subjects are shown in *Fig.* 5.2. (p. 99) in which are plotted the times of successive awakenings. Clearly the time of waking in each of these subjects became progressively later with each successive day indicating a period for the rhythm of sleep and wakefulness in excess of 24 hours. Results from these cave experiments also will be discussed in more detail in Chapter 5.

In those experiments in which the free-running period has been slightly in excess of 24 hours, it has usually been found that all measured overt rhythms free-ran with the same period, and the same period as that for the rhythm of sleep and wakefulness. When internal rhythms thus remain synchronized we term this *internal synchronization*. When internal synchronization is exhibited, caution must be taken about interpretation of results. For example, although a given rhythm may free-run with a period deviating from an exact 24 hours this alone is not indisputable evidence that the rhythm is endogenous. That is, if the rhythm remains in synchrony with the rhythm of sleep and wakefulness, it is possible that this sleep–wakefulness rhythm, associated with which there is a new rhythm of activity and inactivity and a change of mealtimes, might be producing all the other rhythms in the individual. For example, each period of activity effects an increase in body temperature at any time and independent of rhythmicity. An activity rhythm, therefore, is most likely reflected by a body temperature rhythm running in parallel. However, evidence can be gained in free-running experiments that a rhythm is oscillating independently of the sleep–wakefulness rhythm. First, there can be an altered phase relationship between the two rhythms; if the one rhythm were simply a consequence of the rhythm of sleep and wakefulness, then the two rhythms should always bear the same phase relationship no matter the period. An example of a changed phase relationship is shown in *Fig.* 2.3 (p. 28), from which it can be seen that from days 1 to 14 the maximum rectal temperature is found at the beginning of the period of wakefulness rather than at its usual position in the middle or later half in subjects living a nychthemeral existence (*see* Chapter 2, s. 2). More conclusive evidence for the independence of a rhythm from that of sleep and wakefulness is indicated by the phenomenon of *internal desynchronization,* also shown in *Fig.* 2.3. In this phenomenon a rhythm may suddenly free-run with a period different from that for the sleep–wakefulness rhythm so that these two rhythms no longer remain synchronized. This is shown in *Fig.* 2.3 where it can be seen that after day 15 the rhythms of rectal temperature and of sleep–wakefulness oscillate independently of one another. (It should also be noted that this phenomenon of internal desynchronization has implications for the number of oscillators controlling rhythms; this will be described further in Chapter 12, s. 2.)

Before leaving this section it should be mentioned that the existence of free-running rhythms with periods deviating from an exact 24 hours has not been interpreted by all workers as indicating the existence of an innate oscillator. The most ardent objector is F. A. Brown who, over the past 30 years, has repeatedly advocated an alternative hypothesis, namely that all rhythms are externally timed. The evidence on which this hypothesis is based includes observations in some plants and lower animal species that rhythms continue oscillations with remarkable accuracy (accurate to some geophysical or other environmental periodicity) in conditions in which the major zeitgeber are excluded. Brown's explanation for those occasions where rhythms have free-run with a period deviating from an exact environmental periodicity is the phenomenon of 'auto-phasing' (Palmer, 1976). This phenomenon is based on the differential sensitivity of an organism to zeitgeber at different times of the day, the phase-response curve (*see* Chapter 12, s. 4.2). In the authors' view, however, any mechanism involving a phase-response curve must assume the existence of some innate oscillator. Readers are referred to a recent review of the arguments (Brown, 1976).

Experiments with altered time

If, as will be shown, the endogenous oscillator controlling circadian rhythms is stable and resists change in zeitgeber, then evidence for the endogeneity of rhythms may be derived from experiments in which the zeitgeber are rearranged so that they do not conform to their customary pattern. In practice, the period or the phase of the zeitgeber may suddenly be changed. Following such a zeitgeber change, endogenous rhythms would not adapt immediately to the altered time.

Experiments with non-24-hour days: The period of zeitgeber may be changed by, for example, switching the lights on in an isolation unit and so signalling morning, not every 24 hours but rather after a longer or shorter period than this. In addition, subjects can be given clocks or watches which run fast or slow such that they live on a day shorter or longer, respectively, than 24 hours. Subjects have thus been studied and various variables measured on artificial daylengths of 18, 21, 22, 22·67, 26·67, 28 and 32 hours. The results from these experiments will be discussed in detail in the course of Chapters 2–7. For the present, however, it is sufficient to generalize the types of result which were obtained and their interpretation.

Though there was much inter-individual variability and variability between the different measured variables, in general one of three possible results has been obtained in these experiments with artificial daylengths.

First, the observed rhythm never shows any adaptation to the artificial daylength but rather adheres to a period of 24 hours or near-24 hours. Examples of this behaviour are shown for the temperature rhythm in *Fig.* 3.3 (p. 46) and the urinary potassium rhythm in *Fig.* 4.7 (p. 29). This finding provides convincing evidence of an endogenous clock and that the

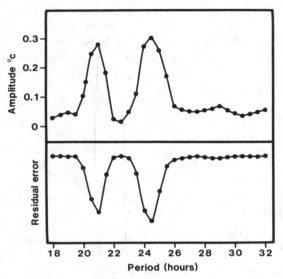

Fig. 1.4. Least-squares spectrum of the rectal temperature rhythm of a subject living on a 21-hour day. The upper panel indicates the amplitude and the lower panel the residual error (sum-squared deviations of observed values from predicted values) of best-fitting cosine curves with periods indicated by the abscissa. Minima in residual error (and maxima in amplitude) are seen when cosine curves with periods of 21 and 24·5 hours were fitted indicating the simultaneous presence of both periods in the temperature rhythm. (Data of Minors, 1975.)

rhythm is endogenous. It should be noted, however, that since there are no 24-hour zeitgeber in these experiments, one would not expect the period of these endogenous rhythms to be exactly 24 hours; rather, they should take up their free-running period. Secondly, a rhythm may immediately take up a period equal to the length of the artificial day. Examples of their behaviour are also shown in *Figs.* 3.3 (heart rate) and 4.7 (urinary flow). Though this may indicate that such rhythms are wholly exogenous, the possibility that such rhythms may have a small endogenous component which is swamped by a large exogenous component cannot be excluded. The third, and more usual, type of result to be obtained from these experiments is that the rhythm gives evidence of two periodic influences being simultaneously present due to the interaction of endogenous and exogenous influences; one, the endogenous component, with a period of about 24 hours and the other, the exogenous component, with a period equal to the length of the artificial day. Such simultaneous presence of two periodicities may be demonstrated by performing a least-squares spectrum analysis (*see* Appendix); an example is shown in *Fig.* 1.4 from an experiment in which a subject lived on a day of 21 hours. In addition,

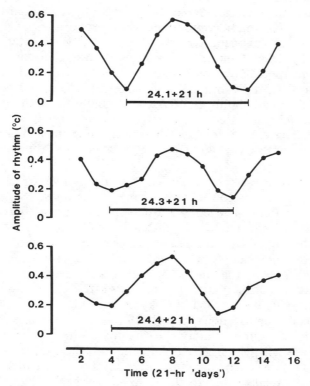

Fig. 1.5. The amplitude of cosine curves fitted to the daily temperature data of 3 subjects who spent 16 days living by a 21-hour clock. The phenomenon of beats is clearly seen. Horizontal bars indicate the period of beats derived from the simultaneous presence of the two periods indicated. (*From* Mills et al., 1977, Fig. 2.)

as a result of the simultaneous presence of two periodicities with near-equal periods, such rhythms usually give evidence of 'beats' (Lewis and Lobban, 1957). Beats are indicated when the amplitude of rhythm waxes and wanes and results from the two periodic influences being at first in phase (when a large amplitude results) and then progressively moving out of phase (when a small amplitude results). An example of this beat phenomenon is shown in *Fig.* 1.5, which shows the daily amplitudes (derived from cosinor analysis) of the rectal temperature rhythm of 3 subjects who spent 16 (artificial) days living by a 21-hour clock. The frequency of these beats can be calculated as $\frac{a \cdot b}{a \sim b}$ where a and b are the two periods simultaneously present. This calculation has been performed for the data in *Fig.* 1.5 and the predicted beat frequency is shown.

In these experiments in which rhythms give evidence of two periodic components, information as to the extent of the endogenous component can also be gained. To do this a circadian amplitude ratio (CAR) is calculated. This ratio is calculated by expressing the amplitude of the near-24-hour (endogenous) component as a proportion of that of the artificial day length (exogenous) component. A ratio greater than unity would thus indicate a stronger endogenous than exogenous component. The results from the calculation of such CARs for different variables will be described in Chapter 4 and are summarized in *Table* 4.1 (p. 83). The amplitudes used in calculating the CAR are usually those of cosine curves with periods which indicated minima in residual error in a least-squares spectrum or, better, those derived when a multiple linear regression of a function containing two cosine terms is performed (*see* Appendix).

Experiments with abrupt time shifts: An alternative to changing the period of the zeitgeber is to abruptly change their phase. Such an abrupt phase-shift is experienced whenever a person travels a sufficient distance round the world into a new time zone. Alternatively, a time-zone transition can be simulated by studying a subject in an isolation unit with all the zeitgeber, for example, the light—dark cycle and general temporal structure of the day in the unit, put forward or back by a number of hours. The relative merits of studying subjects after real or simulated time-zone transitions will be discussed in Chapter 9 (s. 1).

Evidence for the endogeneity of a rhythm is given in these experiments if the rhythm does not immediately shift its phase fully so that temporarily it is out of phase with the external zeitgeber. Rather, the rhythm should progressively phase-shift so that only after several days will it again have the normal phase relationship to the external zeitgeber. Examples of the changes in phase of rhythms following phase-shifts of the zeitgeber are shown in *Figs.* 4.5 (p. 76) and 9.1—9.5 (pp. 189—94), from which it can be seen that most rhythms studied do not shift their phase so as to be immediately appropriate with the new phase of the zeitgeber, indicating an endogenous component. It can further be seen that the rate of shift of phase to the new zeitgeber phase varies between different rhythms and is also dependent on whether the phase of the zeitgeber was advanced or retarded. The inferences from this and the interpretative problems associated with time-zone transitions will be discussed in more detail in Chapter 9.

As was the case with rapid adaptation of a rhythm to a non-24-hour day, the rapid shift of phase of a rhythm following a phase-shift of the zeitgeber is not evidence against the endogeneity of that rhythm. Rather, it might be that the phase of the new environment (times of sleep, meals, activity etc.) is exerting a direct effect upon the rhythm to such an extent as to obscure any endogenous influences. This will be referred to again in Chapter 9 (s. 4.2).

Return to the accustomed environment: Another piece of evidence to substantiate the endogeneity of a rhythm, which can be derived from these experiments in which subjects live on altered time, is to observe the behaviour of the rhythm when the subjects are returning to their accustomed time. Thus, if after several days the rhythm seems to have adapted to the altered time, no matter whether this be a change in zeitgeber phase or period, when the subject is returned to his former time zone or to a customary 24-hour day, an endogenous rhythm will not immediately revert to the former phase or a 24-hour period.

The results obtained from experiments in which people have been returned to their accustomed environment after a phase or period change of zeitgeber have been quite different, depending on whether the period or the phase of the zeitgeber has been changed. Thus, following a change in zeitgeber period, although many rhythms show a period equal to the length of the artificial day, rarely have the periods of rhythms failed to revert immediately to an exact 24 hours when individuals are returned to a customary nychthemeral environment. In only one experiment (Kleitman, 1963) has there been a report of subjects having difficulty in adjusting times of sleep to the customary solar day after life with a non-24-hour clock. However, one cannot conclude that these rhythms which adapt rapidly to changes in period of zeitgeber are purely or largely externally timed; it is possible that they are controlled by a rather labile oscillator or, under nychthemeral conditions, 24-hour influences might swamp any others. This point will be discussed further in Chapter 5 (s. 3.1) and Chapter 12 (s. 4.1).

By contrast with the results obtained after life on a non-24-hour day, the results obtained when subjects are returned to their former time zone, after rhythms appear to have adapted to a phase-shift of zeitgeber, have indicated that rhythms which took some time to adapt to the first time shift similarly did not immediately adapt to the return to the initial time zone (*see* Chapter 9). This is convincing evidence that these rhythms are endogenous in nature. It should be noted, however, that the rate of adaptation to each time shift is not always identical; the rate of adaptation being usually more rapid when the zeitgeber are delayed than when they are advanced. This point will be taken up again in Chapter 9.

The use of constant routines

A problem that has recurred in the foregoing discussion is that, whether the subject was free-running, following a non-24-day or adapting to a time-zone shift, in all cases a new rhythm for sleep and wakefulness is adopted. Sleep is known to exert a direct effect on many variables (Mills

et al., 1978a) and, associated with the new rhythm of sleep and wakefulness, there is a new timing of meals which might also directly influence
several rhythms (such as urinary excretory rhythms). Therefore, it is
unclear to what extent any rhythm observed to be following that of
sleep and wakefulness is simply a reflection of the new timing of sleep and
wakefulness which may be obscuring any endogenous component of
the rhythm under investigation. This phenomenon has been termed
'masking' by Aschoff (1978). This problem can be avoided by placing a
subject on a constant routine; that is, the subject can be kept awake for
24 hours or more, in constant light, sedentary or lightly active throughout,
and asked to take hourly small identical snacks. In this way most exogenous influences are removed and thus any underlying endogenous
component is unmasked. This is the method used by Mills et al. (1978b)
and will be discussed in Chapter 9 (s. 4.2).

7. Summary

The foregoing section has shown that very rarely can we describe a rhythm
as wholly endogenous or exogenous. Rather, rhythms usually result from
an interaction of endogenous and exogenous influences though there are
variations in the strengths of these components between different rhythms.
In Chapter 2 we will describe the rhythm in deep body temperature which
has been shown, using the methods described above, to have a stronger
endogenous component. Chapter 3 will then describe those rhythms
where the exogenous component is strong and in Chapter 4 we will
describe urinary excretory rhythms which fall between. In Chapters 5–8
the concepts introduced in this Chapter will further be used to describe
the rhythms in sleep and wakefulness, psychometric performance, the
endocrine system and the development of rhythms in infancy. Chapters
9–11 will then describe some practical applications of the study of
circadian rhythms and Chapter 12 will deal with theoretical problems
raised in this Chapter and throughout the body of the text.

References

Aschoff J. (1954) Zeitgeber der tierischen Tagesperiodik. *Naturwiss* **41**, 49–56.
Aschoff J. (1958) Tiersische Periodik unter dem Einfluss von Zeitgebern. *Z. Tierpsychol.* **15**, 1–30.
Aschoff J. (1960) Exogenous and endogenous components in circadian rhythms. *Cold Spring Harbor Symp. Quant. Biol.* **25**, 11–26.
Aschoff J. (1963) Comparative physiology: Diurnal rhythms. *Ann. Rev. Physiol.* **25**, 581–600.
Aschoff J. (1978) Features of circadian rhythms relevant for the design of shift schedules. *Ergonomics* **21**, 739–54.
Brown F. A. (1976) Evidence for external timing of biological clocks. In: Palmer J. D. (ed.) *An Introduction to Biological Rhythms.* New York, Academic Press, pp. 209–79.
Bruce V. G. and Pittendrigh C. S. (1957) Endogenous rhythms in insects and micro-organisms. *Am. Nat.* **91**, 179–95.

Bunning E. (1973) *The Physiological Clock*. London, English University Press.

Cannon W. B. (1939) *The Wisdom of the Body*, 2nd ed. New York, Norton.

Cloudsley-Thompson J. L. (1980) *Biological Clocks*. London, Wiedenfeld & Nicholson.

Conroy R. T. W. L. and Mills J. N. (1970) *Human Circardian Rhythms*. London, Churchill.

Czeisler C. A. (1978) Human circadian physiology: internal organization of temperature, sleep—wake and neuroendocrine rhythms monitored in an environment free of time cues. PhD Thesis, Stanford University.

De Candolle A. P. (1832) *Physiologie Vegetale*, Vol. 2. Paris, Bechet Jeune, pp 854—62.

Halberg F. (1959) Physiologic 24-hour periodicity: general and procedural considerations with reference to the adrenal cycle. *Z. Vitamin- Hormon- Fermentforsch.* **10**, 225—96.

Halberg F. (1965) Some aspects of biologic data analysis; longitudinal and transverse profiles of rhythms. In: Aschoff J. (ed.) *Circadian Clocks*. Amsterdam, North Holland, pp. 13—22.

Halberg F., Carandente F., Cornelissen G. et al. (1977) Glossary of chronobiology. *Chronobiologia* **4**, Suppl. 1.

Halberg F. and Katinas G. S. (1973) Chronobiologic glossary. *Int. J. Chronobiol.* **1**, 31—63.

Halberg F., Siffre M., Engeli M. et al. (1965) Étude en libre-cours des rythmes circadien du pouls, de l'alternance veille-sommeil et de l'estimation du temps pendant les deux mois de sejour souterrain d'un homme adult jeune. *C.R. Acad. Sci. [D] (Paris)* **260**, 1259—62.

Halberg F., Tong Y. L. and Johnson E. A. (1967) Circadian system phase—an aspect of temporal morphology; procedures and illustrative examples. In: von Mayersbach H. (ed.) *The Cellular Aspects of Biorhythms*. Berlin, Springer-Verlag, pp. 20—43.

Halberg F., Visscher M. B. and Bittner J. J. (1954) Relation of visual factors to eosinophil rhythm in mice. *Am. J. Physiol.* **179**, 229—35.

Kleitman N. (1963) *Sleep and Wakefulness*, 2nd ed. Chicago, University of Chicago Press.

Lewis P. R. and Lobban M. C. (1957) The effects of prolonged periods of life on abnormal time routines upon excretory rhythms in human subjects. *Q. J. Exp. Physiol.* **42**, 356—71.

de Mairan J. (1729) Observation Botanique. *Histoire de l'Academie Royale des Sciences.* Paris, p. 35.

Mills J. N. (1966) Human circadian rhythms. *Physiol. Rev.* **46**, 128—71.

Mills J. N., Minors D. S. and Waterhouse J. M. (1977) The physiological rhythms of subjects living on a day of abnormal length. *J. Physiol.* **268**, 803—26.

Mills J. N., Minors D. S. and Waterhouse J. M. (1978a) The effect of sleep upon human circadian rhythms. *Chronobiologia* **5**, 14—27.

Mills J. N., Minors D. S. and Waterhouse J. M. (1978b) Adaptation to abrupt time shifts of the oscillator(s) controlling human circadian rhythms. *J. Physiol.* **285**, 455—70.

Minors D. S. (1975) PhD Thesis, University of Manchester.

Palmer J. D. (1976) *An Introduction to Biological Rhythms*. San Francisco, Academic Press.

Pittendrigh C. S. (1960) Circadian rhythms and the circadian organization of living systems. *Cold Spring Harbor Symp. Quant. Biol.* **25**, 159—84.

Reinberg A. (1974) Aspects of circannual rhythms in man. In: Pengelley E. T.(ed.) *Circannual Biological Rhythms*. New York, Academic Press, pp. 423—505.

Saunders D. S. (1977) *An Introduction to Biological Rhythms*. New York, Wiley.

Siffre M. (1963) *Hors du Temps*. Paris, Tuillard.

Siffre M. (1965) *Beyond Time*. (Trans. by Briffault H.). London, Chatto & Windus.

Wever R. (1979) *The Circadian System of Man. Results of Experiments under Temporal Isolation*. Berlin, Springer-Verlag.

chapter 2 *The Circadian Rhythm of Deep Body Temperature*

1. Methods of Measurement

The rhythm of deep body temperature has been the subject of many studies, some of which were performed over a century ago. Two factors contribute to the large amount of experimentation. First, with the development of reliable mercury-in-glass thermometers, accurate measurement of oral temperature became convenient; and secondly, the frequency with which measurement can be made, at least during waking hours, is limited only by the cooperation of the subject.

If a large number of readings are to be taken—and this is an important consideration when one is concerned with an accurate assessment of the rhythm—a compromise has often to be reached between the requirements of the experimenter and what is acceptable to the subject. Rectal temperature is believed to be an accurate measure of deep body temperature and is the preferred method, but it is often unpopular with the subject and so, in practice, alternative sites of measurement have often been used (Sasaki, 1972).

One possibility that is widely used is oral temperature, but this suffers from the disadvantage that the oral mucosa will not reflect deep body temperature accurately if drinks have been taken or the subject has been talking recently. In conditions such as exercise when breathing through the mouth is likely, an alternative site must be found. Another possibility, described by Brooke et al. (1973), is to measure the temperature of mid-stream urine by a rapidly registering thermometer; *Fig.* 2.1 shows that it is possible by this method to obtain good agreement between oral and urine temperatures provided that the volume of urine that has been voided is not too low. It also illustrates the general finding that oral temperature slightly underestimates deep body temperature. Generally, skin temperature is a poor indicator for deep body temperature, not only because the results are lower (no doubt due to convection of air past the measuring site) but also because the phasing of the rhythm at this site differs from that of deep body temperature. Nevertheless, skin temperature over the breast has been used as an indicator of breast carcinoma (*see* Chapter 11). The axilla as a site for temperature measurement can be difficult to interpret in terms of its relation to either 'skin' or 'deep body' temperature since the extent to which the position is insulated from air movement and to which this temperature would then reflect deep body temperature is not known with certainty in any particular case.

The development of methods whereby the temperature can be recorded automatically has enabled data for the whole 24 hours to be obtained without the need to wake the subject. One advantage of this is that with data points uniformly spaced throughout the nychthemeron, the variety and sophistication of possible mathematical treatment increases markedly

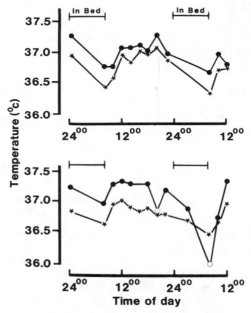

Fig. 2.1. Comparison of temperature measured in the urine stream (●) with oral temperature (★) in two subjects. Note the close correspondence between the two rhythms except on occasions when the volume of urine passed was less than 70 ml (○). (Data of Minors, 1975.)

(*see* Chapter 1). The recent development of continuous recording systems that are compact enough to be conveniently carried by the subject promises to be a powerful tool, especially in field studies in which subjects are required to continue their normal routine (*see,* for example, Smith et al., 1980).

2. The Nychthemeral Rhythm

The earliest records of the diurnal variation of temperature are generally attributed to Gierse (1842) or Davy (1845). They observed that oral temperature was low on waking, rose rapidly until about mid-morning and then remained fairly stable until the evening, when it started to fall. If records are kept while the subject is asleep, a minimum value at about 04⁰⁰ is usually found. Thus, the evening fall continues during the first part of sleep, and the rise after waking is a continuation of a process that started in the latter half of sleep. A typical result from a group of 8 subjects is shown for rectal temperature in *Fig.* 2.2 and shows a daily range of temperature of about 1·0 °C.

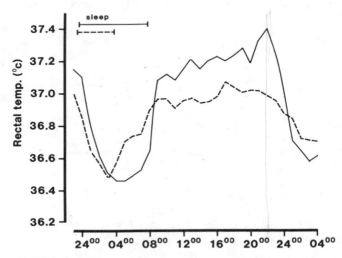

Fig. 2.2. Mean circadian changes in rectal temperature measured hourly in 8 subjects living a normal nychthemeral existence (solid line) and in the same subjects awoken at 04⁰⁰ and spending the subsequent 24 hours awake in constant light and taking hourly small identical snacks (dashed lines). (Unpublished data of Minors and Waterhouse.)

3. The Endogenous Component

A number of experimental approaches indicate that the rhythm of deep body temperature has a large endogenous component; these approaches will now be considered.

3.1. Experiments in which exogenous influences are decreased

Aschoff et al. (1974) measured the rhythm of rectal temperature in 6 males confined to an isolation unit. Three protocols were used:

1. Subjects lived on a normal routine, sleeping uninterrupted in the dark and being awake in the light.

2. Subjects were awakened briefly for test sessions on two occasions during the night but otherwise the lighting regimen was unchanged.

3. Subjects lived in continuous darkness with or without interruptions during the sleep period.

The temperature rhythms both for individuals and for the group as a whole were remarkably similar under all circumstances. Thus the rhythm was little influenced by lighting conditions and brief interruptions of sleep. Similarly, the circadian rhythm of temperature has been found to persist in subjects who have fasted or remained in bed throughout the nychthemeron. (For reviews, *see* Mills, 1966; Conroy and Mills, 1970; Czeisler, 1978).

Fig. 2.2 also shows data obtained from 8 subjects during a 24-hour constant routine (in which posture, feeding, activity and external temperature, light and humidity were all constant). The rhythm shows a smaller amplitude but similar times of minimum and maximum to those on a normal routine. Slight differences compared with nychthemeral conditions—the slower fall about midnight, the smaller amplitude and the gentler rise about 08^{00}—can be interpreted to show that sleep and wakefulness exert some effect upon body temperature, a point that will be considered below (s. 4).

3.2. Free-running experiments

When subjects have been isolated from external time cues by being placed either in caves (Reinberg et al., 1966; Colin et al., 1968) or in specially constructed isolation units (Mills et al., 1974; Czeisler, 1978; Wever, 1979), the rhythm of deep body temperature does not disappear but continues with an amplitude that is only slightly diminished and with a period of about 25 hours. It is observed too that under such circumstances the temperature rhythm varies less from day to day than those of cardiovascular and respiratory variables (Chapter 3), of the excretion of urinary constituents (Chapter 4), of sleep—wakefulness (Chapter 5), of performance in psychometric tests (Chapter 6) or of many hormones (Chapter 7).

In most subjects in such free-running experiments there is internal synchronization between different variables measured; that is, all variables have the same period though often there is a change in phase relationship. An example of this is shown in *Fig.* 2.3 before day 15 where it can be seen that the minimum temperature (downward-pointing arrow) occurs at about the time of retiring rather than about the time of midsleep as is usually observed under nychthemeral conditions. A putative explanation of this result will be given in Chapter 12.

In these cases of internal synchronization it cannot be decided whether the observed rhythms are being driven by an internal clock or whether they are produced by other factors such as rhythms of sleep—wakefulness or feeding (*see* Chapter 1). However, in about one-third of the free-running experiments that have been performed by Aschoff's group (summarized in Wever, 1979) the phenomenon of 'spontaneous internal desynchronization' has been observed. *Fig.* 2.3 also indicates an example of this phenomenon. Initially, before internal desynchronization occurs, there is a constant

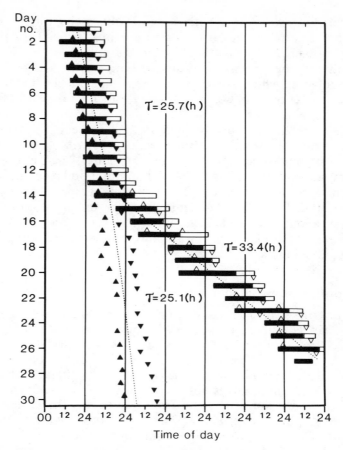

Fig. 2.3. Circadian rhythms of wakefulness and sleep (black and white bars respectively) and of rectal temperature (▲ maxima, ▼ minima) in a subject who lived alone in an isolation unit without indication of the time of day. Successive days are plotted from above downwards. From days 1 through to 14 the subject is internally synchronized with the two rhythms showing similar periods (25·7 hours). At day 15 spontaneous desynchronization takes place so that thereafter the two rhythms show different periods (rectal temperature 25·1 hours; sleep—wakefulness 33·4 hours). Open triangles show temporally corrected positions of temperature, maxima and minima represented by corresponding black triangles. (*From* Wever, 1975, Fig. 7.)

phase relationship between the rhythms of temperature and sleep—wakefulness, both of which showed a period of 25·7 hours. About day 15, internal desynchronization took place, the temperature rhythm continuing with a period similar to that before the desynchronization (25·1 hours) but sleep and wakefulness (and feeding) now showing a much longer period of 33·4 hours. Clearly there is no constant phase relationship between temperature and sleep—wakefulness under these conditions. Internal desynchronization is a phenomenon with a number of important implications in circadian rhythms and will be mentioned again in later chapters. However its importance at the moment is that it indicates most strongly that the rhythm of deep body temperature need not result from rhythms in sleep—wakefulness or feeding. Further, since the period of the temperature rhythm remains significantly greater than 24 hours, it cannot be attributed to influences from normal nychthemeral factors imposing themselves upon the subject. Other evidence dissociating temperature rhythms from the pattern of sleep and wakefulness comes from experiments in which subjects have been isolated in caves for 3 (Reinberg et al., 1966) or 6 (Colin et al., 1968) months. The sleep—wakefulness pattern has become irregular whereas the temperature rhythm has retained a far more stable period of slightly more than 24 hours (*see* Chapter 5).

3.3. Experiments with phase-shifts

The Arctic summer is associated with light and temperature conditions that are almost constant throughout the 24 hours. Sharp (1961) took advantage of this lack of zeitgeber to investigate the effect upon axillary temperature of inversion of the sleep—wakefulness routine. Subjects slept blindfolded to mimic darkness and 'night time' and were exposed to the light during their 'daytime'. The experiment was divided into three parts; in the first and third, subjects slept from 22^{30} until 07^{00} and in the second from 10^{30} until 19^{00}. The temperature rhythm did not adjust to the changed routine immediately but took about 3 days. After reverting to the original routine, another 3 days were required before the temperature rhythm returned to normal. Both results are most easily interpreted to suggest that adaptation to the changed routines involved the gradual adjustment of an endogenous oscillator. The potential difficulty expressed earlier, namely that axillary temperature might not accurately reflect core temperature, does not alter the validity of the results. The important point is the time taken by the temperature rhythms to return to control values after inversion of the sleep—wakefulness routine, whatever these control patterns might be. In fact, the result that the highest temperatures recorded were in the early part of 'daytime' and reached only about $36 \cdot 4\,^{\circ}\mathrm{C}$ suggests that they might not have been identical to deep body temperature.

Similar results have been obtained from studies of real transmeridian flights or simulated time-zone transitions in isolation units (Elliott et al., 1972). Although the exact time-course of adaptation varies (*see* Chapter 9 for details) there is general agreement that a delay of some days is experienced in both types of experiment. By contrast, if flights did not involve time-zone transition (northwards or southwards), no change in the temperature rhythm was observed (Hauty and Adams, 1966). These results again suggest that slow adaptation of an internal clock is involved. Such an interpretation is confirmed by the further observation that if the return journey was made after a stay of some days in a different time zone, then a second period of adjustment was required. Thus it was not the flight *per se* but rather living in a new time zone which resulted in slow adaptation of the endogenous component of the temperature rhythm.

In night-shift workers too, adaptation of the temperature rhythm to night work is certainly not rapid, but there are difficulties in the interpretation of this result. First, it is difficult to assess the extent to which adaptation has taken place, and second it is observed that adaptation back to a normal routine is much faster than the rate of adaptation to night work (these problems will be considered in Chapter 10). Nevertheless, the fact that adaptation is not immediate again argues for an endogenous component of the temperature rhythm.

3.4. Experiments with non-24-hour days

Experiments in which temperature rhythms have been measured in subjects living 'days' that bear a simple relationship to 24 hours (for example, 6, 12 or 48 hours in length) have indicated that the 24-hour rhythm persists in spite of these disturbances in the sleep—wakefulness pattern. An example is the study of Meddis (1968) in which 3 subjects lived on a 48-hour 'day' for at least 2 months. Oral temperature showed two peaks per 'day' corresponding closely with times on a normal 24-hour day. As has been pointed out by Conroy and Mills (1970), on such schedules to a certain extent subjects are still in phase with normal routine and normal zeitgeber might contribute to the stability of rhythms that is observed (*see also* Chapter 12).

This potential objection does not apply in other experiments that have been performed. Thus in a series of experiments upon student volunteers leading a sedentary existence in an isolation unit and subject to 21- or 27-hour days (Mills et al., 1977), rhythms of rectal temperature and of a number of urinary constituents were investigated. In these experiments two periodic components could often be seen in the data—one with a period of about 24 hours and the other equal to the length of the artificial day. The near-24-hour (endogenous) component, when compared with the simultaneously present 21- or 27-hour (exogenous) component, was greatest for temperature; thus the amplitude of the cosine curve best-fitting the near-24-hour component was greater than that of the cosine curve best-fitting the 21- or 27-hour component (*see Table* 4.1).

In a further study performed by Kleitman and Kleitman (1953) in the constant light of an Arctic summer, 3 subjects used watches that had been adjusted to show the passage of 24 'hours' when 18 or 28 real hours had elapsed. The oral temperature rhythm of one subject did not adjust to either abnormal day length, showing instead peak values separated by about 24 (real) hours (*see Fig.* 3.3) On the other hand, the temperature rhythms of the other 2 subjects showed a number of peaks more in accord with the number of activity periods, that is a peak every 18 or 28 hours. The explanation for the difference between subjects is not known, but the result does suggest that for two of the subjects the greater activity shown in the Arctic experiment, when compared with that in the isolation unit experiment, leads to an accentuation of the exogenous component.

4. Exogenous Effects upon the Temperature Rhythm

Even though the endogenous component of the temperature rhythm is marked, an exogenous component is present too, as is indicated by the observation that the temperature rhythms of two of the subjects adapted to 18- and 28-hour days in the Arctic experiment just described. In the isolation unit experiments too, an exogenous component was present in the temperature rhythm even though it was less marked than endogenous influences.

The exogenous component is generally assumed to result from the pattern of sleep and wakefulness, that is the alternation between inactivity and activity. The fall in temperature produced by sleep and the rapid rise on waking have been referred to already (*see Fig.* 2.2). Consequently, the phase of the temperature rhythm varies less when expressed relative to sleep rather than to clock time, as a result of which it is possible to compare more easily the rhythms of subjects on different schedules (Halberg et al., 1969). A similar finding has been obtained in a study by Mills and Waterhouse (1973) using temperature data from a single subject who spent a year in social isolation and whose hours of sleep were influenced by seasonal changes in the length of daylight. It was found that the variation in acrophase of the circadian rhythm of temperature on different days of the year was less when referred to mid-sleep than when referred to midnight.

The effect of sleep can be seen also in schedules in which the hours of sleep are changed so that, for any hour, a comparison can be made between the temperatures when the subject was awake and when asleep. *Fig.* 2.4 shows the temperature data in a subject who lived on a 12-hour day. This figure shows that each sleep period is associated with a fall in temperature even if the sleep is taken during the daytime. These falls are superimposed upon the endogenous component of the temperature rhythm such that the fall associated with daytime sleep is not so great as that associated with sleep at conventional hours. Further, more recently it has been demonstrated that the effect of sleep in lowering temperature is itself subject to circadian variation (Mills et al., 1978) though the reason for this is unknown.

The result that sleep exerts a direct effect upon the temperature rhythm has led to the view that the sleep—wakefulness cycle normally

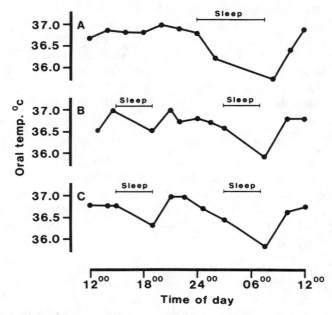

Fig. 2.4. The oral temperature rhythm in a single subject measured over 3 successive days. In A the subject slept for 8 hours at a conventional time. In B and C the subject adhered to a 12-hour day, sleeping for 4 hours separated by 8 hours of wakefulness. Note each sleep period is associated with a fall in temperature, this being greater, however, when sleep is at a conventional time. (Data of Mills, 1968.)

entrains the endogenous oscillator controlling the temperature rhythm to a period of 24 hours. But a distinction must be drawn between a direct (masking) effect of some factor and its role as a zeitgeber, a problem that is considered again in Chapters 5 and 12. Some evidence that is relevant to this problem comes from a study by the authors (Minors and Waterhouse, 1980a). They showed that the rhythms of subjects on irregular patterns of sleep and waking 'free-ran' (*Fig.* 2.5A) but that the rhythms could be stabilized to a 24-hour period if 4 hours of the customary 8-hour sleep were taken at a regular time each day (*Fig.* 2.5B). With respect to the temperature rhythm, the result might have been solely due to the regular decrement in temperature that would have been produced by the 4-hour sleep taken at the same time each day. That this was not the whole explanation is indicated by later experiments (Minors and Waterhouse, 1980b) in which the temperature rhythm immediately after such a procedure was assessed during a 'constant routine' (*see* Chapter 1). On this regimen, when the exogenous influences had been minimized, the temperature rhythm seemed unchanged from that during a control constant routine, that is the endogenous component too had been entrained to a

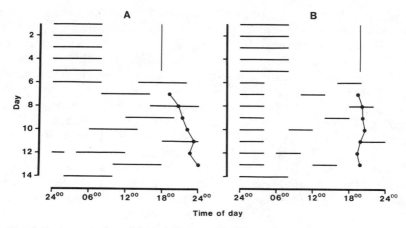

Fig. 2.5. Acrophases of the rectal temperature rhythm of subjects studied in an isolation unit. Horizontal bars indicate times in bed. Experiments began with a control phase during which sleep was taken at conventional times. Vertical line represents acrophase of rhythm during the control phase. Thereafter the subject slept for 8 hours at irregular times (A) or for two 4-hour periods one of which was regularly taken from 24⁰⁰–04⁰⁰ (B). (Data of Minors and Waterhouse, 1980a.)

24-hour period by the regular 4-hour sleep. The inference from this is that the 4-hour sleep was acting as a zeitgeber as well as exerting a direct exogenous influence upon the temperature rhythm.

It is worth stressing at this point that it is not possible to state which component(s) of sleep—changes in posture, light flux to the retina, food and water intake or alterations in neurophysiological activity in the brain—is (are) responsible for either the masking or zeitgeber influence that is exerted by the pattern of sleep and wakefulness. Further consideration of this is given in Chapter 12.

5. The Interaction between Exogenous and Endogenous Influences

The temperature rhythm at any instant will therefore be derived from both endogenous and exogenous components. Normally these are in phase with each other, but when subjects live on non-24-hour days or show spontaneous internal desynchronization these two influences will continually change with respect to one another. Sometimes they will be in phase (as in normal conditions) with a resultant rhythm of high amplitude; at other times the two components will be out of phase, the rhythm now being of low amplitude. Mills et al. (1977) observed this phenomenon in their experiments with 21- and 27-hour days (*see Fig.* 1.5). (It is interesting to note that the beat frequency observed suggested that the endogenous component had a period of slightly more than 24 hours, that is, it was free-running.) Such a phenomenon was clearly seen also in some cases of internal desynchronization (Aschoff et al., 1967).

Further, whereas the entrained rhythm during a nychthemeral routine tended to show a daytime plateau and rapid drop with sleep onset (*see* Fig. 2.2), the rhythm during free-running conditions resembled more a sinusoidal shape (Czeisler, 1978). In conditions of internal desynchronization, Czeisler found also that the temperature rhythm, when submitted to Fourier spectral analysis, contained two periods, one with a value equal to about 25 hours and the other equal to the period of the sleep–wakefulness cycle. Eduction of the temperature waveform at both periods (*see* Chapter 1) produced curves of different shape; that at 25 hours had a sinusoidal, 'free-running' shape whereas that at the sleep–wakefulness period had a plateau-plus-dip, 'entrained' shape. That is, under conditions of internal desynchronization the temperature rhythm had two components, one endogenous and free-running, the other a direct effect of sleep and entrained to the sleep–wakefulness cycle.

6. Origin of the Rhythm

The deep body temperature at any moment results from the balance between heat gain and heat loss, though the thermal capacity of the human body is such that changes in body temperature take place rather slowly when heat imbalances do occur.

6.1. Heat gain or heat loss?

The evidence is against the view that the rhythm results from changes in the rate of heat production since the rhythm persists almost unchanged under the conditions where changes in metabolic rate are minimal, as on a constant routine or when the subject is confined to bed throughout the nychthemeron. Such studies indicate that the variations in metabolic rate seen in these circumstances are far too small to affect heat gain and thus body temperature by anything like the required amount (*see also* Chapter 3); indeed, it is likely that the slight rhythm of metabolic rate is a consequence, not a cause, of the temperature rhythm. Further, moderate changes in metabolic rate produced by waking activities have little effect upon temperature rhythms. Therefore it is a circadian variation in heat loss that must account for the circadian rhythm in temperature. Measurements of skin temperature and blood flow, summarized by Aschoff and Heise (1972) and by Hildebrandt (1974), show that for the extremities (but not the trunk, proximal limbs and forehead) there is a circadian rhythm that is phased differently from that for rectal temperature (*Fig.* 2.6). The observations that the minimum cutaneous temperature in hand, foot and finger regions coincides with the maximum rate of rise of deep body temperature and that these cutaneous temperatures are highest in the evening suggest that there might be a causal link between the two.

Aschoff and Heise (1972) have calculated that the changes in heat loss from these sites account for at least 75 per cent of the circadian variation in deep body temperature. More recently, Tokura et al. (1979) have

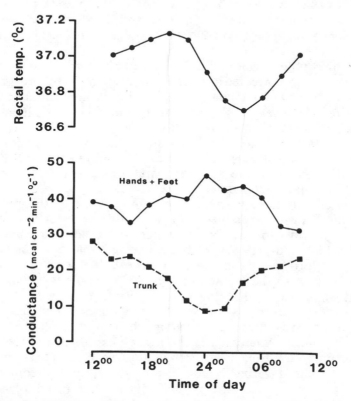

Fig. 2.6. Rectal temperature and thermal conductance of the trunk and hands-plus-feet as a function of time of day. Average values from 9 subjects, resting in the nude at 32°C ambient temperature. Trunk is average from two measurements (chest and abdomen) and 'hands and feet' refer to data averaged from four measurements (back of one hand, one finger, back of one foot and one toe). (*From* Aschoff and Heise, 1972, Fig 6.)

measured the rate of insensible perspiration in subjects under conditions of constant humidity (55 per cent) and temperature (32°C). The subjects were sedentary during the daytime and slept at night with little cover from bedclothes. The rate of sweating was highest in the late evening and early night—at a time when core temperature was falling—and lowest during the latter part of the night and early morning—when deep body temperature was rising.

6.2. The hypothalamus and changes in the 'set-point'

From the comments above it would be possible that the rhythm in core temperature is a consequence of rhythms in vasomotor and insensible sweating, but there are several observations that argue against this. These observations suggest rather that the circadian rhythms in the activities of the heat loss mechanisms are simultaneous and that they are manifestations of a changing 'set-point' of the thermoregulatory mechanisms. These observations are:

1. The amplitude, mean and acrophase of the temperature rhythm are very similar (when corrected for different sleep—wakefulness patterns) in countries with different climates, that is, in circumstances in which the factors affecting heat loss—ambient humidity and temperature—are vastly different (Halberg et al., 1969). Further, changing the climatic environment in which an individual lives does not affect his circadian rhythms. Such observations are interpreted far more easily if it is accepted that it is the control of the body temperature rather than the loss of heat that varies in a circadian manner.

2. Measurements have been made of the response of the sweating and shivering mechanisms to a thermal load at different times of the nychthemeron. Attempts to raise body temperature by drinking tea or taking exercise (Hildebrandt, 1974) or by exposing subjects to heat (Timbal et al., 1975) produce more vigorous sweating during the night than during the day. Also the latency of onset of sweating is less at night and Timbal et al. consider that the core temperature at which sweating is initiated is nearer to resting core temperature during the night. The largest sweating response is during the evening when the temperature rhythm is in its falling phase, whereas it is often not possible to produce sweating in the morning at a time when temperature is rising at its greatest rate (*Fig.* 2.7). When shivering is considered, the reverse circumstances are met, shivering being easiest to elicit during the rising phase of the circadian rhythm and most difficult during the evening (Hildebrandt, 1974).

3. Hildebrandt (1974) has demonstrated also that vasoconstrictor responses to thermal loads show a circadian variation. The skin flow can be estimated by measuring how long it takes the skin to warm up after being cooled in a water-bath. In the evening the hand takes less time to warm up (i.e. its blood flow is greater) after cold immersion than during the morning (*Fig.* 2.7).

4. Cabanac et al. (1976) investigated the preferred skin temperature of subjects who had been artificially warmed or cooled. They concluded from their experiments that subjects preferred cool cutaneous temperatures if their deep body temperature was above the 'set-point' and warm cutaneous temperatures if their core temperature was below the 'set-point' at that time. Their further finding that subjects tolerated warm environ-

ments far better during the morning (the rising phase of the circadian rhythm) and cool environments during the evening (the falling phase) accords with the view that the circadian rhythm of deep body temperature results from the 'set-point' being raised in the morning and lowered in the evening.

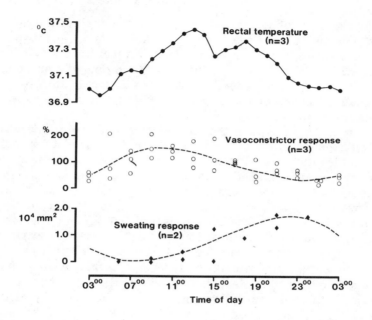

Fig. 2.7. Circadian variations in rectal temperature, duration of vaso-constrictor response after a cold hand-bath and planimetric magnitude of the sweating reaction of the forehead after drinking diaphoretic tea. (*From* Hildebrandt, 1974, Fig. 4.)

6.3. The effects of posture upon heat regulation

Recent experiments by Czeisler (1978) have cast some light on the obser-vation made earlier that core temperature drops very rapidly at the onset of sleep and rises very rapidly on waking (*see Fig.* 2.2A). He has found that the rapid fall in core temperature that occurs on lying down and going to sleep is accompanied by a large increase in wrist temperature, and hence blood flow, and that this cannot be attributed to insulation from bedclothes. No doubt the decreased sympathetic outflow to the cutaneous vessels that mediates these changes in flow is controlled in part by the temperature-regulating centre, as has just been described; but it seems likely that, in addition, the changes in posture and brain activity associated with lying down and going to sleep would produce further decreases in sympathetic tone. Also the rapid increase in core temperature on waking and rising probably derives in part from the increased sympathetic discharge that is produced by the start of a new day for the individual.

7. Implications and Usefulness

From this account of the circadian rhythm of deep body temperature in man, it emerges that the rhythm is notable for its comparative immunity from disturbances from the external environment. This stability comes about because of the large endogenous component that the rhythm possesses and, as will be argued in Chapter 12, this in turn is a reflection of its nearness to one of the oscillators that is held to be responsible for circadian rhythmicity.

If one is asked to explain the physiological and evolutionary advantage of the circadian rhythm in temperature, one is forced to speculate. The advantage of a high body temperature must be offset against its metabolic cost. It could then be argued that during the sleep of an animal the cost of maintaining body temperature at its waking value is unacceptably high. The decrease associated with sleep would then reduce the energy requirements (at a time when the animal is not eating) but the modest size of this fall would not jeopardize too much its chance to escape predators (at a time when it is normally hidden from them). The argument could then continue that the rise of body temperature towards the end of sleep, together with other factors that will be discussed later in Chapters 3–7, ensures that the newly woken animal is immediately effective for the rigours of a new period of activity—a process of pre-adaptation (Cloudsley-Thompson, 1980). Alternatively, the changes in 'set-point' might be just one manifestation of the multifarious changes associated with rhythms of brain activity. The observation that nocturnal animals have peak temperatures at night argues for the usefulness of higher waking and lower sleeping temperatures but does not enable a decision to be made in favour of either of the above speculations.

Later chapters will consider the role that temperature plays in affecting other rhythms, most notably that of performance (Chapter 6) and the implications of this relationship in subjects who have undergone time-zone shifts or perform shift work (Chapters 9 and 10). Further, usefulness of the temperature rhythm in clinical diagnosis and its modification in breast tumours will be considered in Chapter 11. Finally, the changing amplitude of the temperature rhythm during the passage of an individual from birth through maturity to senescence will be seen to provide clues as to the causal nexus that is believed to exist between different circadian rhythms (Chapter 8) and the temperature rhythm in general will be used to indicate some basic properties of the endogenous clock and the way in which it can be entrained by the external environment (Chapter 12).

References

Aschoff J., Fatranska M., Gerecke U. et al. (1974) Twenty-four-hour rhythms of rectal temperature in humans: effects of sleep-interruptions and of test sessions. *Pflügers Arch.* **346**, 215–22.

Aschoff J., Gerecke U. and Wever R. (1967) Desynchronization of human circadian rhythms. *Jap. J. Physiol.* **17**, 450–7.

Aschoff J. and Heise A. (1972) Thermal conductance in man: its dependence on time of day and on ambient temperature. In: Itoh S., Ogata K. and Yoskimura H. (ed.) *Advances in Climatic Physiology.* Tokyo, Igaku Shoin, pp. 334–48.

Brooke O. G., Collins J. C., Fox R. H. et al. (1973) Evaluation of a method for measuring urine temperature. *J. Physiol.* **231**, 91P.

Cabanac M., Hildebrandt G., Massonnet B. et al. (1976) A study of the nychthemeral cycle of behavioural temperature regulation in man. *J. Physiol.* **257**, 275–91.

Cloudsley-Thompson J. L. (1980) *Biological Clocks. Their Functions in Nature.* London, Weidenfeld & Nicholson.

Colin J., Timbal J., Boutelier C. et al. (1968) Rhythm of the rectal temperature during a 6-month free-running experiment. *J. Appl. Physiol.* **25**, 170–6.

Conroy R. T. W. L. and Mills J. N. (1970) *Human Circadian Rhythms.* London, Churchill.

Czeisler C. A. (1978) Human circadian physiology: internal organization of temperature, sleep–wake and neuroendocrine rhythms monitored in an environment free of time cues. PhD Thesis, Stanford University.

Davy J. (1845) On the temperature of man. *Phil. Trans.* **2**, 319–33.

Elliott A. L., Mills J. N., Minors D. S. et al. (1972) The effect of real and simulated time zone shifts upon the circadian rhythms of body temperature, plasma 11-hydroxycorticosteroids and renal excretion in human subjects. *J. Physiol.* **221**, 227–57.

Gierse A. (1842) Quaemiam sit ratio caloris organici. Dissertation, Halle.

Halberg F., Reinhardt J., Bartter F. C. et al. (1969) Agreement in endpoints from circadian rhythmometry on healthy human beings living on different continents. *Experientia* **25**, 106–12.

Hauty G. T. and Adams T. (1966) Phase shifts of the human circadian system and performance deficit during periods of transition: III. North–south flight. *Aerospace Med.* **37**, 1257–62.

Hildebrandt G. (1974) Circadian variations of thermoregulatory response in man. In: Scheving L. E., Halberg F. and Pauly J. E. (ed.) *Chronobiology.* Tokyo, Igaku Shoin, pp. 234–40.

Kleitman N. and Kleitman E. (1953) Effect of non-twenty-four-hour routines of living on oral temperature and heart rate. *J. Appl. Physiol.* **6**, 283–91.

Meddis R. (1968) Human circadian rhythms and the 48-hour day. *Nature* **229**, 964–5.

Mills J. N. (1966) Human circadian rhythms. *Physiol. Rev.* **46**, 128–71.

Mills J. N. (1968) Temperature and potassium excretion in a class experiment in circadian rhythmicity. *J. Physiol.* **194**, 19P.

Mills J. N., Minors D. S. and Waterhouse J. M. (1974) The circadian rhythms of human subjects without timepieces or indication of the alternation of day and night. *J. Physiol.* **240**, 567–94.

Mills J. N., Minors D. S. and Waterhouse J. M. (1977) The physiological rhythms of subjects living on a day of abnormal length. *J. Physiol.* **268**, 803–26.

Mills J. N., Minors D. S. and Waterhouse J. M. (1978) The effect of sleep upon human circadian rhythms. *Chronobiologia* **5**, 14–27.

Mills J. N. and Waterhouse J. M. (1973) Circadian rhythms over the course of a year in a man living alone. *Int. J. Chronobiol.* **1**, 73–9.

Minors D. S. (1975) PhD Thesis, University of Manchester.

Minors D. S. and Waterhouse J. M. (1980a) Anchor sleep as a synchronizer of rhythms on abnormal schedules. *Int. J. Chronobiol.* **7**. In the press.

Minors D. S. and Waterhouse J. M. (1980b) Does 'anchor sleep' entrain the internal oscillator that controls circadian rhythms? *J. Physiol.* **308**, 92P–93P.

Reinberg A., Halberg F., Ghata J. et al. (1966) Spectre thermique (rythmes de la température rectale) d'une femme adulte avant, pendant et après son isolement souterrain de trois mois. *CR Acad. Sci. [D]Paris* **262**, 782–85.

Sasaki T. (1972) Circadian rhythm in body temperature. In: Itoh S., Ogata K. and Yoshimura H. (ed.) Tokyo, Igaku Shoin. pp. 319–33.

Sharp G. W. G. (1961) Reversal of diurnal temperature rhythms in man. *Nature* **190**, 146–8.

Smith P., Davies G. and Christie M. J. (1980) Continuous field monitoring of deep body temperature from the skin surface using subject-borne portable equipment. Some preliminary observations. *Ergonomics* **23**, 85–6.

Timbal J., Colin J. and Boutelier C. (1975) Circadian variations in the sweating mechanism. *J. Appl. Physiol.* **39**, 226–30.

Tokura H., Ohta T. and Shimomoto M. (1979) Circadian change of sweating rate measured locally by the resistance hygrometry method in man. *Experientia* **35**, 615–6.

Wever R. (1975) The circadian multi-oscillator system of man. *Int. J. Chronobiol.* **3**, 19–55.

Wever R. A. (1979) *The Circadian System of Man, Results of Experiments under Temporal Isolation.* Berlin, Springer-Verlag.

chapter 3 *Cardiovascular, Respiratory, Metabolic and Gastrointestinal Rhythms*

In the previous chapter it was emphasized that the rhythm of deep body temperature possessed a large endogenous component. The rhythms to be dealt with in this chapter lie very much toward the other end of the spectrum since they possess a large exogenous component and show very clearly the direct effect of the environment and activity of the individual; in fact, there has sometimes been difficulty in demonstrating that a particular rhythm has any endogenous component at all!

Accordingly, having described the rhythm as normally observed in nychthemeral conditions, the evidence that there is also an endogenous component will be considered. The general criteria by which such a conclusion can be reached have been outlined in Chapter 1 and the way in which these criteria can be applied in the case of the rhythm of deep body temperature has been illustrated in Chapter 2. In the case of most of the variables to be considered in this chapter, evidence in favour of an endogenous component is very limited. This lack of data often does not reflect the difficulty in recording observations (for examples, heart rate and blood pressure) but rather indicates that in rhythms with large exogenous components the recordings tell us little about the internal clock but rather reflect the subject's activity and the environmental exigencies at the time of measurement. For this reason also, few measurements of the rhythms of these variables have been obtained during shift work or after time-zone transitions.

I. CARDIOVASCULAR SYSTEM
I.1. Continuous Monitoring

The frequent sampling of many cardiovascular variables is not a difficult task and continuous monitoring of patients under intensive care (for example, Wilson et al., 1977) and astronauts in flight (for example, Halberg et al., 1970) has become normal practice, but the extent to which such results reflect the rhythms in healthy man in a normal environment is of course uncertain.

The problem of measuring accurately the circadian rhythm of a variable with a large exogenous component is illustrated by the example of blood pressure. This variable is much influenced by the circumstances under

41

which it is measured even when factors such as posture are controlled. Thus blood pressure has been found to be higher when measured in the strange confines of the clinic rather than in the comfort of one's home (Gordon and Mortimer, 1973) and the presence of a physician has been observed to raise blood pressure (Richardson et al., 1964). Such results would argue for self-measurement (autorhythmometry) of blood pressure as the preferred method (see Chapter 11 and Halberg et al., 1972) and may have implications for the diagnosis of hypertension.

Autorhythmometry may be performed manually or by the automatic recording of data; this latter has the advantages that not only is the subject more able to continue a normal routine but also the recordings can be made when the subject is asleep or otherwise unaware of what is happening. Two variables which have been monitored successfully in this way are heart rate and blood pressure. The former can be measured from the electrocardiogram (Millar-Craig et al., 1978) or the echocardiogram (Halberg et al., 1970). Blood pressure can be measured directly by cannulation (for example, Goldberg et al., 1976; Mitchell et al., 1979) or indirectly by automatically inflating a sphygmomanometric cuff and estimating systolic and diastolic pressures from the echocardiogram (Wertheimer et al., 1974) or by an ultrasound flow meter (Smith and Malyj, 1974). Further, the more recent development of portable devices for storing all this information will make possible the collection of large amounts of data from ambulatory subjects (see, for example, Millar-Craig et al., 1978; Mitchell et al., 1979).

I.2. Blood Pressure

It has long been known that blood pressure is lower during sleep and this is the case even if sleep is taken at unusual hours. (For an account of this earlier work, see Conroy and Mills, 1970.) When measurements are made automatically throughout the sleep period (Richardson et al., 1964), minimum values are found between midnight and 04[00], often declining to surprisingly low values, for example 70/40 mmHg (see also Delea, 1979). Higher values observed by other workers during the night (after the subjects had just been woken) might emphasize the rise in blood pressure in response to factors such as the presence of another person and the 'shock' of being woken. There is disagreement too when the diurnal values are considered; thus some workers have found maximum values just after waking with a slight fall throughout the rest of the day (Millar-Craig et al., 1978), whereas others have found maximum values in the later afternoon or evening (for examples, Richardson et al., 1964; Mitchell et al., 1979). In a recent review, Smolensky et al. (1976) have applied cosinor analysis to data from many studies performed upon subjects of different states of health and undergoing different regimens. They found general agreement that the acrophase is about 12–13 hours after mid-sleep. Such a result

applied to both systolic and diastolic blood pressure and to both normotensive and hypertensive subjects. The further observation that a rhythm
of pulse pressure has the same acrophase indicates the amplitude of the
rhythm of systolic to be greater than that of diastolic blood pressure.

The agreement between different subjects and protocols is perhaps
surprising in view of the large exogenous component in the rhythm. The
referral of the acrophases to midsleep rather than clock time once again
indicates the importance of sleep and wakefulness in the timing of this
rhythm. However, the result that the acrophase differs from midsleep by
12–13 hours might result from the marked nocturnal nadir rather than
reflect the diurnal time course. This result, which is similar to body
temperature in subjects on a nychthemeral existence, would then be
compatible with a wider range of times of peak of the blood pressure (*see
above*) than the cosinor analysis indicates.

Accepting that there is this large exogenous effect, is there an endogenous component? The evidence here is fragmentary, but some approaches
to this problem are described below.

I.2.1. Experiments in which some exogenous influences have been removed

Wertheimer et al. (1974) studied 10 patients who showed no evidence of
cardiovascular disease. The patients were confined to bed, semi-recumbent,
throughout the nychthemeron, were allowed to sleep at night (in the
dark) and ate meals at normal times. A rhythm in systolic and diastolic
blood pressures persisted, though with what appeared to be a decreased
amplitude. This result indicates that changes in posture are not solely
responsible for the rhythms seen nychthemerally. A similar experiment
has been performed by Reinberg et al. (1970) with the added refinements
that meals were controlled and taken at 4-hourly intervals and that the
data were analysed by cosine curve fitting. The rhythm persisted in these
circumstances also with the same acrophase but with a decreased amplitude
compared with control subjects on unrestricted diets and diurnal activity;
thus the rhythms do not depend upon diurnal food intake either.

I.2.2. Do changes independent of sleep take place?

The above experiments do not eliminate the possible role of sleep in
influencing the circadian rhythm. This is particularly apposite in the case
of blood pressure since the rapid changes associated with being woken at
night have already been described. An important consideration when an
endogenous component of the rhythm is being sought is whether changes
in blood pressure occur before waking or sleeping or rather simultaneously
with these events.

Millar-Craig et al. (1978), using normal subjects and hypertensive patients living on a normal nychthemeral routine, have claimed that there is a rise in blood pressure before waking from a minimum at about 03⁰⁰ to a maximum at about 10⁰⁰ and then a slower fall thereafter (*Fig.* 3.1). In the study by Richardson et al. (1964), maximum rates of rise and fall were found at 04⁰⁰ to 08⁰⁰ and 20⁰⁰ to 24⁰⁰ respectively and highest

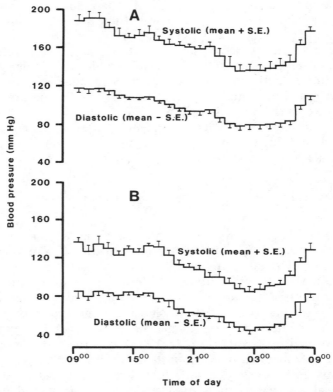

Fig. 3.1. Hourly mean systolic and diastolic blood pressures throughout the 24 hours: A, from 20 untreated hypertensive patients; B, from 5 normotensive patients. (*From* Millar-Craig et al., 1978, Figs. 1 and 3.)

values at 16⁰⁰ to 20⁰⁰. Even though there is disagreement as to time of peak, rapid changes preceding waking and sleeping are indicated by these authors, the inference being that an endogenous factor is operating.

Similar studies by other groups have produced different results. Littler and Watson (1978) and Floras et al. (1978) found only a slight rise before waking and a large and rapid rise immediately after, implying a large exogenous effect of sleep; to correct for the different times of waking of their subjects they expressed the results with reference to time of waking (*Fig.* 3.2A). When the same data were expressed in clock hours (*Fig.* 3.2B),

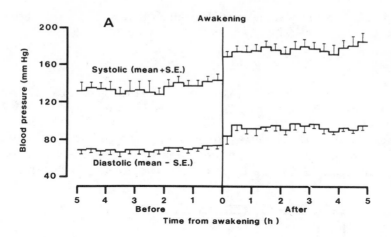

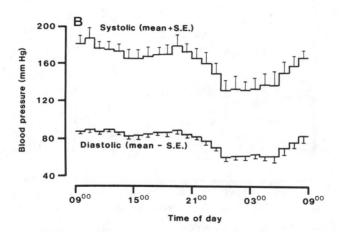

Fig. 3.2. Mean systolic and diastolic blood pressures (± 1s.e.) in 14 un-treated hypertensive subjects recorded at 20-min intervals. A, Blood pressures represented with reference to the subjects' waking times; B, the same data but expressed in clock hours. (*From* Floras et al., 1978, Fig. 1.)

since times of waking showed a spread between different subjects, these results looked remarkably like those of Millar-Craig and his colleagues. Even though this can be interpreted to indicate that irregularity of waking times produced the spurious 'pre-waking' rise found in the study shown in *Fig.* 3.1, factors other than waking alone are implicated. Thus, Raftery and Millar-Craig (1978) assert that blood pressure rose before *any* of the group had woken, and falls of blood pressure during the late evening before sleep have been observed, especially in the study of Richardson et al. (1964). A resolution of this problem is still awaited. In a recent study of the circadian rhythm of blood pressure (Mann et al., 1979), Millar-Craig's group stress the pre-waking rise less but conclude that 'the rapid post-waking rise is not entirely dependent on physical activity'.

I.3. Pulse Rate

Slightly more data relating to pulse rate are available and, like the data for the other cardiovascular variables, these have been reviewed by Smolensky et al. (1976).

Pulse rate is lower at night than during the day in diurnally active subjects and assessment of the rhythm by cosinor analysis indicates acrophases about 13 hours after midsleep. The use of midsleep rather than

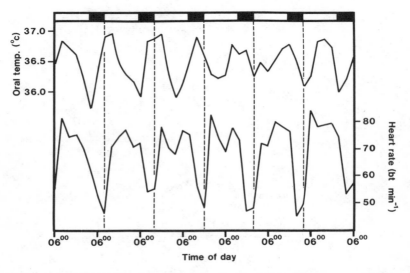

Fig. 3.3. Oral temperature and heart rate of a subject living on a 28-hour routine. Data represent 4-hourly means from three 7-day cycles of the 28-hour routine. The abscissa is divided into real days and the dashed vertical lines represent the ends of 28-hour periods. Black bars represent the times when the subject was asleep. Note a peak in oral temperature occurs for every real day whilst for heart rate a peak occurs on each artificial day. (*From* Kleitman and Kleitman, 1953, Fig. 4).

clock time as a reference point once again indicates the importance of the sleep process (Halberg et al., 1969; Smolensky et al., 1976).

Some of the strongest evidence in favour of there being a dominant exogenous component in the rhythm of pulse rate comes from the study by Kleitman and Kleitman (1953), the temperature data from which have already been discussed (Chapter 2). The three subjects lived in the constant light of the Arctic summer on 18- or 28-hour routines. *Fig*. 3.3 shows some results from the subject whose temperature rhythm did not adjust well to the non-24-hour day. There is a clear dissociation between temperature (with a number of peaks equal to the number of 'real' days) and pulse rate, this latter following the 28-hour routine. Similarly, a rhythm with a period of less than 24 hours was found in one astronaut whose activity rhythm too had a similar period of under 24 hours (Rummel, 1974).

However, other evidence exists that there is also at least some endogenous influence upon the circadian rhythm in heart rate:

1. In the experiments by Wertheimer et al. (1974) and Reinberg et al. (1970) (in which posture and meal times were regulated, *see above*, s. I.2.1) the results for heart rate, like those for blood pressure, showed the same time of peak but a decreased amplitude and so similar inferences can be drawn. Additionally, Halberg et al. (1970) have compared two groups of healthy subjects, both confined to bed for the nychthemeron, one of whom was allowed diurnal isometric exercise; no significant difference between the circadian rhythms in the two groups was found. In other words, the circadian rhythm cannot be attributed to rhythms of food intake or activity.

2. The general finding in free-running experiments is that the pulse rhythm continues with a period in excess of 24 hours and is indistinguishable from those of the sleep—wakefulness cycle and deep body temperature (Halberg et al., 1965; Siffre et al., 1966; Schaefer et al., 1967). The inability to distinguish in these data between an independent endogenous rhythm and one driven by the rhythm of sleep—wakefulness has been mentioned in Chapter 2. Data for when internal desynchronization has taken place or the pattern of sleep—wakefulness becomes irregular (*see* Chapter 5) have not yet been reported. Even then, it would be necessary to show that the rhythm was not due to that in core temperature before an independent internal oscillator for pulse rate was established. Similarly, the difficulty of demonstrating that the rhythm of pulse rate is not due to that of sleep—wakefulness or temperature exists when results from shift work (Schaefer et al., 1979) or time-zone transitions (Hauty and Adams, 1966a, b) are considered.

By none of these approaches can the endogenous rhythm in pulse rate yet be distinguished from that in some other variable (for example, temperature) which possesses an endogenous component and might reasonably be expected to exert an influence.

3. Millar-Craig et al. (1978) and Floras et al. (1978) have both recorded heart rate as well as blood pressure in ambulatory subjects. As for blood pressure, a rise in heart rate before waking is claimed by the former group and denied by the latter. This disagreement also awaits resolution.

4. A result from cardiac transplant patients also suggests that some factor other than an exogenous influence can affect heart rate. Smolensky et al. (1975) describe a case where a recipient's heart developed a period of considerably more than 24 hours while the donor heart continued with

a 24-hour rhythm. The explanation of this difference is uncertain. (The authors speculated that the donor heart was responding to blood-borne catecholamines whereas the recipient's heart might have been more affected by the many drugs administered.) The important point for the present is that the rhythm of the recipient's heart was presumed to possess a period unlike that of any other variable and so this rhythm would be 'endogenous' and not driven by another variable.

I.4. Other Cardiovascular Variables

The data here are very limited and again have been reviewed by Smolensky et al. (1976). The experiments have rarely been performed in circumstances that enable an estimate to be made of the size of any endogenous component.

In brief, therefore, other measures of heart function (for example length of systole, stroke volume, left ventricular ejection time and QRS interval of the electrocardiogram) all show circadian rhythms under nychthemeral conditions which indicate peak activity at or just after the time of mid-activity. Some confirmation of this comes from measurements of cardiac output summarized by Smolensky et al. (1976) and in the recent work of Miller and Helander (1979) using thoracic impedance cardiograms. In these studies the time of maximum output was generally about the late afternoon with considerably lower values during the night. Another important cardiovascular variable is blood flow. In Chapter 2 the importance of distal vasculatures in circadian changes of deep body temperature was described. It was also pointed out that these changes were not observed in blood flows to the trunk, arms and legs. Measurement of blood flow through the forearm by venous occlusion plethysmography (Kaneko et al., 1968; Smith and Malyj, 1974) has indicated peak values in the late afternoon. Since the flow through the trunk, arms and legs is greater than that through the extremities, *in toto,* blood flow is maximum at a time coincident with the acrophases of most other cardiovascular variables.

I.5. The Causal Nexus between Cardiovascular Variables

The preceding sections have indicated that in nychthemeral conditions rhythms of the cardiovascular variables are coincident. This naturally leads to the speculation that this coincidence results not by chance but because the different variables are driven by some common mechanism. Stronger evidence for this view would come from a demonstration of simultaneous changes in response to changed routines such as time-zone shifts, non-24-hour days etc. Evidence such as this is rare; Hauty and Adams (1966a, b) measured many variables, including heart rate and blood pressure, before and after a series of time-zone transitions. Unfortunately the irregularity of the data does not permit it to be ascertained whether the two rhythms adjusted at the same rate. Even if they had, it would then not be possible to distinguish between one rhythm causing the other and both being caused by some other factor or factors.

There has been speculation as to what these common factors might be:

I.5.1. The autonomic nervous system

Since the autonomic nervous system exerts a profound effect upon the cardiovascular system it is very reasonable to consider what role it might play in its circadian rhythms. Not surprisingly, recordings of the autonomic nerve outflow in humans have not been reported and so a direct comparison between the rhythms and nervous system is not possible. However, Gautherie (1973) has classified subjects as 'sympathicotonics', 'vagotonics' or 'equilibrates' on the basis of their response to adrenaline, atropine and pilocarpine, the first group having the greatest amount of sympathetic tone, and so on. When systolic and diastolic blood pressures are considered, there are small differences in acrophase between the different groups (earliest for 'sympathicotonics', latest for 'vagotonics') and between the mean and amplitude of the best-fitting cosine curves (those for the 'sympathicotonics' are highest, those for the 'vagotonics' lowest). It is argued that since the acrophase for 'sympathicotonic' subjects is earliest, the morning rise in blood pressure is brought about mainly by increasing sympathetic discharge. Some evidence that might be adduced in favour of this view is the observation that the diurnal rise of blood pressure in hypertensives can be reduced by administering a β-blocker (de Leeuw et al., 1977); data for normotensives do not seem to be available.

An indirect way of assessing autonomic nervous system function is to measure the concentration of catecholamines in plasma or urine. The results of such measurements will be considered in detail in Chapter 6; in brief, catecholamine concentrations are generally highest at about noon, but reports that place the maximum values at both earlier and later times have appeared.

Two problems are raised when the possible significance of the simultaneity between cardiovascular rhythms and catecholamine levels is considered. First, one cannot decide whether the cardiovascular system is responding to the circulating levels of catecholamines (for example, Reinberg et al. (1970) have proposed that circulating adrenaline is responsible for the diurnal rise in blood flow) or whether the cardiovascular and catecholamine rhythms are two independent manifestations of some other rhythmic process. Secondly, the simultaneity is not exact since not all cardiovascular variables peak at identical times; for example, the claim that blood pressure is highest immediately after waking and falls thereafter would be difficult to correlate with most estimates of the time of peak of catecholamine concentration. This last difficulty, together with the observation that blood pressure rises more rapidly than heart rate after waking (Raftery and Millar-Craig, 1978), suggests that catecholamine concentrations alone cannot be involved.

I.5.2. The temperature rhythm and general 'arousal'

A direct effect of the blood temperature upon at least the sino-atrial node of the heart is to be expected. Kleitman and Ramsaroop (1948) have shown a close parallelism between heart rate and oral temperature and the heart rate of a patient is often used as part of the assessment as to whether or not a fever is present. This parallelism is generally taken to imply some causal connection between the two. A similar conclusion can be drawn from the observation that heart rate paralleled the core temperature in an exercising human who was pharmacologically denervated (Mills, 1973, p. 44). But the dissociation between pulse and temperature found in the Arctic studies of Kleitman and Kleitman (1953) (see Fig. 3.3) indicates that the cardiovascular rhythms are normally less influenced by temperature than by activity rhythms, even though these, in turn, are rarely independent (see Chapter 5).

The effect that the temperature rhythm might exert upon other cardiovascular variables, for examples, cardiac output and blood flow, is not known but a more general problem is raised when the effects of this rhythm are considered. Undoubtedly the temperature rhythm will exert some effect upon the brain itself and there seems to be no reason to exclude an effect upon the autonomic nervous system. Conversely it is not known to what extent the change in 'set-point' of the thermoregulatory centre in the hypothalamus is influenced by the autonomic nervous system or even to what extent both the thermoregulatory and autonomic nervous systems are affected by rhythms in 'arousal' originating from the reticular formation. Such an interrelationship between the reticular formation, the hypothalamus and the autonomic nervous system would be exceedingly difficult to unravel. By way of an example, consider the three observations: (a) that the rhythms of sleep—wakefulness, temperature and heart rate develop in the neonate at about the same age (see Chapter 8); (b) that patients suffering from drug-induced coma showed changed rhythms in rectal temperature, heart rate and systolic blood pressure (Reinberg, Gervais et al., 1973); (c) that the rise in pulse rate associated with assuming an upright posture shows a circadian rhythm (Klein et al., 1968; Menzel, 1942). These could all be used to support theories that attribute the endogenous components of circadian rhythms in the cardiovascular system to the level of catecholamines, to core temperature or to general 'arousal'. Equally these pieces of evidence do not enable distinctions to be made between different models of the causal nexus that exists between these three factors. Finally, there are some observations of a temporary dip in systolic blood pressure about 14^{00} (Zülch and

Hossmann, 1967) at the same time as a 'post-prandial dip' in psychometric performance. As will be argued in Chapter 6, this dip in performance is believed to be a further example of changes in 'arousal'.

II. RESPIRATORY SYSTEM

In general, circadian rhythms in the respiratory system have been studied even less than those of the cardiovascular system. The reason seems to be a combination of two factors. First, since the respiratory system is much influenced by external factors, the rhythms often do little more than reflect these exogenous influences, and secondly, unlike those required for the cardiovascular system, the measurements or samples required are often inconvenient to obtain outside the laboratory and rather time-consuming.

II.1. Oxygen Consumption, Respiratory Frequency and Blood Gas Partial Pressures

The rate of oxygen consumption in a resting subject under nychthemeral conditions shows a circadian change of about 10 per cent with a minimum at about 04^{00}. In part such a result stems from all the accompaniments of sleep and lack of feeding during the night, and wakefulness and meal times during the daytime; but, as *Fig.* 3.4 shows, a rhythm with a similar timing of the minimum is still found when fasting subjects sleep at an unusual time (Bornstein and Völker, 1926). *Fig.* 3.4 indicates also that this sleep produces a direct decrease in oxygen consumption. As was argued in Chapter 2, the rhythm in metabolic rate is believed to reflect the rhythm in deep body temperature rather than to be an important contribution to it (Mills, 1973). In connection with this, Klein et al. (1972) studied a group of subjects undergoing flights across six time-zones in both eastward and westward directions. A fall in metabolic rate following the flight was found in the daytime that was coincident with pre-shift sleep time; it took about 4 days for a nocturnal minimum to reappear. This was a time course similar to that required for a re-establishment of an appropriately phased temperature rhythm and again suggests a link between the two rhythms of temperature and metabolic rate.

Investigations of respiratory frequency under both nychthemeral and free-running conditions (Schaefer et al., 1967) show a similar time course for this rhythm and that of core temperature. This parallelism applies in particular to the nocturnal values with the minimum for both rhythms appearing about the middle of sleep during nychthemeral conditions and at sleep onset during free-running circumstances. However, later in this experiment there was some evidence for dissociation between the two rhythms, that in respiratory frequency (but not core temperature) perhaps surprisingly diverging from the pattern of sleep—wakefulness. Experiments in which time-zone transitions have been carried out (Hauty and Adams,

1966a, b) have indicated similar time courses of adaptation to the new time zone for both temperature and respiratory frequency. However the sleep–wakefulness cycle plays an important role too, as is demonstrated by other studies by Schaefer et al. (1979) upon shift workers living on an 18-hour day (12 hours awake, 6 hours asleep). As time on this system

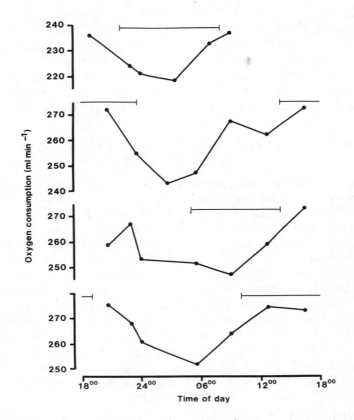

Fig. 3.4. Oxygen consumption of a subject fasting and recumbent, but sleeping at different hours on different occasions. Horizontal bars indicate the times of sleep. (*From* Mills, 1973, Fig. 2.4; data of Bornstein and Völker, 1926.)

passed, the frequency of respiration showed a decreased rhythm with a period of 24 hours and an increase in 18-hour periodicity.

Night time is accompanied by a fall in ventilation that is greater than the decrease in metabolic rate, the result of which is that alveolar P_{CO_2} rises and P_{O_2} falls (Conroy and Mills, 1970). This result cannot be attributed wholly to the effect of sleep, since Mills (1953) demonstrated that a small rise in P_{CO_2} takes place even if the subject stays awake.

II. 2. Mechanical Factors

Some data exist from subjects following nychthemeral routines for measurements of airway resistance (peak expiratory flow rate, FEV_1), vital capacity and lung compliance (Reindl et al., 1969; Conroy and Mills, 1970; Reinberg and Gervais, 1972; Gaultier et al., 1977). Unfortunately, there is not always agreement as to the times of peak when measured by the different groups. This probably reflects not only the few sampling times in some of these studies but also the susceptibility of these rhythms

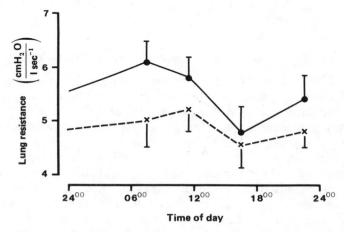

Fig. 3.5. Circadian variations in lung resistance measured in 7 children (means ± 1 s.e.). Lung resistance measurements made before (●) and 10 minutes after (X) inhalation of a β-agonist (2 mg orciprenaline). (*From* Gaultier et al., 1977, Fig. 1.)

to exogenous factors. Airway resistance, whether assessed by FEV_1 (Reinberg and Gervais, 1972) or by peak expiratory flow rate (Reindl et al., 1969; Gaultier et al., 1977) has been found to decline during the hours after waking. The cause of this rhythm in pulmonary resistance is uncertain but a role for the autonomic nervous system has been suggested. Evidence in favour of this view is that adrenaline is known to act as a bronchodilator and administration of a β-agonist or vagolytic agent reduces airway resistance to a value that is approximately equal to minimum values found during the nychthemeral rhythm, as is shown in *Fig.* 3.5 (Gaultier et al., 1977).

If the autonomic nervous system along were responsible for the normal rhythm of airway resistance then a peak in activity of this system after noon rather than at about the time of waking would seem most appropriate (but *see* Chapter 6). Since airway resistance is raised in asthmatics and there appear to be more or worse attacks around the hours of midnight (Clark and Hetzel, 1977), at a time when endogenous cortisol levels are lowest (*see* Chapter 7), a possible link between the two rhythms has been postulated (Reinberg et al., 1963). It is tempting to speculate that the rhythm in airway resistance is influenced by both the rhythmicity of the autonomic nervous system and by substances such as histamine, the release of which is in turn controlled by the rhythm of the natural anti-inflammatory agent, cortisol (but *see* Barnes et al., 1980).

Experiments upon asthmatic subjects (Hetzel and Clark, 1979) have indicated that the normal nocturnal changes in airway resistance can be modified, but not overridden, by spells of waking or exercise. The inference is that the rhythm of airway resistance has both endogenous and exogenous components.

II.3. The Effect of Exercise at Different Times of Day

Klein and Wegmann (1979) have recently reviewed the subject. As their review indicates, the findings are complex, depending upon the physical fitness of the subject as well as the severity and duration of the exercise itself. During light exercise, the heart rate and rate of oxygen consumption are higher by day and lower by night, times of maximum and minimum response being about 12^{00} to 18^{00} and 03^{00} to 05^{00} respectively. With higher workloads the amplitude of the rhythms diminishes because nocturnal values fall below diurnal ones by a smaller amount. Under maximal exercise conditions, the heart rate rhythm is absent and that of oxygen consumption might even invert, nocturnal use now being greater than that during the daytime. This finding relating to the respiratory response accords with the view that at night the body is less 'efficient', has a smaller work capacity and a greater need of oxygen to fulfil this capacity (*see also* Hildebrandt and Engel, 1972). Subjectively also, exercise at night is hardest (Costa et al., 1979; Ilmarinen et al., 1980).

The results with light exercise loads are potentially of more interest since they enable the sensitivity of the cardiovascular and respiratory systems to be estimated. (High exercise loads are less useful due to the intrusion of complicating factors such as stress and motivation.) The circadian rhythms in the responses to exercise are often found to be far less marked if, instead of considering the exercise values directly (as above), the response is assessed as the difference from resting or zero-load (pedalling but doing no external work) values (Davies and Sargeant, 1975). However, Cohen and Muehl (1977) stated that exercise amplified the resting circadian rhythms, that is, the *difference* between exercise and resting values showed a circadian rhythm with a daytime peak. This is shown most convincingly in a study cited by Hildebrandt and Engel (1972) who measured the increment over resting values in pulse rate that was produced by different work loads. Some of their results are shown in *Fig.* 3.6. At the moment, whether the difference between resting (or zero-load) and exercise conditions is independent or not of the time of

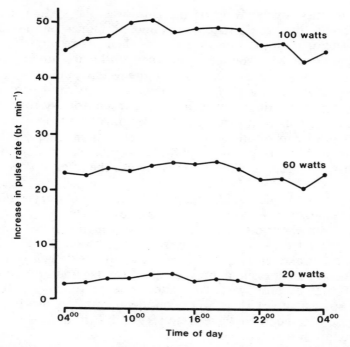

Fig. 3.6. Circadian variation in the increase in pulse rate over the resting level for three levels of leg work performed on a bicycle ergometer. (Data of Voigt et al., 1968.)

day is not yet clear; the result might well be influenced by factors such as the severity of the exercise (*see Fig.* 3.6). One implication of this controversy will be considered in the next section.

II.4. The Causal Nexus

From this brief account of the respiratory system, the similarity between its rhythms and those of the cardiovascular system is obvious enough but the lack of data has made it possible only to suggest that there might be some endogenous component. This component is not invariably present; thus in nychthemeral conditions a rhythm in the ventilation/perfusion ratio of the lung has been measured but this rhythm is absent in conditions of continuous recumbency and so is believed to result from changes in posture alone (Abernethy et al., 1967).

When attempts are made to understand the causes of the endogenous components of respiratory rhythms, it seems that, as before, rhythms of deep body temperature, the autonomic nervous system and reticular formation must be considered. These endogenous components could be influenced by exogenous factors as well as by the direct effect of the sleep—wakefulness cycle. Whereas the cardiovascular system is obviously much affected by the efferent nerves and hormones of the autonomic nervous system, these influences are less when the respiratory system is considered; possible exceptions are their influence upon mechanical aspects such as airway resistance and the more recently described possibility that the sensitivity of the peripheral chemoreceptors can be controlled by efferent sympathetic fibres. Instead, for the respiratory system, it is likely that the endogenous components of the rhythms arise from the combined effects of core temperature acting upon the brain and the activity in the reticular formation itself. As with the cardiovascular system it is quite impossible with present evidence to distinguish the roles of these different factors.

In support of the view that sensitivity of the respiratory system shows rhythmic changes are the following pieces of evidence:

1. There is a variation in the respiratory response to carbon dioxide (Bulow, 1963). This was assessed by requiring the subject to inhale CO_2 while he was 'awake tensed', 'awake relaxed', 'drowsy', 'lightly asleep' and 'deeply asleep', all measured by standard EEG techniques. A number of indices of the response were calculated, one of which was the carbon dioxide response curve. There were considerable differences in the size of the response obtained and these were related by the author to the degree of wakefulness or depth of sleep.

2. If subjects remain lying but awake during the night, there is a slight rise in alveolar P_{CO_2} (Mills, 1953); this is most easily explained on the assumption that the sensitivity of the respiratory system to carbon dioxide is decreased at night.

3. The observations of rhythmicity in the cardiovascular and respiratory responses to exercise have just been described. If the views of Cohen and Muehl (1977) and Hildebrandt and Engel (1972) are correct (*see Fig.* 3.6), then again an increased diurnal sensitivity of the respiratory system seems a likely explanation.

In conclusion, perhaps it is worth pointing out that by attributing the endogenous components of the cardiovascular and respiratory systems to a combination of rhythms—in deep body temperature, the autonomic nervous system and 'arousal' mediated by the reticular formation—one is in no way forced to site the internal clock in one or more of these areas: what is being acknowledged is that these areas are important components in the nexus between the timing mechanism (wherever it might be and whatever its nature) and the observed rhythms (*see also Fig.* 12.1, p. 273).

III. METABOLIC AND GASTROINTESTINAL RHYTHMS

Metabolic and gastrointestinal rhythms, like those of the cardiovascular and respiratory systems, are much affected by external events, in this case the rhythmic intake of food. Experiments in which food intake is modified in an attempt to investigate the endogenous components of the rhythms lead to two difficulties in interpretation. First, changes of dietary intake, such as fasting or taking meals at abnormal times, will often induce further changes, for example, a switch in metabolism from carbohydrate to fat substrates. The result is that with restricted diets or complete fasting, any rhythm is superimposed upon a baseline that is steadily changing (Reinberg, 1974). Secondly, since there is a 'cephalic phase' of digestion, knowledge that it is really 'dinner time' might well produce a response of the gastro-intestinal tract that is independent of actual food intake. Further, there is evidence that the metabolic rhythms are influenced by the nature of the diet. As a result of all these complicating factors, as the following account will indicate, it is rarely possible to infer much about endogenous factors from the present body of knowledge. Changed metabolic rhythms in certain metabolic disorders will be referred to in Chapter 11.

III.1. Metabolic Rhythms
III.1.1. Glucose and insulin

In healthy fasting subjects, plasma glucose levels remain very constant (Jarrett, 1979) and this is attributed to an effective homeostatic system for 'sparing' glucose and making use of alternative sources of energy; on the other hand, insulin levels in fasting subjects, measured by radioimmuno-assay, are slightly higher in the morning than in the afternoon. In another study by Reinberg, Apfelbaum et al. (1973) a large number of obese patients were placed on a low calorie protein diet for 2 days and ate four isocaloric meals equally spaced throughout the nychthemeron. Average levels of insulin fell compared with controls, but the time of peak, 14^{00} to 16^{00}, was unchanged. These studies suggest that the rhythm of plasma insulin (as well as that of many other variables that were measured in this study) does not require carbohydrate intake, but the possible roles of sleep and wakefulness and other exogenous factors have not been eliminated and so the results do not establish whether the rhythm is truly endogenous.

After a meal it is to be expected that glucose and insulin levels will rise. In accord with this are the results of a study by Goetz et al. (1976), who compared the insulin rhythms in a group of subjects studied first when all food was taken at 07^{00} and then at 17^{00}. The acrophase of the rhythm of plasma insulin was changed by 9½ hours, which suggests a very strong exogenous effect of meal times upon this rhythm. Similar findings have been made by Haus (1976) and Graeber et al. (1978). However, other studies have indicated that the rhythms of both substances in the plasma are not wholly dependent upon meal times. In one study by Ahmed et al. (1976), a 'typical diet' was divided into three isocaloric portions and eaten at normal breakfast, lunch and dinner times. Each meal produced a rapid rise in plasma glucose and insulin levels but the rate of decline of glucose after each meal was less as the day progressed. In the other study, Terpstra et al. (1978) have investigated the effect of a carbohydrate-rich diet divided equally between meals taken at conventional times, at eight equally spaced times or at 2-hour intervals between 09^{00} and 23^{00}. On all regimens glucose showed a post-prandial rise, but there was also a tendency for the average level to rise to a maximum about 20^{00}. Both studies suggest that the body is less able to deal with glucose as the day progresses but detailed comparison of the results is complicated since different diets were used.

A more closely controlled investigation of glucose and insulin changes is possible when the effects of glucose intake alone are considered; this is the basis of the glucose tolerance test. The results of this test indicate a poorer tolerance to glucose later in the day (see, for example, Jarrett and Keen, 1970; Sensi, 1974; Gibson et al., 1975). Since this effect is observed also when the glucose is administered intravenously it cannot be attributed to changed uptake characteristics from the gut. The result seems to arise from two components. The first is that for any rise of plasma glucose, more insulin is released in the morning (compare with the observation of a raised morning insulin level in fasting subjects) and the second is that a given dose of insulin administered intravenously is more effective in the morning, even though its rate of clearance from plasma does not seem to depend upon the time of day (Sensi and Capani, 1976; Coscelli et al., 1978; Jarrett, 1979).

The factors that might be responsible for these two components are not known. Effects directly due to the diabetogenic hormones (adrenaline, cortisol, growth hormone, glucagon) seem unlikely. Thus growth hormone is rarely released until sleep (see Chapter 7) and plasma glucagon is reported not to show circadian rhythmicity (Reinberg, Apfelbaum et al., 1973; Ahmed et al., 1976; Jarrett, 1979). However, plasma cortisol levels fall throughout the day (Chapter 7) and they might exert some influence. This speculation does not seem to have been tested, just as a possible role of the autonomic nervous system in contributing to the circadian release of insulin has not yet been investigated (Ungar and Dobbs, 1978).

In an experiment upon a group of subjects living without timepieces in the Arctic, the subjects' sleep—wakefulness cycle spontaneously lengthened. When, as a result, the routine of sleep and wakefulness had become reversed, glucose tolerance was reduced at 08^{00} real time (when the subjects had been awake 10 hours) when compared with 16^{00} real time (when the subjects had just woken) (Campbell et al., 1975). The argument that in some way the pattern of sleep and wakefulness had determined the degree of glucose tolerance is acceptable but not necessarily true; other rhythms had not been measured and any that had changed in phase with the pattern of sleep and wakefulness, assuming internal synchronization, could have been responsible for the changed glucose tolerance.

III.1.2. Cholesterol, lipoproteins, free fatty acids and triglycerides

Rhythms in the concentrations of plasma cholesterol and lipoprotein are probably small in amplitude or insignificant (Conroy and Mills, 1970; Schlierf, 1978; Terpstra et al., 1978). Rhythms in both serum free fatty acids (FFA) and triglycerides have been claimed and there is evidence that, as with glucose, the type of diet influences the rhythmicity that is observed.

Free fatty acids

1. A group of 14 healthy males was given 50 g glucose at 24^{00} or 09^{00}, fasted for 8 hours and then given a standard intravenous dose of insulin. The fasting FFA level immediately before the insulin injection was higher at 17^{00} than at 08^{00} but at both times the levels were decreased to similar minima with similar time courses by the insulin injections (Gibson et al., 1975).

2. In a series of experiments already described (Terpstra et al., 1978), in which the subjects were on a carbohydrate-rich diet and different patterns of meal times were compared, the FFA levels were described as 'variable' whatever the number and distribution of meals. In later experiments in which a fat-rich diet was used instead (Van Gent et al., 1979), diurnal peaks accompanied diurnal feeding regimens, but no circadian rhythm was found when evenly spaced meals were taken.

3. Three-hourly isocaloric sucrose meals were associated with no rhythm in FFA, but with overnight fasting there was a nocturnal maximum (Barter et al., 1971).

4. FFA have been found to be higher at night than in the daytime, but the size of the diurnal fall was less if the diet was rich in fat rather than carbohydrate (Schlierf, 1978).

Triglycerides

1. Claims have been made for higher values at night than in the daytime when on low-fat diets (Schlierf, 1978). However, other studies have indicated higher diurnal values when on carbohydrate-rich diets (Maruhama et al., 1967; Barter et al., 1971; Terpstra et al., 1978), or fat-rich diets (Van Gent et al., 1979). These last results cannot be attributed solely to the effect of meals, since the rise commenced before the first meal and was not prevented by the last meal.

2. Diet might affect the amplitude of the rhythm also. Thus Kuo and Carson (1959) compared triglyceride rhythms in subjects on a normal diet, one rich in corn oil and one of rice and fruit. Meals were taken at normal times. All diets (both fat-poor and fat-rich) showed post-prandial rises and later dips, with low nocturnal values, but the diurnal rises were smallest in the case of the diet of rice and fruit.

From these results it is clear that both the timing and composition of meals influences the lipid rhythms in a complex way. Hardly any comment can be made relating to the extent to which these rhythms can be said to have any endogenous component since investigations of subjects on abnormal routines etc. do not seem to have been performed. Even if some endogenous component does exist, then we are quite ignorant of the role played by different hormones and the autonomic nervous system in its production.

III.1.3. Amino acids

Feigin et al. (1968) considered the rhythms of the many amino acids in healthy volunteers on a normal diet eaten at conventional times. Other work has substantially confirmed their findings (for example, Wurtman et al., 1968), namely that there is a minimum of concentration at about 04^{00} with maximum values in the late afternoon. Some studies have claimed post-prandial rises but other factors must be involved as the evening meal does not prevent the evening decline (Fernstrom, 1979). Further the ratio of aromatic to large neutral amino acids falls as the total amount of protein in the diet is increased (Fernstrom et al., 1979), the possible significance of which will be considered in Chapter 12.

The effect of dietary content upon amino acid rhythms, especially tyrosine and tryptophan, has been considered in some detail and is reviewed by Fernstrom (1979). He has studied subjects taking three equal meals per day at 08^{00}, 12^{00}, 17^{00}. The diet was balanced for fat and carbohydrate but had high (150 g), fairly low (75 g) or very low (0 g) protein intake. On the high intake (normal intake is 100+ g) the 'normal' diurnal peak was seen; on the very low intake the rhythm was inverted; on the intermediate intake it was absent (*Fig.* 3.7). The level of an amino acid is determined by the balance between influx into the plasma pool (due to the diet) and loss into cells (due to the action of insulin released by ingestion of food). It is proposed that this latter factor will be relatively more important on the low protein intake but less so with high protein intake. This is because, although food intake will stimulate insulin release and amino acid removal in both cases, there will be a greater amino acid influx with the high-protein diet. Such a result might explain the findings of Grundig et al. (1978) in which nocturnal peak times were reported and, as Fernstrom (1979) points out, this theory predicts that diets high in carbohydrate would be expected to decrease diurnal plasma amino acid concentrations.

From the data available it is not known whether an endogenous rhythm of plasma amino acid concentrations exists. If it did then one possible cause would be rhythms in metabolism of amino acids by the liver and other tissues, probably under the influence of hormonal rhythms. In rats, there is evidence that the catabolism of amino acids by the liver shows a

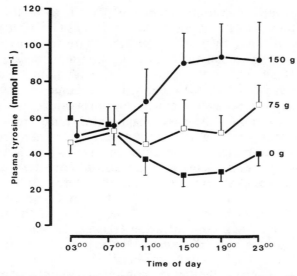

Fig. 3.7. Relation between dietary protein content and plasma tyrosine concentration in normal human subjects. Subjects consumed three meals per day at 08⁰⁰, 12⁰⁰ and 17⁰⁰, each containing 0, 25 or 50 g of protein. Each diet was ingested for 4 consecutive days. Plasma samples were collected on the third and fourth days and analysed for tyrosine. Data represent means ± standard deviations. (*From* Fernstrom, 1979, Fig. 1. Reproduced from Krieger D. T. (ed.) *Endocrine Rhythms,* 1979, by kind permission of Raven Press, New York.)

circadian rhythm, but evidence in the human does not yet exist. We are also ignorant of the extent to which circadian rhythms in gluconeogenic and anabolic hormones might influence amino acid levels in the plasma.

III.2. Gastrointestinal Rhythms

Since gastrointestinal function is strongly linked to food intake, under normal circumstances a circadian rhythm with a diurnal maximum is to be expected. The extreme inconvenience associated with most techniques for sampling gut activity has effectively prevented anything but the most rudimentary observations in circadian functions being made. Not surprisingly, there are no observations after manoeuvres designed to separate the endogenous and exogenous components of these rhythms. The meagre amount of data (*see* Conroy and Mills for a review up to 1970) generally supports the view that activity is greatest after a meal. More recent data agree with such a view. Thus, Emonts et al. (1979) used a non-reabsorbed marker to measure the flow of fluid along the small intestine in healthy subjects with normal eating habits and Jorde et al. (1980) measured plasma gastric inhibitory polypeptide concentrations in similar conditions; in both studies marked post-prandial changes were observed.

A possible exception are the circadian rhythms of saliva. Both whole saliva and the separate secretions from the parotid and submandibular glands can be obtained with comparative ease. Dawes (1974) has reviewed the field up to that date. In brief, it appears that, under nychthemeral conditions, there are circadian rhythms in both stimulated and unstimulated secretions of whole, parotid gland and submandibular gland salivas (*see also* Ferguson and Botchway, 1979a, b, 1980). Some of these rhythms are of flow, protein and potassium concentrations (which peak in the late afternoon) and of sodium, chloride and magnesium concentrations (which peak on waking or during the night if samples were taken then), results that are confirmed by cosinor analysis. The cause of the rhythms is not known. Dexamethasone suppression of cortisol secretion did not alter the circadian rhythm of protein concentration in either stimulated or unstimulated parotid saliva (Dawes and Ong, 1973), but another adrenal steroid, aldosterone, is believed to be important in the genesis of the rhythms of both sodium and potassium concentration (Dawes, 1974). The potential use of salivary rhythms as a diagnostic aid in both systemic and oral disorders has been described (Dawes, 1974, 1975).

If one attempts to assess whether or not an endogenous component is present in gastrointestinal rhythms, it is necessary to rely upon indirect evidence. Cagnoni et al. (1974) assayed the amount of gastric juice secreted at 2-hour intervals in control subjects and in patients who had undergone vagotomy for duodenal ulcers; meals were taken at conventional times. In both groups, acid secretion showed a diurnal rhythm with peak values about 20^{00}. The observation that the same rhythm of secretion was present in the vagotomized group indicates that it was not mediated by neural mechanisms alone. Moore and Englert (1970) had observed a similar time of peak gastric juice secretion in an earlier study. In this work subjects did not eat or smell food and received an intravenous infusion of 5 per cent dextrose. The cause of such a rhythm could not be mechanical or chemical and the authors considered that thinking about food would have produced a peak earlier than 20^{00}. Hormone levels were not measured in this study.

When these results, together with the claim that there is a circadian rhythm of gastrin release (Feurle et al., 1972) are considered, it is possible to consider that the circadian rhythms of gastrointestinal activity have endogenous and exogenous components, the latter being affected by mechanical, neural and humoral influences. The observation that a group of patients with pancreatic cannulas receiving 15–30 ml water per hour by mouth showed rhythms of low amplitude in volume, and in enzyme and bicarbonate secretion, but that the amplitude was accentuated when light meals were taken, accords with this view (Lines and Ranger, 1969).

A study of gastric contractions in 7 subjects during sleep (Lavie et al., 1978) indicated an ultradian rhythm with a period of about 100 minutes (but one which bore no obvious relation to the REM–non-REM cycle, *see* Chapter 5). Similar ultradian rhythmicity has been found in fasting subjects when awake (Hiatt and Kripke, 1975). In Chapter 12 the possible link between ultradian and circadian components will be considered.

In summary, therefore, factors such as meal times are dominant when metabolic and gastrointestinal rhythms are considered, but a small endogenous component to these rhythms might exist. Like the cardiorespiratory system, the gastrointestinal system is much affected by the autonomic nervous system but hormones and humoral factors also affect the gut and are of great importance in metabolism. The extent to which these factors might be involved in producing the small endogenous component is not known, and it would seem that the solution to problems as complex and as difficult to investigate as these is still far off. Nevertheless, as Chapters 9 and 10 will indicate, gastrointestinal disorders are a frequent concomitant of changed routines. Not only should this result be a spur to investigations in this field, it also suggests that the rhythms are not wholly exogenous since adaptation to these changed regimens is not immediate.

References

Abernethy J. D., Maurizi J. J. and Farhi L. E. (1967) Diurnal variations in urinary-alveolar N_2 difference and effects of recumbency. *J. Appl. Physiol.* **23**, 875–9.

Ahmed M., Gannon M. C. and Nuttall F. Q. (1976) Post-prandial plasma glucose, insulin, glucagon and triglyceride responses to a standard diet in normal subjects. *Diabetologia* **12**, 61–7.

Barnes P., Fitzgerald G., Brown M. et al. (1980) Nocturnal asthma and changes in circulating epinephrine, histamine, and cortisol. *N. Engl. J. Med.* **303**, 263–7.

Barter P. J., Carroll K. F. and Nestel P. J. (1971) Diurnal fluctuations in triglyceride, free fatty acids and insulin during sucrose consumption and insulin infusion in man. *J. Clin. Invest.* **50**, 583–91.

Bornstein A. and Völker H. (1926) Über die Schwankungen des Grundumsatzes. *Z. Ges. Exp. Med.* **53**, 439–50.

Bulow K. (1963) Respiration and wakefulness in man. *Acta Physiol. Scand.* Suppl. **209**, 1–110.

Cagnoni M., Tarquini B., Orzalesi R. et al. (1974) Gastric function in a group of vagotomized patients with duodenal ulcer and in a control group: a temporal study. In: Scheving L. E., Halberg F. and Pauly J. E. (ed.) *Chronobiology.* Tokyo, Igaku Shoin, pp. 98–9.

Campbell I. T., Jarrett R. J. and Keen H. (1975) Diurnal and seasonal variation in oral glucose tolerance: studies in the Antarctic. *Diabetologia* **11**, 139–45.

Clark T. J. H. and Hetzel M. R. (1977) Diurnal variation of asthma. *Br. J. Dis. Chest* **71**, 87–92.

Cohen C. J. and Muehl G. E. (1977) Human circadian rhythms in resting and exercise pulse rates. *Ergonomics* **20**, 475–79.

Conroy R. W. T. L. and Mills J. N. (1970) *Human Circadian Rhythms.* London, Churchill.

Coscelli C., Alpi O. and Butturini U. (1978) The response of glucose, insulin and FFA to intravenous glucose and glucagon in elderly subjects in the course of morning and afternoon tests. *Chronobiologia* **5**, 387–90.

Costa G., Gaffuri E., Perfranceschi G. et al. (1979) Re-entrainment of diurnal variation of psychological and physiological performance at the end of a slowly rotated shift system in hospital workers. *Int. Arch. Occup. Environ. Health* **44**, 165–75.

Davies C. T. M. and Sargeant A. J. (1975) Circadian variation in physiological responses to exercise on a stationary bicycle ergometer. *Br. J. Indust. Med.* **32**, 110–14.

Dawes C. (1974) Rhythms in salivary flow rate and composition. *Int. J. Chronobiol.* **2**, 253–79.

Dawes C. (1975) Circadian rhythms in the flow rate and composition of unstimulated and stimulated human submandibular saliva. *J. Physiol.* **244**, 535–48.

Dawes C. and Ong B. Y. (1973) Circadian rhythms in the concentrations of protein and the main electrolytes in human unstimulated parotid saliva. *Arch. Oral Biol.* **18**, 1233–42.

Delea C. S. (1979) Chronobiology of blood pressure. *Nephron* **23**, 91–7.

Emonts P., Vidon N., Bernier J.-J. et al. (1979) Étude sur 24 heures des flux liquidiens intestinaux chez l'homme normal par la technique de la perfusion lente d'un marqueur non absorbable. *Gastroenterol. Clin. Biol.* **3**, 139–46.

Feigin R. K., Klainer A. S. and Beisel W. R. (1968) Factors affecting circadian periodicity of blood amino acids in man. *Metabolism* **17**, 764–75.

Ferguson D. B. and Botchway C. A. (1979a) Circadian variations in flow rate and composition of human stimulated submandibular saliva. *Arch. Oral Biol.* **24**, 433–7.

Ferguson D. B. and Botchway C. A. (1979b) Circadian variations in the flow rate and composition of whole saliva stimulated by mastication. *Arch. Oral Biol.* **24**, 877–81.

Ferguson D. B. and Botchway C. A. (1980) A comparison of circadian variation in the flow rate and composition of stimulated human parotid, submandibular and whole salivas from the same individuals. *Arch. Oral Biol.* **25**, 559–68.

Fernstrom J. D. (1979) The influence of circadian variations in plasma amino acid concentrations on monoamine synthesis in the brain. In: Krieger D. T. (ed.) *Endocrine Rhythms.* New York, Raven Press, pp. 89–122.

Fernstrom J. D., Wurtman R. J., Rand W. M. et al. (1979) Diurnal variations in plasma concentrations of tryptophan, tyrosine and other neutral amino acids: effect of dietary protein intake. *Am. J. Clin. Nutr.* **32**, 1912–22.

Feurle G., Ketterer H., Becker H. et al. (1972) Circadian serum gastrin concentrations in control persons and patients with ulcer disease. *Scand. J. Gastroenterol.* **7**, 177–83.

Floras J. S., Jones J. V., Johnston J. A. et al. (1978) Arousal and the circadian rhythm of blood pressure. *Clin. Sci. Mol. Med.* **55**, Suppl. 4, 395S–397S.

Gaultier C., Reinberg A. and Girard F. (1977) Circadian rhythms in lung resistance and dynamic lung compliance of healthy children. Effect of two bronchodilators. *Resp. Physiol.* **31**, 169–82.

Gautherie M. (1973) Circadian rhythm in the vasomotor oscillations of skin temperature in man. *Int. J. Chronobiol.* **1**, 103–39.

Gibson T., Stimmler L., Jarrett R. J. et al. (1975) Diurnal variation in the effects of insulin on blood glucose, plasma non-esterified fatty acids and growth hormone. *Diabetologia* **11**, 83–8.

Goetz F., Bishop J., Halberg F. et al. (1976) Timing of single daily meal influences relations among human circadian rhythms in urinary cyclic AMP and hemic glucagon insulin and iron. *Experientia* **32**, 1081–4.

Goldberg A. D., Raftery E. B. and Green H. L. (1976) The Oxford continuous blood-pressure recorder—technical and clinical evaluation. *Postgrad. Med. J.* **52**, Suppl. 7, 104–9.

Gordon R. D. and Mortimer R. H. (1973) Time related aspects of arterial blood pressure, plasma volume, and electrolyte homeostasis in man. *Int. J. Chronobiol.* **1**, 25–30.

Graeber R. C., Gatty R., Halberg F. et al. (1978) *Human Eating Behaviour: Preferences, Consumption Patterns, and Biorhythms.* Food Sciences Laboratory, US Army Natick Research and Development Command, Natick, Massachusetts.

Grundig V. E., Kollner U. and Simanyi M. (1978) Tagesschwankungen der Aminosäurenkonzentration im Blutserum des Menschen. *Acta Med. Austriaca* **5**, 8–11.

Halberg F., Johnson E. A., Nelson W. et al. (1972) Autorhythmometry—procedures for physiologic self-measurements and their analysis. *The Physiology Teacher* 1, 1–11.

Halberg F., Reinhardt J., Bartter F. C. et al. (1969) Agreement in endpoints from circadian rhythmometry on healthy human beings living on different continents. *Experientia* 25, 106–12.

Halberg F., Siffre M., Engeli M. et al. (1965) Étude en libre-cours des rythmes circadiens du pouls, de l'alternance veille-sommeil et de l'estimation du temps pendant les deux mois de sejour souterrain d'un homme adulte jeune. *C. R. Acad. Sci. [D] Paris* 260, 1259–62.

Halberg F., Vallbona C., Dietlin L. F. et al. (1970) Human circadian circulatory rhythms during weightlessness in extraterrestrial flight or bedrest with and without exercise. *Space Life Sci.* 2, 18–32.

Haus E. (1976) Pharmacological and toxicological correlates of circadian synchronization and desynchronization. In: Rentos P. G. and Shephard R. D. (ed.) *Shift Work and Health.* Washington, DC, US Department of Health, Education and Welfare, pp. 87–117.

Hauty G. T. and Adams T. (1966a) Phase shifts of the human circadian system and performance deficit during the periods of transition: I. East—West flight. *Aerospace Med.* 37, 668–74.

Hauty G. T. and Adams T. (1966b) Phase shifts of the human circadian system and performance deficit during the periods of transition: II West—East flight. *Aerospace Med.* 37, 1027–33.

Hetzel M. R. and Clark T. J. H. (1979) Does sleep cause nocturnal asthma? *Thorax* 34, 749–54.

Hiatt J. F. and Kripke D. F. (1975) Ultradian rhythms in waking gastric activity. *Psychosom. Med.* 37, 320–5.

Hildebrandt G. and Engel P. (1972) The relation between diurnal variation in psychic and physical performance. In: Colquhoun W. P. (ed.) *Aspects of Human Efficiency —Diurnal Rhythm and Loss of Sleep.* London, The English Universities Press, pp. 231–240.

Ilmarinen J., Ilmarinen R., Korhonen O. et al. (1980) Circadian variation of physiological functions related to physical work capacity. *Scand. J. Work Environ. Hlth* 6, 112–22.

Jarrett R. J. (1979) Rhythms in insulin and glucose. In: Krieger D. T. (ed.) *Endocrine Rhythms.* New York, Raven, pp. 247–58.

Jarrett R. J. and Keen H. (1970) Further observations on the diurnal variation in oral glucose tolerance. *Br. Med. J.* 4, 334–5.

Jorde R., Burhol P. G., Waldum H. L. et al. (1980) Diurnal variations of plasma gastric inhibitory polypeptide in man. *Scand. J. Gastroenterol.* 15, 617–19.

Kaneko M., Zechman F. W. and Smith R. E. (1968) Circadian variation in human peripheral blood flow levels and exercise responses. *J. Appl. Physiol.* 25, 109–14.

Klein K. E. and Wegmann H. M. (1979) Circadian rhythms of human performance and resistance: operational aspects. In: *Sleep, Wakefulness and Circadian Rhythm.* NATO-AGARD Lecture Series No. 105, London, Paris, Toronto, NATO-AGARD, Ch. 2.

Klein K. E., Wegmann H. M. and Brüner H. (1968) Circadian rhythm in indices of human performance, physical fitness and stress resistance. *Aerospace Med.* 39, 512–18.

Klein K. E., Wegmann H. M. and Hunt B. I. (1972) Desynchronization of body temperature and performance circadian rhythm as a result of outgoing and home-going transmeridian flights. *Aerospace Med.* 43, 119–32.

Kleitman N. and Kleitman E. (1953) Effect of non-24-hour routines of living on oral temperature and heart rate. *J. Appl. Physiol.* 6, 283–91.

Kleitman N. and Ramsaroop A. (1948) Periodicity in body temperature and heart rate. *Endocrinology* **43**, 1–20.

Kuo P. T. and Carson J. C. (1959) Dietary fats and the diurnal serum triglyceride levels in man. *J. Clin. Invest.* **38**, 1384–93.

Lavie P., Kripke D. F., Hiatt J. F. et al. (1978) Gastric rhythms during sleep. *Behav. Biol.* **23**, 526–30.

de Leeuw P. W., Falke H. E., Kho T. L. et al. (1977) Effects of beta-adrenergic blockade on diurnal variability of blood pressure and plasma noradrenaline levels. *Acta Med. Scand.* **202**, 389–92.

Lines J. G. and Ranger I. (1969) Diurnal rhythms of pancreatic function. *J. Physiol.* **200**, 57P–58P.

Littler W. A. and Watson R. D. S. (1978) Circadian variation in blood pressure. *Lancet* **1**, 995–6.

Mann S., Millar-Craig M. W., Melville D. I. et al. (1979) Physical activity and the circadian rhythm of blood pressure. *Clin. Sci.* **57**, Suppl. 5, 291S–294S.

Maruhama Y., Goto Y. and Yamagata S. (1967) Diabetic treatment and the diurnal plasma triglyceride. *Metabolism* **16**, 985–95.

Menzel W. (1942) Der 24-Stunden-Rhythmus des menschlichen Blutkreislaufes. *Ergeb. Inn. Med. Kinderheilk.* **61**, 1–53.

Millar-Craig M. W., Bishop C. N. and Raftery E. B. (1978) Circadian variation of blood pressure. *Lancet* **1**, 795–7.

Miller J. C. and Helander M. (1979) The 24-hour cycle and nocturnal depression of human cardiac output. *Aviat. Space Environ. Med.* **50**, 1139–44.

Mills J. N. (1953) Changes in alveolar carbon dioxide tension by night and during sleep. *J. Physiol.* **122**, 66–80.

Mills J. N. (1973) Transmission processes between clock and manifestations. In: Mills J. N. (ed.) *Biological Aspects of Human Circadian Rhythms.* New York, Plenum, pp. 27–84.

Mitchell R. H., Ruff S. C. and Murnaghan G. A. (1979) Computer-aided study of circadian variation of blood pressure in pregnancy. *Irish J. Med. Sci.* **148**, 113–14.

Moore J. G. and Englert E. (1970) Circadian rhythm of gastric acid secretion in man. *Nature* **226**, 1261–2.

Raftery E. B. and Millar-Craig M. W. (1978) Circadian variation in blood pressure. *Lancet* **1**, 996.

Reinberg A. (1974) Chronobiology and nutrition. *Chronobiologia* **1**, 22–7.

Reinberg A., Apfelbaum M. and Assan R. (1973) Chronophysiologic effects of a restricted diet (220 cal/24 h as casein) in young healthy but obese women. *Int. J. Chronobiol.* **1**, 391–404.

Reinberg A. and Gervais P. (1972) Circadian rhythms in respiratory functions, with special reference to human chronophysiology and chronopharmacology. *Bull. Physiopath. Resp.* **8**, 663–75.

Reinberg A., Gervais P., Pollack E. et al. (1973) Circadian rhythms during drug-induced coma (transverse study of rectal temperature, heart rate, systolic blood pressure, urinary water and potassium). *Int. J. Chronobiol.* **1**, 157–62.

Reinberg A., Ghata J., Halberg F. et al. (1970) Rythmes circadiens du pouls, de la pression artérielle, des excrétions urinaires en 17-hydroxycorticosteroides catéchol-amines et potassium chez l'homme adulte sain, actif et au repos. *Ann. Endocrinol. (Paris)* **31**, 277–87.

Reinberg A., Ghata J. and Sidi E. (1963) Nocturnal asthma attacks: their relationship to the circadian adrenal cycle. *J. Allergy* **34**, 323–30.

Reindl K., Falliers C., Halberg F. et al. (1969) Circadian acrophase in peak expiratory flow rate and urinary electrolyte excretion of asthmatic children: phase shifting of rhythms by prednisone given in different circadian system phases. *Rass. Neurol. Veg.* **23**, 5–26.

Richardson D. W., Honour A. J., Fenton G. W. et al. (1964) Variation in arterial pressure throughout the day and night. *Clin. Sci* **26**, 445–60.

Rummel J. A. (1974) Rhythmic variation in heart rate and respiration rate during space flight—Apollo 15. In: Scheving L. E., Halberg F. and Pauly J. E. (ed.) *Chronobiology*. Tokyo, Igaku Shoin, pp. 435–40.

Schaefer K. E., Clegg B. R., Carey C. R. et al. (1967) Effect of isolation in a constant environment on periodicity of physiological functions and performance levels. *Aerospace Med.* **38**, 1002–18.

Schaefer K. E., Kerr C. M., Buss D. et al. (1979) Effect of 18-h watch schedules on circadian cycles of physiological functions during submarine patrols. *Undersea Biomed. Res.* (Submarine Supplement) S81–S90.

Schlierf G. (1978) Diurnal variations in plasma substrate concentration. *Eur. J. Clin. Invest.* **8**, 59–60.

Sensi S. (1974) Some aspects of circadian variations of carbohydrate metabolism and related hormones in man. *Chronobiologia* **1**, 396–404.

Sensi S. and Capani F. (1976) Circadian rhythm of insulin-induced hypoglycaemia in man. *J. Clin. Endocrinol. Metabol.* **43**, 462–5.

Siffre M., Reinberg A., Halberg F. et al. (1966) L'isolement souterrain prolongé. Étude de deux sujets adultes sains avant, pendant et après cet isolement. *Presse Méd.* **18**, 915–19.

Smith R. E. and Malyj W. (1974) Studies of peripheral circulatory rhythms in resting and exercising humans. In: Scheving L. E., Halberg F. and Pauly J. E. (ed.) *Chronobiology*. Tokyo, Igaku Shoin, pp. 733–6.

Smolensky M. H., Kraft I. A., Sothern R. B. et al. (1975) Circadian rhythmicity in heart rate and duration of cardiac cycle (P–P interval) in donor and recipient cardiac tissue of heart homograft patients. *Texas Rep. Biol. Med.* **33**, 533–48.

Smolensky M. H., Tatar S. E., Bergman S. A. et al. (1976) Circadian rhythmic aspects of human cardiovascular function: a review by chronobiologic statistical methods. *Chronobiologia* **3**, 337–71.

Terpstra J., Hessel L. W., Seepers J. et al. (1978) The influence of meal frequency on diurnal lipid, glucose and cortisol levels in normal subjects. *Eur. J. Clin. Invest.* **8**, 61–6.

Ungar R. H. and Dobbs R. E. (1978) Insulin, glucagon, and somatostatin secretion in the regulation of metabolism. *Ann. Rev. Physiol.* **40**, 307–43.

Van Gent C. M., Pagano Mirani-Oostdijk C., Van Reine P. H. et al. (1979) Influence of meal frequency on diurnal lipid, glucose and insulin levels in normal subjects on a high fat diet: comparison with data obtained on a high carbohydrate diet. *Eur. J. Clin. Invest.* **9**, 443–6.

Voigt E.-D., Engel P. and Klein H. (1968) Tagesrhythmische Schwankungen des Leistungspulsindex. *Germ. Med. Mth.* **17**, 394–5.

Wertheimer L., Hassen A., Delman A. et al. (1974) Cardiovascular circadian rhythm in man. In: Scheving L. E., Halberg F. and Pauly J. E. (ed.) *Chronobiology*. Tokyo, Igaku Shoin, pp. 742–7.

Wilson D. M., Kripke D. F., McClure D. K. et al. (1977) Ultradian cardiac rhythms in surgical intensive care unit patients. *Psychosom. Med.* **39**, 432–5.

Wurtman R. J., Rose C. M., Chou C. et al. (1968) Daily rhythms in the concentration of amino acids in human plasma. *N. Engl. J. Med.* **279**, 171–5.

Zülch K. J. and Hossmann V. (1967) 24-hour rhythm of human blood pressure. *Ger. Med. Mth.* **12**, 513–18.

chapter 4 *The Kidney and Hormones Affecting it*

The collection of urine samples is a straightforward task, there being little difficulty in persuading subjects to void urine every hour. Even more frequent sampling is possible if fluid intake is high. A bonus associated with frequent sampling is that, by measuring mid-flow urine temperature, it is possible to obtain data on deep body temperature as has been referred to in Chapter 2. Urine samples differ from temperature and heart rate data, for example, in that they describe the average value since the last time of voiding urine, that is, they are continuous rather than discrete data. The advantage and disadvantage of this can be appreciated by consideration of the inferences that can be drawn from a urine sample collected on rising; such a sample gives some information about nocturnal renal activity since it is an average value, but it will underestimate the minimum or maximum rates of excretion if they occur nocturnally.

Because the kidneys deal with so many substances, a study of urinary rhythms enables one to compare the effectiveness of exogenous and endogenous factors in these different substances. Further, in accord with the kidneys' homeostatic and excretory functions, the average rate of excretion of a substance on different days will change and the effect of this upon rhythmicity can be assessed. Finally, since the process of urine formation has so many stages, each of which might manifest circadian rhythmicity, a study of the rhythmicity shown by these different stages can give some idea of the causal nexus that exists within the system.

In other accounts, consideration of these points is given to each constituent in turn (Wesson, 1964; Conroy and Mills, 1970; Reinberg, 1971; Mills, 1973; Bartter et al., 1979). Whilst this approach has obvious merits, in the context of the present book it would seem more appropriate to consider rather the types of experiments by which the balance of exogenous and endogenous components is assessed and then the experimental approaches by which it is hoped the causal nexus can be unravelled. Examples from a variety of constituents will illustrate the argument, but for any constituent in particular, the reader is referred to the reviews cited above.

1. Endogenous and Exogenous Components

1.1. Nychthemeral conditions

Wesson (1964) studied subjects on normal diets and under nychthemeral conditions; limited diurnal activity and nocturnal sleep were allowed and urine samples were collected every 3 or 4 hours throughout the nych-themeron. These were analysed for sodium, chloride, potassium, phosphate, calcium and magnesium and glomerular filtration and renal plasma flow rates (GFR and RPF respectively) were calculated from clearance data. The excretion of most substances was greater during the day than during the night, the GFR and RPF also being highest diurnally; peak times were at about 16^{00}. Exceptions to these generalizations were phosphate (the excretion of which was phased differently) and calcium and magnesium (whose rhythms and times of peak seemed to vary more, both between individuals and between days). Simultaneous collection of plasma samples indicated that only in the case of phosphate did concentrations vary rhythmically throughout the 24 hours. For other urinary constituents that have been investigated (*see* Mills, 1966; Conroy and Mills, 1970), times of maximum excretion are normally coincident with mid-activity, further exceptions being urea (which peaks in the evening), and titratable acidity (which is the inverse of the potassium rhythm). In some subjects the rhythms have two diurnal peaks, the second generally occurring in the evening, as a result of which the acrophase of fitted cosine curves tends to fall between the two peaks (*see* Appendix) (Minors et al., 1976).

A more recent assessment by cosine curve fitting of urinary data from subjects under nychthemeral conditions has confirmed many of these findings (Minors et al., 1976). Acrophases in all constituents except phosphate were between 14^{00} and 18^{00}. The phosphate acrophase was just before midnight, this being determined in part by low excretory rates about 10^{00} and in part by high nocturnal rates.

The mean rate of excretion of any constituent will vary from day to day due to the excretory and homeostatic functions of the kidney. *Fig.* 4.1, from previously unpublished data by the authors, indicates that there is a direct relationship between the amplitude and mesor of the cosine curve fitted to each 24 hours of data. Such a relationship has been observed before in urine rhythms (Mills and Waterhouse, 1973; Mann et al., 1976) even though in the latter study, the amplitude and mean were estimated without cosine curve fitting. It is also evident in those data of Wesson (1964), whose subjects were given saline infusions or placed on low-salt diets and so their mean rates of excretion varied. This relationship has been observed in other variables and species (for example, *see* Sollberger, 1955; Mann et al., 1972; Aschoff and von Saint Paul, 1973) and gives some justification to the practice of expressing rhythmic data in the form 'amplitude/mean'. As yet, the mechanism by which this relationship arises is not known though there has been some speculation on this point (Mann et al., 1976).

As Wesson (1964) found, not all rhythms show the same degree of reproducibility even when some control over activity and diet is exerted. One of the most reproducible of electrolyte excretory rhythms is that of potassium; by contrast, calcium shows a more erratic rhythm and other

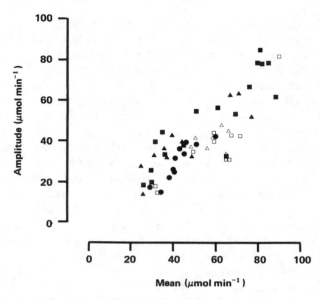

Fig. 4.1. Relationship between the amplitude and the mean level of the urinary potassium excretory rhythm for 5 subjects. Amplitude and mean level derived from best-fitting cosine curve. Symbols represent data from different subjects. (Minors and Waterhouse, unpublished data.)

substances, such as sodium, chloride and phosphate, have rhythms that are intermediate. This can be assessed more formally in a number of ways: (*a*) by assessing the proportion of occasions when the rate of excretion at night is greater than that for the 24-hour mean (Minors et al., 1976); (*b*) by comparing the variance in acrophases either between different subjects or between different days in the same subject (Mills, 1974); (*c*) by comparing the confidence intervals of acrophases (Minors et al., 1976). In all cases a similar sequence of urinary constituents is obtained (*see above*), though it seems that, even so, the most reproducible, potassium, is less reliable than core temperature.

1.2. Renal function when diet and activity are controlled

These differences in reproducibility are believed to reflect the balance between endogenous and exogenous influences for different variables. Diet, changes in activity and posture are all known to affect renal function. Some idea of the relative importance of exogenous and endogenous influences can be obtained by comparing renal rhythms when these external factors are changed.

Lobban and Tredre (1964) found that the amplitudes of renal rhythms were less than normal when patients were subjected to complete bedrest

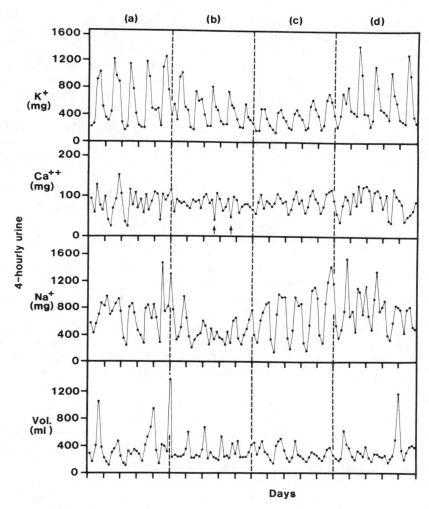

Fig. 4.2. Urinary excretion of potassium, calcium, sodium and water every 4 hours in a single subject during periods of different diets: (a) on free diet; (b) on 4-hourly bread and modified milk; (c) as (b) but subject confined to bed; (d) on free diet and activity. ↑ indicate periods of heavy exercise. (*From* Loutit, 1965, Fig. 3.)

or severe limitation of exercise. The change in amplitude was less marked for potassium than for water, sodium or chloride. Diet was not controlled in this study, but in one by Reinberg et al. (1969), isocaloric meals were taken throughout the experimental period while subjects were confined to bed but allowed to sleep. For both potassium and 17-hydroxycorticosteroid excretion, the circadian rhythms were similar on this regimen to results from subjects living a conventional existence.

Two studies, by Loutit (1965) and Moore-Ede et al. (1972), have been particularly informative since they have investigated the effects of both changes in diet and activity upon a number of urinary rhythms; *Fig.* 4.2 shows some of the data of Loutit. From *Fig.* 4.2 the stability of the potassium rhythm is evident with similar times of peak on different days and on different regimens. (The gradual fall in excretion in (b) and (c) presumably reflects a low potassium intake on the controlled diet.) On the other hand, there are considerable differences in the data for calcium when diet and activity are changed. As *Fig.* 4.2 indicates, calcium shows regular circadian rhythms only when both diet and activity are controlled (c). The fall in excretion produced by marked activity is shown (↑) in (b); the general decrease in regularity when (b) and (c) are compared presumably reflects the lack of control of activity during (b); when neither diet nor activity is controlled, as in (a) and (d), the excretion becomes quite irregular. Sodium excretion and urine volume give results intermediate between those for potassium and calcium.

Fig. 4.3 shows some of the data from the study by Moore-Ede et al. (1972). Similar conclusions can be drawn. Thus:

1. Under normal conditions (N) the inter-individual variation for the calcium rhythm is greater than that for potassium.

2. The potassium rhythm is comparatively little affected by changes in regimen. This applies whether one considers the time of peak of the rhythm, the time at which excretion first rises above the mean, the general shape of the rhythm, the amplitude of the rhythm or (not shown here) the mean daily rate of excretion.

3. For calcium the picture is rather different. There are considerable differences between different regimens whichever of the measures is considered; the normal regimen is associated with a range of up to 12 hours in peak times in calcium excretion (N). With regular meals and constant bedrest (BD), the endogenous component of the excretory rhythm in calcium is uncovered. The effect of activity (AD) is to decrease the excretion of calcium and a comparison of AD with AF suggests that the peaks are more regularly placed in the absence of food intake. If one accepts that the nychthemeral routine modifies the endogenous component of calcium excretion by two mechanisms—a rise produced by feeding and a fall by activity—then it is not surprising that the nychthemeral routine can be so irregular. However, the data indicate that, even though sleep and meal times considerably influence the calcium excretory rhythm, they cannot wholly account for it; the residuum is the endogenous component of the rhythm. *Fig.* 4.3 indicates that this component is more marked for potassium than calcium. In the strictest sense it is necessary to eliminate further environmental influences such as noise and social interaction before one can attribute the rhythm unambiguously to an endogenous oscillator. This has been achieved in experiments performed in an isolation unit and confirms the presence

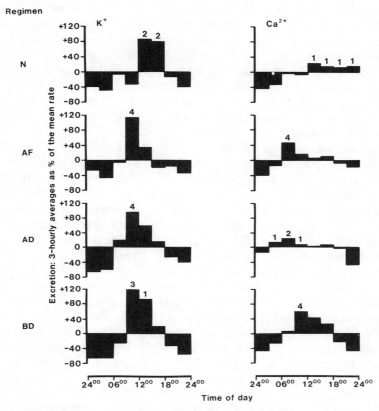

Fig. 4.3. Average excretion rates of potassium and calcium for 4 subjects on 4 activity and diet patterns: N, normal activity and diet; AF, increased activity and fasting; AD, increased activity and regular meals; BD, bedrest and regular meals. Numbers above the plots indicate the number of subjects with peak excretion rates at each time. (*From* Moore-Ede et al., 1972, Fig. 1.)

of a small endogenous component in the urinary calcium rhythm (Mills et al., 1978).

These data from experiments with changed feeding and activity patterns give rise to the concept of a spectrum of possibilities, from that for constituents much influenced by external factors (for example, calcium) to that for those little influenced by the same (for example, potassium). Such a concept gains support from other types of experiment that will now be considered.

1.3. Free-running experiments

It has already been mentioned that in these conditions internal desynchronization often does not occur, all variables showing a free-running period of about 25 hours, but that, in passing from the entrained to the free-running stage, there is a change in the phase relationship that exists between the pattern of sleep–wakefulness and the rhythm of a variable (*see also* Chapters 2 and 3). Wever (1979) has argued that in passing from the entrained to the free-running state those variables influenced most by the endogenous oscillator that is important for the temperature rhythm will show the largest advance in phase, whereas those influenced most by the pattern of sleep–wakefulness (or some factor associated with this pattern, for example, meal times) will retain more the phase relationship to sleep and wakefulness shown in the entrained state. (This subject will be elaborated upon in Chapter 12). Rather surprisingly, the large amount of urinary data from the experiments of Aschoff and Wever do not seem to have been treated this way. However, the data of Kriebel (1974), derived from a single subject, suggest that urinary 17-ketosteroids are influenced mainly by activity and that plasma catecholamines and 17-hydroxycorticosteroids, like deep body temperature, possess large endogenous components (*see* Wever, 1979, Fig. 18).

However, as described in Chapter 2, about one-third of the large number of subjects studied by Aschoff and Wever in their bunker have shown internal desynchronization (Wever, 1979). In this phenomenon, activity and temperature rhythms take on different free-running periods, an example of which has already been shown in *Fig.* 2.3. Fourier analysis of the data from another subject showed that the most prominent period for temperature was 24·8 hours and for activity, 33·5 hours. In addition, the urinary data from this experiment were subjected to Fourier analysis and some findings are shown in *Fig.* 4.4. The results, that the dominant period for the potassium rhythm is 24·8 hours, that the dominant period for the calcium rhythm is 33·5 hours and that for volume both periods are equally strong, all accord with the concept that between different constituents there is a range in the relative importance of the endogenous component that controls the temperature rhythm and some factor associated with the sleep–wakefulness cycle.

1.4. Experiments involving time-zone shifts and shift work

Studies have made use of both real and simulated time-zone transitions and shift work. After real flights, rarely have the urine samples been taken often enough or have sufficient days been studied to enable a precise account of the process of adaptation to be given. Those data that have been obtained have been reviewed by Aschoff et al. (1975). It is possible to deduce from these data: (*a*) that potassium and 17-hydrocorticosteroids both adjust slowly but urine flow adjusts more rapidly (Lafontaine et al., 1967); (*b*) that phosphate adapts much more readily than potassium (Conroy and Mills, 1970, p. 142); (*c*) that calcium and chloride adapt more quickly than 17-hydroxycorticosteroids (Halberg et al., 1971); (*d*)

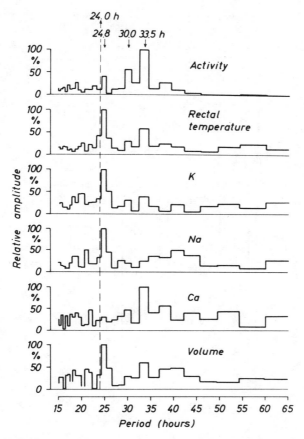

Fig. 4.4. Fourier analysis of the activity, rectal temperature and urinary rhythms in a subject living under constant conditions without time cues for 26 days. Note that the predominant period for the activity rhythm is 33·5 hours and similarly for calcium excretion. Rhythms with a large endogenous component, for example rectal temperature and urinary potassium excretion, have a predominant period of 24·8 hours with only a small 33·5-hour component. (*From* Wever, 1979, Fig. 25/II.)

that catecholamines adjust more rapidly than 17-hydroxycorticosteroids (Wegmann and Klein, 1973). With simulated time-zone shifts in an isolation unit, in which more frequent urine sampling was possible, different rates of adaptation were observed, again the sodium and chloride rhythms adapting more rapidly than that of potassium (Elliott et al., 1972). A clear example is shown in *Fig.* 4.5 which shows that the sodium rhythm of one subject adapted to post-shift time far more rapidly than did his potassium rhythm (*Fig.* 4.5, *above*), as a result of which, on the first day after the 'flight', the sodium rhythm has adapted but the potassium rhythm still follows pre-shift time (*Fig.* 4.5, *below*).

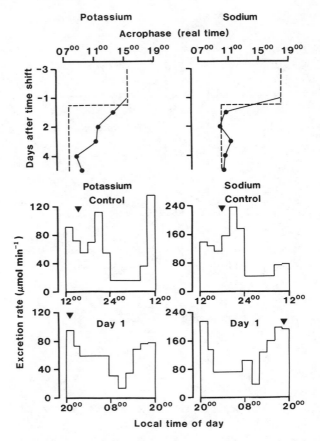

Fig. 4.5. Urinary potassium and sodium excretion before and after a simulated flight through eight time zones in an eastward direction. *Above,* The course of the daily acrophase; the dashed line indicates the course predicted for immediate completed adaptation. *Below,* The excretion patterns during a control day (*middle*) and the first day following the time shift (*bottom*); the arrow indicates the acrophase. (Data of Elliott et al., 1972.)

In studies using shift workers the rates of adaptation of various urinary constituents to changes in routine have been compared. Thus phosphate seems to adapt rapidly and flow and the major cations less quickly (Mills and Thomas, 1957; Conroy et al., 1970). Reinberg et al. (1975) have reported the results of a field study upon 20 healthy shift workers. Subjects worked an 8-hour, weekly-rotating shift system in the sequence 'day', 'night', 'evening' and 'morning'. Urine samples were collected on the first and seventh day of each shift and analysed for (amongst other substances) sodium, 17-ketosteroids, creatinine, potassium and 17-hydroxycorticosteroids. The adjustment of all constituents to the night shift was not equally rapid, the last two appearing to be slower than the other three. (*See also* Chapter 10 for further comment on this study.)

In experiments performed in the constant light and temperature of the midsummer Arctic, a group of subjects inverted their schedule of sleep and wakefulness, wearing blindfolds to simulate darkness during sleeping hours (Martel et al., 1962). 'The urinary constituents measured were ketosteroids, ketogenic steroids, water, sodium and potassium. For water, sodium and potassium, an adapted rhythm showed lowest rates of excretion during the sleep times and highest in activity times and the process of adjustment consisted of a declining influence of old time and an increasing influence of new time. By these criteria, adjustment was very rapid for ketosteroids, took about 4 days for water and sodium, and about 6 for potassium and ketogenic steroids.

In summary, therefore, these experiments confirm that the rhythms of some variables (for example, flow) are affected more than others (for example, potassium) and this is interpreted to indicate that differences exist between constituents in the relative importance of endogenous and exogenous factors.

1.5. Experiments with non-24-hour days

Mills and Stanbury (1952) studied 5 subjects living a 12-hour 'day' for 48 hours. A persistence of the 24-hour rhythm was found for urinary pH, volume, sodium, potassium and chloride, but a 12-hour rhythm in phosphate excretion developed immediately. A rather different approach was used by Wever (1972). He measured the change in phase relationship between a number of variables and the light—dark cycle in subjects who were entrained first to a 24-hour and then to a 26·67-hour 'day'. Results are shown in *Fig.* 4.6. On the longer zeitgeber cycle the phase of all variables is advanced with respect to the zeitgeber; but, as the Figure indicates, for some variables (temperature and potassium) the advance was greater than for others (activity and water). These results were interpreted to confirm the view that temperature and urinary potassium rhythms are influenced far less by exogenous influences than are urinary flow and activity rhythms. (Compare with the data of Kriebel (1974), s. 1.3 *above* and *see also* Chapter 12.)

Many of the data presented so far indicate that rates of adaptation to changed or abnormal routines of sleep and wakefulness vary with different constituents, but it is difficult to ascribe any quantitative value to this comparison; instead one is able to do little more than describe a constituent as being affected mainly by exogenous factors (including the rhythm of sleep—wakefulness) or by endogenous factors. However, experiments have been performed which enable a quantitative comparison of endogenous and exogenous components to be made. Much of this work has been performed in the Arctic under conditions of comparative constancy

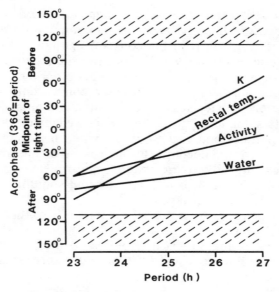

Fig. 4.6. Temporal position of the acrophases of urinary potassium and water excretion, rectal temperature and activity as a function of the period of the light—dark cycle. Acrophases expressed relative to the middle of the light time. Data from 6 subjects with 13 light—dark cycle periods. Hatched area represents dark time. (*From* Wever, 1972, Fig. 5.)

of lighting and temperature (for example, Lewis and Lobban, 1956, 1957a, b; Simpson and Lobban, 1967; Simpson, 1977). Different protocols have been used, all requiring the subjects to live regular routines determined by timepieces that had been made to run fast or slow. One of the most informative protocols, for present purposes, made use of timepieces that indicated the passage of 24 hours when 21 or 27 real hours had elapsed. Accordingly, after 8 apparent days the subjects had in fact lived 7 or 9 real days. In one study (Lewis and Lobban, 1957a, b) urine samples were collected at regular time and analysed for a number of common constituents. *Fig.* 4.7 shows results from one of the subjects on the 27-hour 'day'. There is a clear dissociation between urinary flow, which is much influenced by the sleep—wakefulness cycle and which shows eight peaks (equivalent to 8 apparent days), and the 'endogenous' constituent potassium, which shows nine peaks (equivalent to 9 real days). Similar results were obtained on a later expedition in which the excretion of 17-hydroxycorticosteroids, another 'endogenous' constituent, was studied also (Simpson and Lobban, 1967).

More recently, some of these data have been re-analysed by computer to assess the period of the best-fitting cosine curves (Simpson et al., 1970; Simpson, 1977). This analysis indicated that, on a 21-hour day, as well as the 21-hour component (due to the exogenous effect of the regimen), the

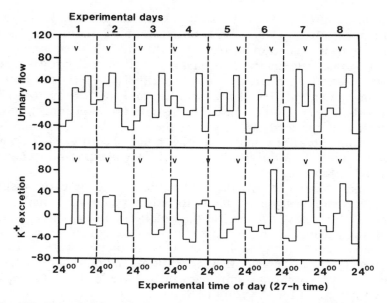

Fig. 4.7. Urinary flow and potassium excretion over 9 days of living on a 27-hour routine. Flow and potassium expressed as percentage deviations from the mean. The abscissa is divided into experimental days and the dashed lines mark the end of each 27-hour day. Arrowheads indicate mid-day by solar time. (*From* Lewis and Lobban, 1957b, Fig. 7.)

endogenous component was better described by a period slightly greater than 24 hours. This accords better with theory since the persistence of an *exact* 24-hour rhythm in the supposed absence of 24-hour synchronizers would be difficult to explain; to consider instead that, in the absence of such synchronizers, the endogenous rhythm free-ran with a period slightly in excess of 24 hours is far more acceptable and also gives some indication of the range of entrainment of the oscillator(s) concerned (*see* Chapters 1 and 12). Further, by comparing the amplitude of the endogenous component to that of the exogenous, 21-hour component, the authors produced a 'circadian amplitude ratio' which would be higher for 'endogenous' constituents and lower for those affected more by the 21-hour routine. The sequence found by this technique was:

(endogenous): 17 OHCS $>$ K$^+$ $>$ Na$^+$ $=$ H$_2$O $>$ Cl$^-$: (exogenous).

The simultaneous presence of two rhythmic influences would be expected to show 'beats' in any stretch of data as the two components were first in phase and then completely out of phase with one another. Such a prediction can be seen in the data for urine volume from subjects D (in Fig. 2) and R (in Fig. 3) in the study of Lewis and Lobban (1957b); in both examples the amplitude is lower in the middle of the 9-day cycle than at either the beginning or end.

A similar approach has been adopted in a series of experiments in an isolation unit (Mills et al., 1977). Both 21- and 27-hour days have been used and a wider range of urinary constituents (flow, sodium, chloride, potassium, urate, calcium and phosphate) as well as deep body temperature has been assessed. Further, the circadian amplitude ratio has been calculated from the amplitudes of the circadian and the 21-hour (or 27-hour) components fitted simultaneously to the data (*see* Appendix). In spite of slight differences in analysis, as well as the more obvious difference in the subjects' comfort, it is remarkable how much accord exists between the circadian amplitude ratio ranking order in the Arctic and isolation unit experiments (*Table* 4.1).

The major difference was the finding in the Arctic experiments that temperature (not subjected later to cosinor analysis) had a strong exogenous component whereas, in the isolation unit experiments, temperature showed a high endogenous component. A possible explanation for the discrepancy is that the subjects in the isolation unit were far less active than those in the inclement field conditions; these circumstances would accentuate any exogenous component in the Arctic study. (Compare this result with those in the study by Kleitman and Kleitman (1953) discussed in Chapters 2 and 3.)

1.6. Summary

In summary, therefore, different urinary constituents show differences in the relative importance of endogenous and exogenous components in determining their rhythmicity. As with other variables, comparisons between rhythms and subjects are more easily made if account of varying exogenous influences is taken. One part of this exogenous influence is the pattern of sleep and wakefulness (and associated changes such as posture, food intake etc.). Once again, comparisons are easier to make if the rhythms are referred to mid-sleep (Halberg et al., 1969) and the variances of acrophases in a subject studied longitudinally were lower when expressed this way rather than with reference to local midnight (Mills and Waterhouse, 1973). Another part of the exogenous influence is eating habits. To the extent that eating habits are normally closely related to the pattern of sleep—wakefulness, then the use of mid-sleep as a reference point will take them into account also.

2. Possible Origins of Renal Rhythmicity

The amount of substance excreted in the urine results from the balance between what is filtered at the glomeruli and what is reabsorbed or secreted by the tubules. The filtered load depends in turn upon two factors, the effective glomerular filtration rate and the plasma concentration of the substance. Since there are known to be mechanical, neural and humoral factors that can influence tubule function, it is clear that there is no lack of possibilities for factors that could be involved in the genesis of renal rhythms!

It is naive to believe that any renal rhythm will be caused by rhythmicity in only one of these factors; in the following account some of the experimental evidence in favour of a role for each factor in renal rhythmicity will be given, together with evidence indicating that it is not the sole cause for a particular urinary constituent. When the hormonal influence upon the kidney is considered, the account will be extended to cover what is known relating to the circadian rhythms of these hormones in general.

2.1. Plasma concentration

A rhythm in the urinary excretion of a substance might result from the rhythm in plasma concentration of that substance. The concept that urine composition reflects that of the plasma would accord with the view of the kidney as a homeostatic organ but would still leave unanswered the question of what causes the rhythms in plasma concentration.

To test this relationship it is necessary to sample plasma concentrations and urinary excretory rates frequently and simultaneously over a sufficiently long period to ensure also that the circadian rhythm can be accurately described. Since changes in activity, food intake and posture are all likely to affect renal function and so obscure any relationship between plasma and urine, these factors must be taken into consideration. One of the clearest results is for the rhythm in urinary phosphate excretion. *Fig.* 4.8 shows data from a group of students under conditions of a 'constant routine', that is remaining active and awake and taking an identical snack each hour throughout the experiment (Mills, 1966). There is a close parallelism between the plasma phosphate concentration and urinary excretion of phosphate. The cause of the changing plasma phosphate is not certain; cortisol is known to lower plasma phosphate levels and promotes the intracellular uptake of this ion, especially by skeletal muscle (Mills and Thomas, 1959). However, one difficulty in accepting a strong link between cortisol and phosphate excretion is that the hormone rhythm is generally very stable and believed to show a large endogenous component (*see* Chapter 7) whereas the rhythm of urinary phosphate excretion is far more influenced by exogenous factors (*Table* 4.1). Lack of data prevents our knowing whether it is the relationship between cortisol and plasma phosphate or that between plasma and urine phosphate that is modified when plasma cortisol and urine phosphate excretion become dissociated.

The relationship between plasma concentration and urine excretion is much less clear for other substances. Two papers (Buchsbaum and Harris, 1971; Meyer et al., 1974) can be cited to illustrate this. In both studies there were simultaneous measurements of plasma levels and urine excretion

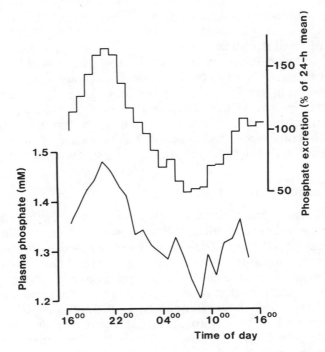

Fig. 4.8. Mean urinary phosphate excretion (*above*) and mean plasma phosphate concentration (*below*) measured hourly in subjects awake and active and with identical hourly food and fluid intake. (*From* Mills, 1966, Fig. 1.)

for a variety of constituents including sodium, potassium and phosphate. The published results show occasions when the plasma concentration and urine excretion behaved in parallel, but equally there are occasions when they changed in opposite directions; Buchsbaum and Harris (1971) concluded that there was no significant correlation between the two. These two studies argue against plasma rhythms being major determinants of urine levels, but slight differences in protocol existed and the experimental conditions were not controlled as closely as in the experiment of Mills (1966). Thus, although activity, electrolyte intake and surroundings were kept constant from day to day, they were not kept constant within the 24 hours; recumbency and sleep at night were allowed and normal meal times were observed. These factors might have confounded any relationship between plasma and urine.

For potassium, another problem is that the ion in the urine is believed to come from tubular secretory processes as well as from the filtered load at the glomeruli. A recent paper by Moore-Ede et al. (1975) has suggested a plausible mechanism for accounting for the potassium rhythm in excretion that invokes this secretory process. The study was based upon three healthy subjects whose activity, food intake and posture were kept as constant as possible. Plasma potassium concentration and urinary excretion of this ion were measured regularly throughout the nych-themeron; a peak in both variables was found at about midday and a minimum 12 hours later. Further, the arteriovenous concentration differ-ence across the forearm was most negative from 09^{00} to 12^{00} and was interpreted to indicate that the maximum net movement of potassium from the intracellular fluid (ICF) to the extracellular fluid (ECF) was at this time. Conversely, at 21^{00}, the difference was positive and maximum

Table 4.1. Rank ordering for circadian amplitude ratio of different constituents measured in Arctic and isolation unit experiments

		Arctic 21-h day	Isolation unit 21-h day	27-h day
Endogenous	17-hydroxycorticosteroids	1	—	—
	Temperature	—	1	1
	Potassium	2	2	2
	Chloride	4	3	3
	Flow	3	4	4
	Phosphate	—	5	5
	Sodium	3	6	6
	Urate	—	7	7
Exogenous	Calcium	—	7	8

—, Not assessed.

and interpreted as a net flux into cells. In agreement with these interpret-ations was the observation that erythrocyte potassium levels were minimum about noon and maximum at about midnight. Finally, the postulated ICF–ECF flux of potassium required to change the ECF concentration by the amount observed was calculated to be similar to the excess rate of urinary potassium loss. The authors concluded that the kidneys were responding to changed ECF potassium concentration in a homeostatic manner which would minimize these changes. They postulated that the renal tubule cells, like other cells, lost potassium under the influence of cortisol but that this was into the tubular fluid rather than into the ECF. The extent to which confirmation of these data and speculations will be obtained by other laboratories remains to be seen, but the basic link between cortisol and potassium efflux would account for the observation that the rhythms have in common a tendency to be less affected than most by exogenous influences (Table 4.1). However, the observation that cortisol and potassium rhythms in night workers are not always in phase with each other (Conroy et al., 1970) indicates that other processes can be involved.

2.2. Glomerular filtration rate

This can be measured by calculating the clearance of inulin or, more conveniently but less accurately, of creatinine. Although most results indicate lower values during the night, the ratio of maximum to minimum rates of excretion varies between subjects and exceptions with higher nocturnal values for clearance are not rare (Mills, 1966). Because the amplitude of the rhythm is low, cosine curves often do not fit the data significantly better than a straight line (Minors et al., 1976).

There is no clear evidence that glomerular filtration rate (GFR) is an important determinant of excretory rhythms. The coincidence that exists between the rhythms of GFR and of excretion of other constituents has been noted by some workers (Wesson, 1964) and can be inferred from other studies (for example, Martel et al., 1962). However, this coincidence does not show that there is a causal link, but rather it might be fortuitous; furthermore it need not apply under all circumstances. For example, Vagnucci et al. (1969) measured the rhythms of inulin clearance and of sodium, potassium and chloride excretion in 5 subjects. When the subjects were recumbent throughout the experimental session, then the rhythms of GFR and electrolyte excretion were coincident. When, in other experiments, changes to an upright posture took place, both GFR and the rates of excretion of electrolytes were decreased; but the decreases were only transient and were less marked (in both size and duration) for GFR and potassium than for sodium and chloride. In other words, because the rhythms of GFR and sodium excretion were dissociated at least temporarily, neither could have been wholly responsible for the other in these circumstances.

2.3. Tubule function

Advances in renal physiology have indicated a plethora of factors that can modify tubular reabsorptive and secretory processes. In many cases the quantitative significance of these factors is uncertain and the role any might play in the circadian rhythms is quite unknown. Nevertheless, since the tubule is the site of uptake and secretion of many different substances, it is to be expected that interaction between some of them will take place; this interaction might give some clues as to the origin of renal circadian rhythms.

Urine flow is obviously determined very much by exogenous influences since water intake changes throughout the nychthemeron, but a rhythm persists under conditions of constant routine with lowest flows at night (Mills, 1966). The levels of antidiuretic hormone are quite high at night in spite of the subject being recumbent and this will play some role in decreasing nocturnal flow. Another factor is the decreased rate of electrolyte loss since that solute which is not reabsorbed will exert an osmotic effect retaining some water in the tubule, one result of which would be a relationship between the rate of cation excretion (sodium and potassium) and urine flow (Mills and Stanbury, 1952). However, the reciprocal relationship might play some part also as there is evidence for a flow-dependent secretion of ions by the distal convoluted tubule. The relationship between solute excretion and water loss is suggested also by the observation of a child who suffered from nocturnal enuresis and had abnormal electrolyte rhythms (Lewis et al., 1970).

The two ions sodium and chloride have generally been considered together probably because it is believed that they are handled similarly by the kidney. In support of this is the finding that variance of the mean difference between the acrophases of the two rhythms is small in nychthemeral circumstances (Mills, 1974). However, the acrophase of the chloride rhythm normally leads that of sodium by about an hour (Minors et al., 1976) and this does not lend itself to a ready explanation, since conventional accounts of tubule handling of the ions, especially by the proximal tubule where most of the ions are reabsorbed, treat chloride movement as a passive consequence of sodium reabsorption. The earlier morning rise in chloride excretion might be attributed to the simultaneous fall of phosphate excretion and the need to balance anion and cation numbers in tubular fluid or to a delayed increase in sodium excretion due to postural changes raising aldosterone levels (Mills, 1973). Recently, another possibility has emerged because, in the rabbit, the ascending limb of the loop of Henle seems to transport chloride actively and sodium follows as a passive consequence (Burg and Green, 1973); however, there is no evidence at the present time for or against the hypothesis that this part of the nephron is responsible for the circadian rhythmicity in sodium or chloride in man. Finally, the observation that there is an increase in the variance of the mean difference between chloride and sodium acrophases after simulated time-zone transitions (Mills, 1974) suggests that the linkage between the two ions is not immutable.

2.4. The influence of hormones upon renal rhythms

A number of hormones are known to affect the renal tubules and it is through these humoral factors that some of the homeostatic functions of the kidney operate. The putative roles of cortisol in renal function have already been considered; three other hormones—antidiuretic hormone (ADH), aldosterone and parathormone (PTH)— must now be considered.

Antidiuretic hormone

ADH levels are determined by exogenous factors that alter plasma osmotic pressure. In addition, in the absence of changes in osmotic pressure, levels of ADH are low when subjects are recumbent and higher when they are upright (Segar and Moore, 1968). This result is attributed to reflexes originating from stretch receptors in the atria, aortic arch and carotid sinus baroreceptor areas.

As a result, any endogenous component in a rhythm of ADH release can be assessed only under rather stringent conditions. George et al. (1975) studied a group of subjects who were recumbent throughout the nychthemeron. Even though water intake was not controlled, blood samples indicated no systematic change in haematocrit or plasma osmotic pressure throughout the 24 hours. However, there was a rhythm in plasma ADH with peak values occurring during the early hours of the night (24^{00} to 04^{00}). As has already been mentioned above (s. 2.3), higher nocturnal values would contribute to the decreased urine flow at night.

The cause of the nocturnal rise in ADH levels is not certain. The nocturnal fall in blood pressure (*see* Chapter 3) and the nocturnal fall in plasma volume (Yoshimura and Morimoto, 1972) have been suggested by Seif and Robinson (1979). However, as Chapter 7 will indicate, many hormone rhythms show peak values during sleep and perhaps ADH is no exception. Even so, Rubin et al. (1978) were unable to find any link between pulses of ADH secretion and any particular sleep stage (a relationship that will be discussed in more detail in Chapter 7).

Aldosterone

In diurnally active man the rhythm of aldosterone is much affected by postural influences, but the rhythm has an endogenous component in that it persists in recumbent subjects with a peak at about 08^{00} as is shown in *Fig.* 4.9 (Williams, Cain et al., 1972). However, the timing of this rhythm is the opposite of that which would be required for there to be a role for this salt-retaining hormone in mediating the nychthemeral rhythm of sodium excretion.

There is some controversy over the cause of this endogenous component of the aldosterone rhythm. The quantitative effects upon the adrenal gland of the changes of plasma potassium concentration resulting from potassium fluxes between the ICF and ECF (*see* s. 2.1) are not known; rather, attention has concentrated upon the renin-angiotensin system and the rhythmic secretion of adrenocorticotrophic hormone (often inferred from plasma cortisol levels), both of which have rhythms that seem appropriately phased (*Fig.* 4.9).

A rhythm in plasma renin with a peak at, or just before, the normal time of waking was found in subjects who remained recumbent and ate evenly spaced meals (Gordon et al., 1966). Since there is a reflex increase in renin release produced by assuming the upright posture, the diurnal routine will accentuate the endogenous rhythm in renin secretion in the hours immediately after waking. Further evidence of an interaction between endogenous and exogenous factors comes from the finding that

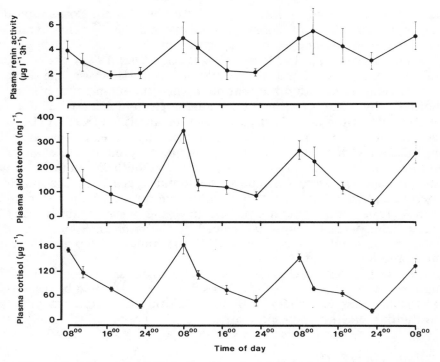

Fig. 4.9. Diurnal variations in plasma renin activity, aldosterone and cortisol in 5 recumbent subjects over 3 successive days. Data are means ± 1 s.e. Subjects consumed 200 mmol sodium and 200 mmol potassium per day in 3 isocaloric meals. Note the similar phasing of the 3 rhythms. Similar results were obtained in subjects whose sodium and potassium intakes were lower. (Data of Williams, Cain et al., 1972.)

remaining upright delays, but does not prevent, the fall in plasma renin levels (Gordon et al., 1966). In the case of adrenocorticotrophic hormone (ACTH), it is known that this can stimulate aldosterone secretion by the adrenal cortex. Details of the rhythmicity of ACTH will appear in Chapter 7, but, as *Fig.* 4.9 shows, the times of peak values of cortisol (and, it is inferred, ACTH) and aldosterone concentrations are close.

A decision in favour of either the renin-angiotensin system or ACTH as the cause of the aldosterone rhythm is not easy. Protagonists for both views (for example, Katz et al., 1972; Williams, Cain et al., 1972; Williams, Tuck et al., 1972; Vagnucci et al., 1974; Armbruster et al., 1975a; Katz et al., 1975) have correlated plasma aldosterone levels with renin and cortisol in subjects in changing postures or with different dietary sodium intakes. All three hormones show an episodic secretion (*see* Chapter 7) that is more frequent during sleep and comparatively slight differences in timing of the peaks seem to account for the differences in correlation between aldosterone and renin or cortisol that have been found. Correlation does not prove causality, of course, but a dissociation between renin and aldosterone on the one hand and cortisol on the other has been claimed (Williams, Cain et al., 1972; Williams, Tuck et al., 1972) and this does indicate something of the causal links involved. However, the evidence that the aldosterone rhythm continues when the renin rhythm has been abolished by administration of an adrenergic β-blocker (Cugini et al., 1977) or when the cortisol rhythm has been abolished (Katz et al., 1975) does not enable a decision in favour of either hypothesis to be made, but only indicates that either mechanism alone can produce the aldosterone rhythm. The view has been put forward that both hormones normally influence the aldosterone rhythm, the sodium balance of the individual determining which of the two is more important (Armbruster et al., 1975a). Thus, when sodium levels are low (and plasma renin is high) the level of aldosterone is more influenced by the renin-angiotensin rhythm; when body sodium is replete, ACTH rhythmicity becomes the more important determinant.

The more recent claims for the presence of 'salt-losing factors' (de Wardener, 1978) might indicate other substances that could be responsible for the rhythms of urinary sodium or chloride excretion. As yet, such possibilities remain purely speculative.

Parathormone (PTH)

Early reports involved infrequent blood sampling but indicated that there was a progressive diurnal rise in PTH levels (Arnaud et al., 1971). This result has been confirmed in studies with more frequent sampling throughout the nychthemeron, the peak midnight value of PTH being succeeded by a nocturnal fall (Sinha et al., 1975). The further observations that the rhythm of plasma calcium was the inverse of that of PTH and that the evening rise of the hormone could be prevented by calcium infusions suggested that the rhythm of PTH could be a response to the rhythm of plasma calcium (Jubiz et al., 1972). However, these authors found that bedrest had no effect upon the noctural PTH rise and yet considerably modified the rhythm of plasma calcium and so other factors must be involved. The relative time courses of the circadian changes in plasma PTH, plasma calcium and urinary excretion of calcium (see s. 1.1) seem to preclude any strong link between, first, plasma calcium and calcium excretion in the urine and, secondly, between PTH and calcium excretion via a hormonal effect upon the nephron. The extent to which the phosphaturic effect of PTH could be a contributory factor to the rhythm of phosphate excretion in the urine is unknown but it is worth noting that, at least on a nychthemeral existence, peak rates of phosphate excretion (about midnight) correspond to the time of highest plasma PTH concentration.

2.5. The influence of nerves upon renal rhythms

Since the release of renin by the kidneys after haemorrhage or changes in posture is mediated in part by renal efferent sympathetic nerves, it is relevant to consider the extent to which the endogenous rhythm of renin release is similarly controlled. The abolition of the renin rhythm after use of an adrenergic β-blocker has already been mentioned and other similar evidence from humans is summarized by Davis and Freeman (1976). Further support for a role of the nerves in affecting the renin rhythm comes from a study upon patients with kidney transplants in whom the rhythm was absent (Armbruster et al., 1975b). These authors add that renin responses to postural changes are present in patients with renal transplants; this result implies that the circadian and homeostatic release mechanisms are controlled independently, a conclusion discussed further in Chapter 7.

Presumably, the release of renin will be affected also by circulating catecholamines from the adrenal medulla. Some comment on the controversy regarding the time course of these substances has been made in the previous chapter and further consideration of it will be given in Chapter 6. At the moment all that need be added is that the relative contributions of the neural and humoral components are not known.

The effect of nerves upon all aspects of renal function can be investigated by considering the rhythms in transplanted kidneys. The results are conflicting. Some have claimed that the rhythms are unchanged (for example, Ginn et al., 1960), while others have claimed that marked changes in rhythms, even their inversion, take place (for example, Berlyne et al., 1968). A more recent review of this topic is included in a paper by Ratte et al. (1974). As with cardiac transplants (*see* Chapter 3) the general health of the patient, the extent of damage to the transplanted tissue and the effects of drugs are all influences that might alter renal rhythms independent of any effect that nerves might exert.

2.6. Causal nexus in renal rhythms

Having described the possible means by which rhythmicity might arise in renal function, it would then be reasonable to suppose that the relative importance of these could be assessed for any urinary constituent. This has been done by other authors (Wesson, 1964; Conroy and Mills, 1970; Reinberg, 1971; Mills, 1973; Bartter et al., 1979), to whose accounts the reader has already been referred. As has been indicated, for all constituents the argument is incomplete and based upon many pieces of evidence, sometimes controversial, often contradictory, derived from many types of experiment or subject. In the case of no constituent is the origin of the rhythmicity fully understood, though, as earlier sections of this chapter have indicated, examples of contributory factors can be given.

Why is the picture so unclear? In part the answer must be that the process of urine formation is controlled at so many stages that the data at present available are quite inadequate to unravel them. For example, an understanding of the cause of a rhythm seems to require simultaneous knowledge for that substance of its plasma concentration, the GFR and all aspects of its tubular handling by the renal tubule. Even with such a mass of information two problems would still remain.

First, as has been stated earlier, it is not known whether, for any one substance, the rhythm derives from one mechanism or (even if it does) whether that mechanism is the same under all circumstances. When other fields of physiology are taken as a model, a whole assemblage of mechanisms is found, amongst which the order of importance can vary according to the circumstances. To believe that a rhythm in urinary excretion can result from more than one possible mechanism and that the balance between them might differ under different circumstances is pure speculation, of course, but then the alternative approach—that a single mechanism accounts for a rhythm—has not yet revealed what this single mechanism might be for any particular substance. Further, the whole concept of a 'hierarchial nexus' (see Chapter 12) argues against there being a single cause.

Secondly, quantitatively, problems are encountered which can be illustrated by the example of sodium treatment by the kidney.

If one assumes that: the rate of excretion is about 50–300 μmol/min; the GFR is 125 ml/min; and the plasma concentration is about 140-μmol/ml; then the filtered load is 17 500 μmol/min. That is, the circadian loss varies between 0·3 per cent and 1·7 per cent of the filtered load and the reabsorptive process necessary to produce the circadian rhythm varies between 99·7 per cent and 98·3 per cent of the filtered load.

Alternatively if one assumes that the reabsorptive processes are constant, then changes in filtered load of 250 μmol/min are adequate to cause the circadian rhythm. That is, changes in the filtered load of less than 2 per cent can account for the circadian rhythm.

Of course such calculations are based upon the naive assumption that the rhythms derive from changes in the filtered load or amount of reabsorption or secretion alone. The evidence of glomerulo-tubular balance shows that such a view is oversimplified and the present writers have argued against a single origin of any renal circadian rhythm. However, these calculations do indicate that the changes that could produce the observed rhythms are very small and that they might be beyond the powers of resolution of our analytical methods at the present time.

Finally, earlier in this chapter it was shown that a 'circadian amplitude ratio'—a measure of the relative effectiveness of endogenous and exogenous components—could be deduced for a variable. An account of the possible nature of these components and the pathways by which they might transmit their effects will be given in Chapter 12, but at the moment it is important to realize that the pathways through which the different components manifest their effects might be identical. For example, it is possible that some of the effects of posture (an exogenous component) and the endogenous component of renin release are both mediated by renal sympathetic nerves. This in no way negates the value of distinguishing exogenous and endogenous elements of a rhythm but, in the present example, would render the interpretation of recordings from the renal nerves that much more complex.

References

Armbruster H., Vetter W., Beckerhoff R. et al. (1975a) Diurnal variations of plasma aldosterone in supine man: relationship to plasma renin activity and plasma cortisol. *Acta Endocrinol.* **80**, 95–103.

Armbruster H., Vetter H., Uhlschmid G. et al. (1975b) Circadian rhythm of plasma renin activity and plasma aldosterone in normal man and in renal allograft recipients. *Proc. Eur. Dialysis Transplant Assoc.* **11**, 268–76.

Arnaud C. D., Tsao H. S. and Littledike T. (1971) Radioimmunoassay of human parathyroid hormone in serum. *J. Clin. Invest.* **50**, 21–34.

Aschoff J. and von Saint Paul U. (1973) Circadian rhythms of brain temperature in the chicken measured at different levels of constant illumination. *Jap. J. Physiol.* **23**, 69–80.

Aschoff J., Hoffmann K., Pohl H. et al. (1975) Re-entrainment of circadian rhythms after phase-shifts of the zeitgeber. *Chronobiologia* **2**, 23–78.

Bartter F. C., Chan J. C. M. and Simpson H. W. (1979) Chronobiological aspects of plasma renin activity, plasma aldosterone, and urinary electrolytes. In: Krieger D. T. (ed.) *Endocrine Rhythms.* New York, Raven Press, pp. 225–45.

Berlyne G. M., Mallick N. P., Seedat Y. K. et al. (1968) Abnormal urinary rhythm after renal transplantation in man. *Lancet* **2**, 435–6.

Buchsbaum M. and Harris E. K. (1971) Diurnal variation in serum and urine electrolytes. *J. Appl. Physiol.* **30**, 27–35.

Burg M. B. and Green N. (1973) Function of the thick ascending limb of Henle's loop. *Am. J. Physiol.* **224**, 659–68.

Conroy R. T. W. L., Elliott A. L. and Mills J. N. (1970) Circadian excretory rhythms in night workers. *Br. J. Indust. Med.* **27**, 356–63.

Conroy R. T. W. L. and Mills J. N. (1970) *Human Circadian Rhythms.* London, Churchill.

Cugini P., Manconi R., Serdoz R. et al. (1977) Influence of propranolol and circadian rhythms of plasma renin, aldosterone and cortisol in healthy supine man. *Boll. Soc. It. Biol. Sper.* **53**, 263–9.

Davis J. O. and Freeman R. H. (1976) Mechanisms regulating renin release. *Physiol. Rev.* **56**, 1–56.

Elliott A., Mills J. N., Minors D. S. et al. (1972) The effect of real and simulated time-zone shifts upon the circadian rhythms of body temperature, plasma 11-hydroxy-corticosteroids and renal excretion in human subjects. *J. Physiol.* **221**, 227–57.

George C. P. L., Messerli F. H., Genest J. et al. (1975) Diurnal variation of plasma vasopressin in man. *J. Clin. Endocrinol. Metabol.* **41**, 332–8.

Ginn H. E., Unger A. M., Hume D. M. et al. (1960) Human renal transplantation: an investigation of the functional status of the denervated kidney after successful homotransplantation in identical twins. *J. Lab. Clin. Med.* **56**, 1–13.

Gordon R. D., Wolfe L. K., Island D. P. et al. (1966) A diurnal rhythm in plasma renin activity in man. *J. Clin. Invest.* **45**, 1587–92.

Halberg F., Nelson W., Runge W. J. et al. (1971) Plans for orbital study of rat biorhythms. Results of interest beyond the biosatellite program. *Space Life Sci.* **2**, 437–71.

Halberg F., Reinhardt J., Bartter F. C. (1969) Agreement in endpoints from circadian rhythmometry on healthy human beings living on different continents. *Experientia* **25**, 106–12.

Jubiz W., Canterbury J. M., Reiss E. et al. (1972) Circadian rhythm in serum parathyroid hormone concentration in human subjects: correlation with serum calcium, phosphate, albumin and growth hormone levels. *J. Clin. Invest.* **51**, 2040–6.

Katz F. H., Romfh P. and Smith J. A. (1972) Episodic secretion of aldosterone in supine man: relationship to cortisol. *J. Clin. Endocrinol. Metabol.* **35**, 179–81.

Katz F. H., Romfh P. and Smith J. A. (1975) Diurnal variation of plasma aldosterone, cortisol and renin activity in supine man. *J. Clin. Endocrinol. Metabol.* **40**, 125–34.

Kriebel J. (1974) Changes in internal phase relationships during isolation. In: Scheving L. E., Halberg F. and Pauly J. E. (ed.) *Chronbiology.* Tokyo, Igaku Shoin, pp. 451–9.

Lafontaine E., Sirot J., Pasquet J. et al. (1967) Influence des voyages aeriens est-ouest et vice versa sur les rythmes circadiens de la diurèse et de l'élimination urinaire du sodium et du potassium. *Rev. Med. Aéronautique* **6**, 11–15.

Lewis P. R. and Lobban M. C. (1956) Patterns of electrolyte excretion in human subjects during a prolonged period of life on a 22-hour day. *J. Physiol.* **133**, 670–80.

Lewis P. R. and Lobban M. C. (1957a) The effects of prolonged periods of life on abnormal time routines upon excretory rhythms in human subjects. *Q. J. Exp. Physiol.* **42**, 356–71.

Lewis P. R. and Lobban M. C. (1957b) Dissociation of diurnal rhythms in human subjects living on abnormal time routines. *Q. J. Exp. Physiol.* **42**, 371–86.

Lewis H. E., Lobban M. C. and Tredre B. E. (1970) Daily rhythms of renal excretion in a child with nocturnal enuresis. *J. Physiol.* **210**, 42P–43P.

Lobban M. C. and Tredre B. E. (1964) Renal diurnal rhythms in human subjects during bed-rest and limited activity. *J. Physiol.* **171**, 26P–27P.

Loutit J. F. (1965) Diurnal variation in urinary excretion of calcium and strontium. *Proc. R. Soc. B.* **162**, 458–72.

Mann H., Rutenfranz J. and Wever R. (1972) Untersuchungen zur Tagesperiodik der Reaktionszeit bei Nachtarbeit. II, Beziehungen zwischen Gleichwert und Schwingungsbreite. *Int. Arch. Arbeitsmed.* **29**, 175–87.

Mann H., Stiller S. and Korz R. (1976) Biological balance of sodium and potassium. A control system with oscillating correcting variable. *Pflügers Arch.* **362**, 136–9.

Martel P. J., Sharp G. W. G., Slorach S. A. et al. (1962) A study of the roles of adrenocortical steroids and glomerular filtration rate in the mechanism of the diurnal rhythm of water and electrolyte excretion. *J. Endocrinol.* **24**, 159–69.

Meyer W. J., Delea C. S., Levine H. et al. (1974) A study of periodicity in a patient with hypertension: relations of blood pressure, hormones and electrolytes. In: Scheving L. E., Halberg F. and Pauly J. E. (ed.) *Chronobiology.* Tokyo, Igaku Shoin, pp. 100–7.

Mills J. N. (1966) Human circadian rhythms. *Physiol. Rev.* **46**, 128–71.

Mills J. N. (1973) Transmission processes between clock and manifestations. In: Mills J. N. (ed.) *Biological Aspects of Human Circadian Rhythms.* London, Plenum, pp. 27–84.

Mills J. N. (1974) Phase relations between components of human circadian rhythms. In: Scheving L. E., Halberg F. and Pauly J. E. (ed.) *Chronobiology.* Tokyo, Igaku Shoin, pp. 560–3.

Mills J. N., Minors D. S. and Waterhouse J. M. (1977) The physiological rhythms of subjects living on a day of abnormal length. *J. Physiol.* **268**, 803–26.

Mills J. N., Minors D. S. and Waterhouse J. M. (1978) Adaptation to abrupt time shifts of the oscillator(s) controlling human circadian rhythms. *J. Physiol.* **285**, 455–70.

Mills J. N. and Stanbury S. W. (1952) Persistent 24-hour renal excretory rhythm on a 12-hour cycle of activity. *J. Physiol.* **117**, 22–37.

Mills J. N. and Thomas S. (1957) Diurnal excretory rhythms in a subject changing from night to day work. *J. Physiol.* **137**, 65P–66P.

Mills J. N. and Thomas S. (1959) The influence of adrenal corticoids on phosphate and glucose exchange in muscle and liver in man. *J. Physiol.* **148**, 227–39.

Mills J. N. and Waterhouse J. M. (1973) Circadian rhythms over the course of a year in a man living alone. *Int. J. Chronobiol.* **1**, 73–9.

Minors D. S., Mills J. N. and Waterhouse J. M. (1976) The circadian variations of the rates of excretion of urinary electrolytes and of deep body temperature. *Int. J. Chronobiol.* **4**, 1–28.

Moore-Ede M. C., Brennan M. F. and Ball M. R. (1975) Circadian variation of inter-compartmental potassium fluxes in man. *J. Appl. Physiol.* **38**, 163–70.

Moore-Ede M. C., Faulkner M. H. and Tredre B. E. (1972) An intrinsic rhythm of urinary calcium excretion and the specific effect of bedrest on the excretory pattern. *Clin. Sci. Mol. Med.* **42**, 433–45.

Ratte J. M., Halberg F., Haus E. et al. (1974) Circadian urinary rhythms in rats with renal grafts. *Chronobiologia* **1**, 62–73.

Reinberg A. (1971) Biological rhythms of potassium metabolism. Proceedings of 8th Colloquium of the International Potash Institute, Uppsala, Sweden.

Reinberg A., Chaumont A.-J. and Laporte A. (1975) Circadian temporal structure of 20 shift workers (8-hour shift-weekly rotation): an autometric field study. In: Colquhoun P., Folkard S., Knauth P. et al. (ed.) *Experimental Studies of Shiftwork.* Opladen, Westdeutscher Verlag, pp. 142–65.

Reinberg A., Ghata J., Halberg F. et al. (1969) Rythmes circadiens du pouls, de la pression artérielle, des excrétions urinaires en 17-hydrocorticosteroides catéchol-amines et potassium chez l'homme adulte sain, actif et au repos. *Ann. Endocrinol. (Paris)* **31**, 277–87.

Rubin R. T., Roland R. E., Gouin P. R. et al. (1978) Secretion of hormones influencing water and electrolyte balance (antidiuretic hormone, aldosterone, prolactin) during sleep in normal adult men. *Psychosom. Med.* **40**, 44–59.

Segar W. E. and Moore W. W. (1968) The regulation of antidiuretic hormone release in man. *J. Clin. Invest.* **47**, 2143–51.

Seif S. M. and Robinson A. G. (1979) Rhythms of the posterior pituitary. In: Krieger D. T. (ed.) *Endocrine Rhythms.* New York, Raven Press, pp. 187–201.

Simpson H. W. (1977) Human 21-h day studies in the high Arctic: a review of analyses carried out on the 1960 Spitsbergen and 1969 Devon Island studies. *Nova Acta Leopoldina* **225**, 407–29.

Simpson H. W. and Lobban M. C. (1967) Effect of a 21-hour day on the human circadian excretory rhythms of 17-hydroxycorticosteroids and electrolytes. *Aerospace Med.* **38**, 1205–13.

Simpson H. W., Lobban M. C. and Halberg F. (1970) Arctic chronobiology. Urinary near-24-hour rhythms in subjects living on a 21-hour routine in the Arctic. *Arctic Anthropology* **7**, 144–64.

Sinha T. K., Miller S., Fleming J. et al. (1975) Demonstration of a diurnal variation in serum parathyroid hormone in primary and secondary hyperparathyroidism. *J. Clin. Endocrinol. Metabol.* **41**, 1009–13.

Sollberger A. (1955) Statistical aspects of diurnal biorhythm. *Acta Anat.* **23**, 97–127.

Vagnucci A. H., McDonald R. H., Drash A. L. et al. (1974) Intradiem changes of plasma aldosterone, cortisol, corticosterone and growth hormone in sodium restriction. *J. Clin. Endocrinol. Metabol.* **38**, 761–76.

Vagnucci A. H., Shapiro A. P. and McDonald R. H. (1969) Effects of upright posture on renal electrolyte cycles. *J. Appl. Physiol.* **26**, 720–31.

de Wardener H. E. (1978) The control of sodium excretion. *Am. J. Physiol.* **235**, F163–73.

Wegmann H. M. and Klein K. E. (1973) Internal dissociation after transmeridian flights. Preprints, XXI International Congress on Aviation and Space Medicine, Munich, 17–21 September 1973, pp. 334–37.

Wesson L. G. (1964) Electrolyte excretion in relation to diurnal cycles of renal function. *Medicine* **43**, 547–92.

Wever R. A. (1972) Mutual relations between different physiological functions in circadian rhythms in man. *J. Interdiscipl. Cycle Res.* **3**, 253–65.

Wever R. A. (1979) *The Circadian System of Man. Results of Experiments under Temporal Isolation.* Berlin, Springer-Verlag.

Williams G. H., Cain J. P., Dluhy R. G. et al. (1972) Studies on the control of plasma aldosterone concentration in normal man. I. Response to posture, acute and chronic volume depletion and sodium loading. *J. Clin. Invest.* **51**, 1731–42.

Williams G. H., Tuck M. L., Rose L. I. et al. (1972) Studies on the control of plasma aldosterone concentration in normal man. III. Response to sodium chloride infusion. *J. Clin. Invest.* **51**, 2645–52.

Yoshimura H. and Morimoto T. (1972) Seasonal and circadian variations in body fluid. In: Itoh S., Ogata K. and Yoshimura H. *Advances in Climatic Physiology.* Tokyo, Igaku Shoin, pp. 381–94.

chapter 5 *Sleep and Wakefulness*

1. Studying the Sleep—Wakefulness Rhythm

The rhythm of sleep and wakefulness differs from those considered in previous chapters in two ways. First, if considered simply as 'awake' or 'asleep', it is an alternation between two states and as such can be described mathematically by a rectangular wave rather than a cosine curve (or indeed any other continuous function). Secondly, some of the properties of the sleep—wakefulness rhythm in man differ from those of other rhythms and so cannot be investigated by the same techniques as will now be discussed.

1.1. Problems

In previous chapters the relative importance of the exogenous and endogenous components of circadian rhythms has been assessed by investigating the extent to which the rhythms are affected by constant routines, time-zone shifts, days of abnormal length and life in the absence of time cues. Not all these approaches can be used in studies of sleep—wakefulness rhythmicity for the following reasons:

1. Provided that he has sufficient motivation, man can considerably modify his hours of sleep and wakefulness. This motivation can derive from necessity (as in the cases of shift workers and intercontinental travellers) or from altruism (as in the cases of experimental subjects who submit to abnormal schedules), for example, those of the 48-hour day (Meddis, 1968; Webb, 1978), the 21- or 27-hour day (Lewis and Lobban 1957a,b; Mills et al., 1977) or the 12-hour day (Mills and Stanbury, 1952).

2. It might be argued that one could investigate the effect of influences such as food, noise or light, by placing subjects in an environment in which these factors vary and then investigating whether or not the subjects were influenced by these external rhythms. In practice the interpretation of the results from such experiments is not straightforward. Thus, if the potential zeitgeber is 'strong' (marked differences in light or noise intensity or in food availability) it will force the subjects to comply; if it is 'weak' (if alternative sources of food or light are available or if there are ear-plugs) then the subjects might choose to ignore such external rhythmicity. In other words, what might be investigated by such protocols is the willing-

95

ness of subjects to put up with any inconvenience that might arise due to desynchronization from the environment rather than some property of endogenous rhythms. Thus, if the subject's sleep—wakefulness pattern remained in phase with the environmental rhythm it would not be possible to be sure whether sleep and wakefulness had become entrained or whether, instead, the subject had chosen to override (that is, to 'mask') an endogenous component. Equally, if the pattern of sleep and wakefulness became desynchronized from the imposed environmental rhythm a distinction between an unentrained endogenous component (or one driven by another free-running rhythm) and a subject tolerant of, or uninterested in, a desynchronized environment could not be made with certainty. Presumably, some information related to this distinction could be obtained by asking the subject why he adopted either procedure, but such experiments do not appear to have been performed.

1.2. Methods

Attempts to investigate the pattern of sleep and wakefulness have been made in two main directions. First, indirect methods of assessing the rhythm have been used and, secondly, subjects have been studied while deprived of all time cues or any rhythmic inputs.

Indirect methods

If one makes the not unreasonable assumption that one sleeps when one feels tired and has difficulty in doing so when one feels alert, then the results of sleep-deprivation experiments act as a marker of the rhythm of sleep and wakefulness. Regular subjective assessments of 'fatigue' and 'alertness' can be made throughout the study (more details of which tests are given in the next chapter). Such assessments can be given in terms of a scale of values (for example 1 is equivalent to 'not at all fatigued'; 10 is equivalent to 'as fatigued as I've ever felt') so that the variable can be described mathematically by continuous functions such as cosine curves. The difficulties are that the relationship between 'fatigued—alert' and 'asleep—awake' is not known and, by their very nature, the experiments can be performed for short durations only.

Another indirect assessment of the sleep—wakefulness cycle is locomotor activity. This frequently makes use of wheel or perch movement in the cages of experimental animals but requires more sophisticated equipment in humans. The isolation unit of Aschoff and Wever (Wever, 1979) has pressure sensors in the floor whereby movement can be monitored. Again the relationship between this and the sleep—wakefulness rhythm is not always clear, as in the example of a subject who spends time relaxing in a chair. However, studies by this group have shown that under most circumstances (the exceptions will be discussed later) there is a close relationship between locomotor activity and the subjective assessment of 'sleep' and 'waking'. As Wever (1979, p. 16) has pointed out, the subjective estimate of 'sleep' will be longer than the true value and the subjective estimate of 'waking' will always be an underestimate.

More recently, locomotor activity rhythms have been measured in man using wrist-mounted accelerometers, the data from which may be stored on a silicon chip. This method of recording locomotor activity is preferred since it allows monitoring of subjects outside the confines of specially constructed units and will detect most activities.

Subjects in isolation

Most experiments have been performed in conditions in which subjects have been removed from time cues or rhythmic influences; such conditions have been met in caves or in specially constructed isolation units. In these experiments, the subjects have decided their own pattern of sleep and wakefulness as well as their own meal times and the ways in which to pass the waking hours. One advantage of such a protocol is that the experiments have not been limited by a lack of sleep (c.f. the indirect methods described above); it is argued that under these circumstances any rhythmicity that remains must originate endogenously (but *see also* the views of Brown, described in Chapter 1).

2. Results of Sleep–Wakefulness Studies

2.1. Sleep deprivation

Fröberg et al. (1972) have studied two groups of soldiers who stayed awake for 75 hours. The subjects scored their fatigue every 3 hours whilst living under conditions that masked nychthemeral cues. The average score for fatigue for the group of 63 subjects as a whole is shown in *Fig.* 5.1. Two main conclusions can be drawn from these data. First, there is a general increase in the sensation of fatigue as the amounts of prior wakefulness and sleep deficit increase. Secondly, superimposed upon this trend, there is a marked circadian rhythm with a peak value during the middle of the normal sleep period and minimum value about the middle of the waking day. This kind of result can no doubt be confirmed by the experience of the reader. In staying up all night, for reasons of work or leisure, a time is reached towards dawn when the feeling of fatigue begins to decrease even though sleep deficit continues to increase (though as well as a circadian influence, a knowledge of real time might be a contributory factor here). Further, during the evening of the following day, a marked increase in fatigue occurs and this results presumably from a combination of sleep deprivation and circadian influence.

In passing it is worth mentioning that after time-zone transitions or during shift work there is evidence (to be considered in more detail in

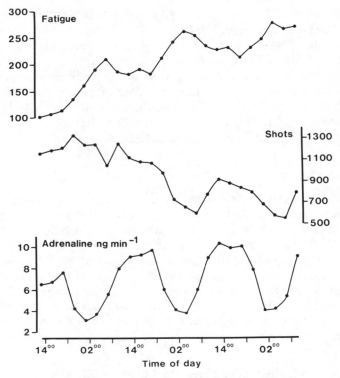

Fig. 5.1. Mean self-rating of fatigue, psychometric performance (number of shots fired by electronic rifle at a target) and urinary excretion of adrenaline measured every 3 hours in 63 subjects over 75 hours during which no sleep was allowed. Note that, despite the overall gradual increase in fatigue and decline in performance, circadian changes can also be seen. (Data of Fröberg et al., 1972.)

Chapters 9 and 10) that individuals are exposed to the combined effects of sleep deprivation and of the circadian variation in the sensation of fatigue. All these results suggest that the sensation of fatigue is influenced at least in part by some endogenous component.

2.2. Subjects isolated from time cues and other rhythmic influences

Experiments performed in caves

The first record of a prolonged sojourn in a cave without a timepiece is that of *Siffre* (Halberg et al., 1965). During his 61 days underground he telephoned to the experimenters at the surface his times of retiring and rising. While awake, Siffre had use of a dim light; he also recorded his pulse rate and the duration of short (1–120 seconds) time intervals given to him by the staff at the surface. The pattern of sleep and wakefulness

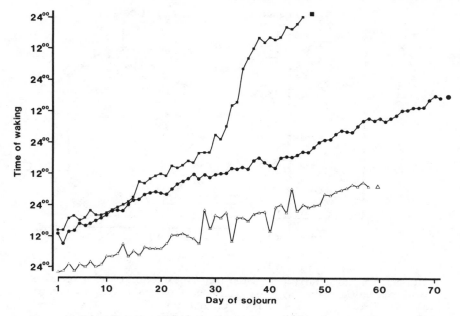

Fig. 5.2. Time of waking over successive days of three subjects isolated from time cues in caves. △ Siffre; ● Workman; ■ Lafferty (latter half of sojourn). (*From* Conroy and Mills, 1970, Fig. 7.2.)

from this first experiment indicated that the subject did not become arrhythmic but rather that his day continued with a frequency of waking and of retiring of about 24·5 hours (*Fig.* 5.2). Very similar periods were found for the rhythms of pulse rate (Chapter 3) and the ability to estimate short intervals of time.

Over the years the periods spent in isolation by subjects have lengthened. The last attempt was in 1972 when Siffre regained his original record, this time by remaining underground for 205 real days. A list of all experiments so far performed is given by Wever (1979, Table 1/I) and reviews are to be found also in Mills (1967), Halberg et al. (1970) and Czeisler (1978).

What physiological data have all these ventures provided? The answer can be only tentative at the moment since, even though later experiments have generally been associated with the collection of a greater range and amount of data, the results of all this effort have not yet been reported fully.

There seems to be general agreement that the original observations—that variables continue rhythmically with a period greater than 24 hours —have been confirmed, at least in the short term (*see Fig.* 5.2). However, as the length of confinement has increased, some circadian rhythms, and especially the relationship between them, have become more erratic. (But

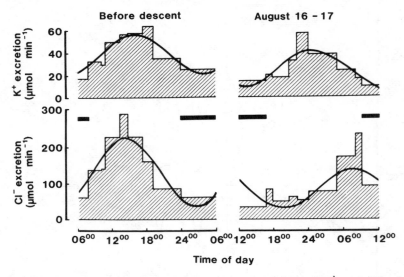

Fig. 5.3. Circadian variations in the urinary excretion of K$^+$ and Cl$^-$ of Workman on (*left*) the day prior to a timeless sojourn in a cave and (*right*) on 16–17 August, after 43 days without a timepiece in the cave. Best-fitting cosine curves are superimposed. Horizontal bars indicate times of sleep. (*From* Mills, 1964, Fig. 2.)

it must be remembered that one is always dealing with undesirably small numbers of subjects.)

In *Workman* (Mills, 1964), the rhythms of sleep–wakefulness and urinary excretion of sodium and potassium seem to have been in phase with each other (internal synchronization) for the first 8 weeks and showed a period of about 24·5 hours. Thereafter, the chloride rhythm, but not that of potassium, first changed its relationship to the activity pattern and then became increasingly irregular (*Fig.* 5.3). Similarly, a changing phase relationship between urinary volume and potassium excretion was seen in *Laures* (Halberg et al., 1970) and a dissociation between urinary electrolytes and activity in *Lafferty* (Mills et al., 1974). In this last subject, plasma samples also were taken every 4 hours for the last 3 days of his isolation. Analysis of these for 11-hydroxycorticosteroids suggested a rhythm with a period of 16 hours, a period found too in the urinary excretion of potassium during this time.

In experiments lasting even longer (4–6 months) even more remarkable changes in the pattern of sleep and wakefulness have been reported. During the first few days of his 182 days in isolation, *Mairetet* showed a normal circadian period (Jouvet, 1968). However, he then spontaneously developed a cycle lasting just over 48 hours—'circa-bi-dian'—and, later still, cycles lasting about three or even four times the 'normal' length (*Fig.* 5.4). This figure also indicates that readaptation to a nychthemeral

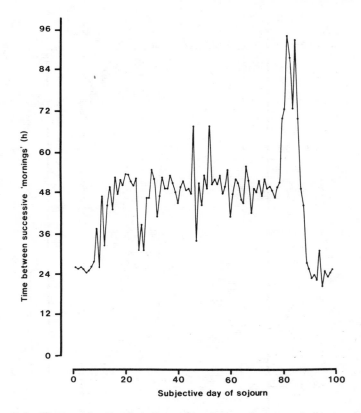

Fig. 5.4. Changes in the duration of real time between 'mornings' subjectively assessed by Mairetet during 6 months' temporal isolation in a cave. Note that for much of the sojourn the subject's assessment of a 'day' corresponded to about 48 hours real time—the 'circa-bi-dian' rhythm. (Data of Jouvet, 1968.)

existence was, as far as can be judged, immediate. A later attempt to get subjects to adjust to a 48-hour 'day' (Jouvet et al., 1975) was successful with only one subject (*Englender*). The other subject (*Chabert*), while not readily adopting a 48-hour 'day', did nevertheless show a large range of patterns of sleep and wakefulness as *Fig.* 5.5 indicates. This result will be discussed further after results from isolation unit experiments have been examined.

Experiments performed in isolation units

Caves are inhospitable environments, generally with low temperatures and high humidity; it is possible that such an environment places a great strain upon the subject and it also has the disadvantage of not being a convenient one for the experimenters to work in. A better arrangement is possible

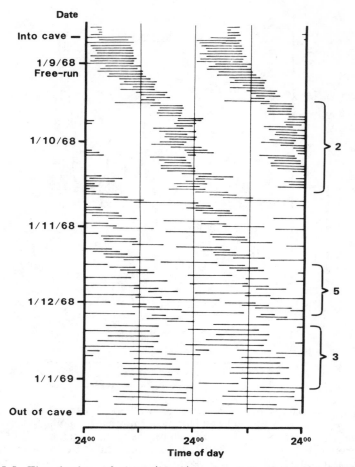

Fig. 5.5. The rhythm of sleep (black) and activity (white) in Chabert during 4 months' temporal isolation in a cave. The periods marked 2, 5 and 3 correspond to the anomalies described in the text. *See text* for further details. (Data of Jouvet et al., 1975.)

in isolation units in which the temperature, humidity and subjects' comfort can all be maintained at a reasonable standard. The majority of experiments have been performed in Germany and have been summarized in Wever's book (1979). Other contributions have come from America (Czeisler, 1978) and England (Mills et al., 1974). Isolation unit experiments have not lasted anything like as long as experiments performed in caves; instead, durations of from 2 to 4 weeks have been more common. Nevertheless, as earlier chapters have indicated, results almost identical to those from Siffre's first sojourn have been obtained; the rhythm of sleep and wakefulness has continued with a period in excess of 24 hours and with other rhythms synchronized to it.

As in cave experiments, however, various 'anomalies' have been observed. These take many forms, but some of them are:

1. A changing phase relationship (internal dissociation) between some variables and the sleep—wakefulness cycle (Mills et al., 1974, Fig. 1).

2. An occasional long or short 'day', the rest of the record showing a regular circadian rhythm (Wever, 1979; Fig. 36).

3. The development of 'circa-bi-dian' (48+ hour) or 'circa-semi-dian' (12+ hour) rhythms (Wever, 1979; Fig. 31 and 33 respectively).

4. The development of internal desynchronization (*Fig.* 2.3).

5. The alternation of one or more short sleeps with one or more long sleeps, these two sequences of sleeps often appearing to be 180° out of phase (Mills et al., 1974, Fig. 4).

Many of these 'anomalies' have been found in Czeisler's study (1978). Further, he has compared data from isolation unit experiments with the patterns of sleep and wakefulness that were observed in the cave dweller Chabert. The similarity which exists between isolation unit experiments and the data of Chabert is indicated on *Fig.* 5.5 by the regions marked (2), (3) and (5) which indicate the coincidence between Chabert's record and the appropriate 'anomaly' described above. Czeisler considers that no study lasting over 2 months has failed to show some form of 'anomaly' in sleep—wakefulness patterns and that the difference between individuals relates to how soon these changes will arise rather than whether or not they will (Czeisler, 1978, p. 136). Indeed Wever (1979) observed internal desynchronization in some subjects as soon as they entered the isolation unit.

3. Inferences from Isolation Experiments

3.1. Are sleep and wakefulness determined by an oscillator?

The observations just described, namely that, in the absence of external rhythmic influences, sleep and wakefulness do not become arrhythmic, suggest that some form of circadian oscillator is involved. Further evidence for this view is as follows.

Intuitively one might suppose that a stochastic relationship between sleep and wakefulness would exist whereby a longer-than-usual period of activity would be followed by a compensatory period of sleep that was longer than usual. However, an alternative has been proposed by Aschoff et al. (1971) that is based upon an oscillator model. Details are to be found in their review but in principle the argument is as follows: the subject goes to sleep when an oscillating function falls below some 'threshold' value and he wakes up when the function rises above this value. If, for any reason, the subject has a span of wakefulness that is longer than usual, that is, his sleep onset is delayed, then he will still tend to wake at the normal time as the oscillation rises above the threshold value; that is, there would be an *inverse* relationship between the lengths of sleep and prior wakefulness. Evidence in favour of this prediction has been presented by Aschoff et al. (1971) that was based upon their studies on isolated subjects who remained internally synchronized throughout; however, such an inverse relationship was not found for all his subjects by Czeisler (1978). Czeisler found also that if periods of prior wakefulness in excess of about 21 hours were considered (as during some 'anomalous' routines, *see above*) then a direct relationship between prior wakefulness and sleep seemed a better description. This result accords better with a 'stochastic' rather than an 'oscillator' model of the sleep–wakefulness rhythm.

In Chapter 4 the changes in acrophase of variables relative to the light–dark cycle produced by alterations in the zeitgeber period were discussed (*see Fig.* 4.6). Such an argument applies to the rhythm of sleep and wakefulness also. As this Figure shows, there is a small advance in the rhythm when the light–dark cycle is lengthened: this would not be observed if the rhythm were determined by the light–dark cycle only. However, it will be noted that the endogenous component is weak and similar to the case of urinary water excretion.

Perhaps the most convincing data favouring an independent oscillator for sleep and wakefulness are those observed during internal desynchronization (*see Fig.* 2.3), a phenomenon that has already been referred to in previous chapters. Why it takes place in about one-third of the subjects in isolation is not understood, though Lund (1974) found a higher degree of 'neuroticism' in subjects showing the phenomenon (for further comment on this *see* Chapter 12, s. 3.1). As an alternative, it was suggested by Mills et al. (1974) that the time of retiring could be delayed by interest or advanced by boredom and that this might produce a sleep–wakefulness rhythm in which an interested subject stayed awake longer and had longer sleeps by way of compensation. (It will be recalled that Czeisler (1978) found that the 'stochastic' rather than the 'oscillator' model better described the sleep–wakefulness rhythm when long periods of wakefulness were involved.) Whether such a suggestion could account for the *regularity* of the desynchronized sleep–wakefulness rhythm or accords with the subjects' own views on whether they were 'interested' or 'bored' has not been investigated.

In summary, there is evidence that the rhythm of sleep–wakefulness can be controlled by an oscillator, albeit a rather weak one. The properties of this oscillator and the extent to which it can act independently of the oscillator that controls the circadian rhythm of temperature will be discussed in Chapter 12.

3.2. Is there a relationship between temperature and sleep–wakefulness rhythms?

Even though the phenomenon of internal desynchronization indicates that the sleep–wakefulness and temperature rhythms can be independent,

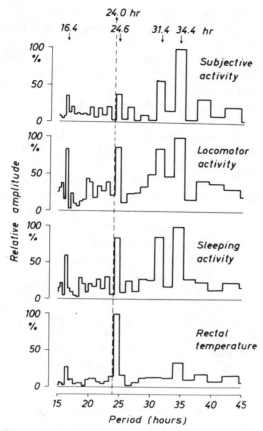

Fig. 5.6. Fourier analysis of the periodic components in the activity and sleep and temperature rhythm of a subject living in temporal isolation in a unit. Subjective activity was recorded by the subject indicating when he was active, about to have a nap or about to go to sleep. Locomotor activity was recorded from floor-mounted motion sensors. Sleeping activity recorded from bed-mounted motion sensors. (*From* Wever, 1979, Fig. 30/III.)

under nychthemeral conditions the two rhythms, together with many others, are phase-locked (hence the practice of referring acrophases to midsleep, Halberg et al., 1969). As a result it cannot be determined whether or not the two rhythms are driven by different oscillators. Under free-running conditions, the generally observed circumstance of internal synchronization produces a result whereby the temperature rhythm is phase-advanced with respect to the sleep—wakefulness cycle and the subject retires at a time of about minimum core temperature (*see Fig.* 2.3 before day 15). In addition, the feeling of fatigue is normally correlated with low temperatures (*see* s. 2.1 and Chapters 6 and 9), and so it becomes plausible to argue that, during free-running conditions with internal synchronization as well as during nychthemeral conditions, the rhythm of core temperature is an important cause of the rhythm of sleep and wakefulness acting via the sensation of fatigue. (Clearly the picture could easily be made far more complex by the inclusion of the reticular formation, as in Chapter 3.) Further evidence in support of a link between the rhythms of temperature and sleep—wakefulness comes from the following experiments:

1. Subjects after time-zone transition or on night work (*see* Chapters 9 and 10) have difficulty in sleeping when their deep body temperature is high and feel tired when it is low.

2. Subjects who, by virtue of living 'circa-bi-dian days' (for example, Mairetet) or of showing internal desynchronization, show an abnormal relationship between temperature and sleep—wakefulness rhythms often take 'naps' at times coincident with low body temperature but rarely do so when body temperature is high (Colin et al., 1968; Wever, 1979, Fig. 82).

In addition, these subjectively assessed 'naps' can last as long as a subjectively assessed 'sleep'. A consequence of these two factors is that, under 'anomalous' conditions, the objective assessment of sleep and wakefulness (made by pressure-sensitive plates in the floor of the isolation unit) will contain a circadian component due to the sleep that is coincident with the minima of the temperature rhythm. By contrast, the subjective assessment of sleep and wakefulness will underestimate the length of 'naps' and have no such component (Wever, 1979); this is illustrated in *Fig.* 5.6.

Thus, the two rhythms of sleep—wakefulness and temperature are normally phase-locked, but occasionally can be separated; however, under these circumstances, each rhythm still exerts some influence upon the other as has been described in Chapter 2 and will be commented upon again in Chapter 12.

3.3. The sleep—wakefulness rhythm as a masking influence

In earlier chapters the effect of the sleep—wakefulness rhythm upon other rhythms has been considered. In many cases a direct action exists (*see* Mills et al., 1978) and this has been termed 'masking' by Aschoff (1978). What connection is there between this masking effect and the idea of an 'exogenous influence' as defined in Chapter 1?

One solution is to consider that the sleep—wakefulness rhythm always acts as a masking influence (because it always exerts a direct effect upon other rhythms), but whether or not it is an exogenous influence depends upon what factor is influencing the pattern of sleep and wakefulness. Thus, on a non-24-hour 'day' when the pattern of sleep and wakefulness is determined by an external factor (adherence to a timepiece) it seems reasonable to assert that the sleep—wakefulness cycle and its direct masking effect are exogenous as they are being determined by external influences. On the other hand, when the process of internal desynchronization is considered, the direct masking effect of the sleep—wakefulness cycle is derived from a second internal oscillator and so it is endogenous. It follows that, under nychthemeral conditions, to the extent that the rhythm of sleep and wakefulness is determined by external factors (knowledge of time, social influences etc.) it is reasonable to consider it to be a direct and exogenous influence upon other rhythms and it is as such that it has been treated in previous chapters. It is worth while to point out that the validity of such an argument is not affected by the nature of the component or components of the sleep—wakefulness rhythm (changes in activity, food or water intake, posture, social awareness, light reaching the retina, posture or neurophysiological activity) that are responsible for the masking effect.

3.4. Does all rhythmicity in a time-free environment ultimately break down?

The observations that 'anomalies' develop on the pattern of sleep and wakefulness during the course of long term experiments performed in caves have already been described; do such 'anomalies' appear in other rhythms under conditions of isolation? Unfortunately the answer to this problem is severely restricted by the fact that many of the data required either have not been measured or the results from them have not yet been published.

Most data come from studies of temperature and so far these indicate that the rhythm is far more stable, retaining a circadian period of about 25 hours even when the sleep—wakefulness rhythm is behaving 'anomalously'. For example, a temperature rhythm with a period of about 25 hours continued: (a) when Mairetet (Colin et al., 1968) or isolation unit subjects (Wever 1979, Fig. 31) were living a 'circa-bi-dian day'; (b) when internal desynchronization had taken place (see Fig. 2.3); (c) when alternating long and short sleeps were being taken (Mills et al., 1974, Fig. 4); (d) when the period of the sleep—wakefulness cycle changed in response to a changing light—dark zeitgeber period (Wever, 1979, Fig. 94).

For urine, the only data from cave experiments so far are those of Workman (Mills, 1964) in which the increasing irregularity of the excretory rhythm of chloride (but less so that of potassium) has already been mentioned (Fig. 5.3).

The more nebulous concept of 'time sense' seems to change markedly in these experiments. The subjective day extends from one sleep until the next and shows a period of about 25 real hours in normal, free-running circumstances. When the sleep—wakefulness cycle behaves 'anomalously', the time sense invariably changes in the same way. Thus subjects showing internal desynchronization, living 'circa-bi-dian days' or being entrained by zeitgeber of varying period length all sense their 'day' length (from one sleep until the next) as being normal; further, they even plan their meals and general routine to accord with their perceived day length. By way of two examples:

1. While living a 'circa-bi-dian day', Mairetet took his 'daily' quota of cigarettes and meals during the course of 48+ rather than 24+ hours (Fraisse et al., 1968).

2. A group of subjects who alternated between a short and long sleep perceived the short sleeps as post-lunch 'naps' which they took fully dressed in easy chairs (Mills et al., 1974, group H).

Finally, two comments can be made that speculatively relate the changes observed in cave experiments to other aspects of human circadian rhythms.

First, the results from Mairetet lead one to wonder whether, in the absence of external rhythmicity, the endogenous rhythmicity gradually 'disintegrates'. How far this process would continue for the sleep—wakefulness cycle is not known nor is the extent to which other, seemingly more stable, rhythms (for example, that of temperature) would undergo the same process in due course, as implied by Czeisler (1978). If progressive deterioration of rhythms did take place, then it would suggest that, in the long term in humans, a rhythmic environment was required for the maintenance of internal rhythmicity. This problem is encountered again in Chapter 8 when the development of circadian rhythms in infancy is considered. A highly speculative link between infancy and cave-dwellers is suggested when *Fig.* 5.5 is examined and compared with *Fig.* 8.2 (p. 168); in some respects the two figures can be considered to be the reverse of one another. Whether this resemblance is fortuitous or indicates some basic similarity between the processes of establishment and 'disintegration' of circadian rhythms is not known but there is further speculation on related aspects in Chapters 8 and 12.

Secondly, the 'disintegration' of rhythmicity is not equally marked for all components, temperature and urinary potassium excretion seeming relatively immune, urinary chloride excretion and sleep—wakefulness relatively susceptible. There might be a link between this result and the relative importance of endogenous and exogenous components in determining the circadian rhythm of a variable; thus variables affected more by exogenous components might 'disintegrate' more readily under conditions of isolation. Perhaps this suggests that the link between the internal clock and different rhythms tends to get weaker in isolation affecting variables with small endogenous components first; perhaps it suggests instead that some oscillators (for example, that responsible for the sleep—wakefulness cycle) are less stable than others (for example, that responsible for the temperature rhythm).

An account by Siffre (1975) of his 205-day sojourn underground indicates that more than just circadian rhythms fall into disarray during such long periods of isolation. 'I am convinced that final results of this experiment will reveal serious problems confronting future long-range space travellers. Whether because of confinement, solitude, or both, my mental processes and manual dexterity deteriorated gravely and inexorably toward the end of my stay in Midnight Cave'. The full analysis of data obtained from this experiment is still awaited and promises to be of great importance; however, some doubt would seem to exist as to whether or not experiments of this duration (or even longer) will be repeated.

4. Sleep Stages

The earlier distinction that has been drawn between the two states of 'sleep' and 'wakefulness' is based upon the profound behavioural and neurological differences that exist between them. However, brain function is not constant throughout either, as is indicated by the changing sensation of fatigue during wakefulness that has already been described. We shall now consider in more detail the rhythmic changes that are associated with each state, during the state of sleep in this chapter and during the waking period in Chapter 6.

The conventional method for assessing sleep stages is by electro-encephalographic (EEG) recording and the method of scoring this is given elsewhere (Rechtshaffen and Kales, 1968). These recordings have to date required bulky recorders and the inconvenience of having to wear many electrodes on the scalp. Consequently considerable amounts of data have been obtained in sleep laboratories but there are fewer and less complete data from 'field' situations. The recent development of a means of recording that can be used conveniently at home is promising (Campbell et al., 1979) but the problem of securing good electrical contact with the scalp will remain and continue to make field studies of sleep stages difficult.

4.1. Sleep stages during the normal night

It is generally found that the recordings from the first night's sleep are slightly abnormal but that thereafter the sequence shown in *Fig.* 5.7 is regularly found. After the onset of sleep, its depth progressively increases from stage 1 through to stages 3 and 4, known together as slow wave sleep (SWS). Thereafter the subject's depth of sleep decreases and normally, after a latency of about 100 minutes from sleep onset, he enters the first episode of 'paradoxical sleep' or rapid eye movement sleep (REM sleep). The sequence back to stage 4 and then to REM sleep is repeated, suggesting that sleep is a cyclic phenomenon (the REM/non-REM cycle). The period between successive periods of REM sleep is usually about 90 minutes such that four or five of these cycles take place during a single night, but they are not identical in composition, the time spent in stage 4

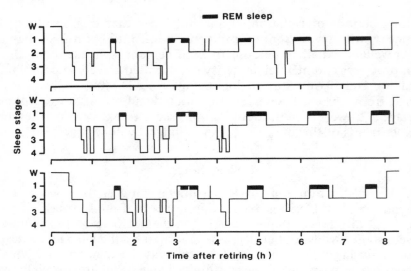

Fig. 5.7. The distribution in sleep stages over 3 consecutive nights recorded from a single subject. (Data of Hume, 1978.)

sleep decreasing and that spent in REM sleep increasing as the night continues (Webb, 1974).

There is evidence that both the individual sleep stages and their distribution within sleep (the sleep profile or 'infrastructure') are important. Thus:

1. The ratio of SWS or REM sleep to total sleep is fairly constant on different nights for any individual (Czeisler, 1978).

2. Abolition of SWS or REM sleep, by waking the subject whenever he passes into that stage during the course of his sleep, becomes a progressively more difficult task since the subject passes ever more readily into the missing stage; at the end of such an experiment, a 'rebound' phenomenon is observed in which large amounts of the missing stage are taken (Kleitman, 1963; Webb, 1971).

3. Attempts to modify the sleep profile within the night (for example by preventing SWS from taking place towards the beginning and REM sleep towards the end of a night's sleep) have met with only moderate success. Thus, the amount of a stage taken at an 'inappropriate' time of the sleep could be raised above normal values but not high enough to equal those found at the 'appropriate' time on control nights (Webb, 1971).

4.2. Sleep stages with changed routines

These results suggest the importance of the sleep infrastructure as observed during a normal night, but why this infrastructure is important is uncertain. Does it indicate that there is a circadian rhythm in the frequency at which different sleep stages occur during the nychthemeron; or does it indicate instead that, for some reason, the body needs SWS more than REM sleep and so tends to take it earlier in the night? A number of approaches to this problem have been made; the results have led to the model that the amount of SWS depends upon the amount of prior wakefulness but that the distribution of REM sleep shows a circadian variation, with most REM occurring just before noon and least before midnight. Some of these approaches and the data they have produced will now be discussed. A report of a workshop covering many aspects of this problem has recently appeared (Czeisler and Guilleminault, 1980).

Experimental changes in sleep time during the normal nychthemeron

An approach that has provided a considerable amount of data is to require subjects to take 'naps' at different times of the nychthemeron in addition to the normal sleep period (Aserinsky, 1973). The results of other studies have been summarized by Taub and Berger (1973). Both studies indicate that there was a similar pattern of sleep stages in 'naps' and in nocturnal sleeps; thus 'naps' taken late in the day (when the subject has experienced considerable prior wakefulness) had a profile similar to the first part of a normal night sleep (that is, high SWS and low REM sleep) whereas 'naps' taken early in the day showed profiles that resembled the last part of a normal nocturnal sleep (that is, low SWS and high REM). If the normal 8 hours of sleep was split into two 4-hour periods, one from 23^{00} to 03^{00} and the second after a waking period of 1, 4 or 12 hours, the amount of SWS correlated positively with the amount of prior wakefulness but did not depend upon the time of day. By contrast, the amount of REM sleep was dependent upon the time of day, being more in the morning and afternoon than late at night (Webb and Agnew, 1971).

The most comprehensive study of the effect of time of naps upon REM sleep and SWS content has been performed by Hume and Mills (1977). Two related protocols were used. In the first (*Fig.* 5.8B), after a control period on a normal routine, subjects took two 4-hour sleeps, one invariably between 24^{00} and 04^{00} and the other systematically 2 hours later each day. The amount of REM sleep and SWS per hour during the moving sleep was measured. As the experiment progressed, the amount of SWS increased whereas the amount of REM sleep rose and then fell to a minimum in the evening. The relative importance of prior wakefulness and circadian rhythmicity in producing these results could be assessed by considering the sleep taken from 24^{00} to 04^{00} since it would be influenced by differing amounts of prior wakefulness only. The results indicated that the amount of SWS depended upon the amount of prior wakefulness but that the amount of REM sleep was hardly affected by this. When any correction necessary for prior wakefulness had been made, then the effect due to time of day was small for SWS but marked for REM sleep, there being highest amounts just before noon and decreasing amounts thereafter. A possible objection to this protocol is that there was a systematic change in the time at which the moving sleep was taken. This is countered by the second protocol (*Fig.* 5.8C) in which the sequence of moving sleeps was randomly arranged; the experimental results were unchanged, however. Mathematical analysis of the pooled data from both protocols indicated a

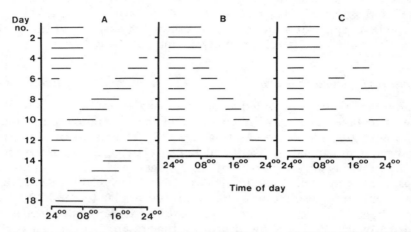

Fig. 5.8. Experimental protocols used by Hume and Mills (1977). Successive days are plotted from above downwards. The times when subjects were asleep are indicated by black bars. (Data of Hume and Mills, 1977.)

significant positive linear correlation between SWS amount and prior wakefulness and a significant fit of a cosine curve to the REM sleep data with an acrophase at about 10^{30}. Presumably the increasing tendency to take REM sleep as sleep progresses during control nights (*Fig.* 5.7) is because, from midnight until 08^{00}, one is on the rising phase of the circadian rhythm.

Taub and Berger (1973) have compared the sleep profiles in 8-hour sleeps taken at a normal time (24^{00} to 08^{00}) and then starting 2 or 4 hours earlier and 2 or 4 hours later. They consider that these changes in hours of sleep mimic more closely the natural changes. As the hour for retiring changed from the earliest (20^{00}) to the latest (04^{00}) time, the subjects showed increased amounts of REM and SWS, in accord with the model described above. In addition, they found a decrease in the length of the REM/non-REM cycle and a marked decrease in the amount of stage 2 sleep. It is generally assumed that this stage, like stage 1, is some form of 'makeweight' in the sleep infrastructure and so 'accommodates' the changing amounts of REM and SWS, but this does not seem to have been investigated at all fully.

Experimental changes in routines and their effect upon sleep stages

Changes in routines have been produced by placing subjects in a time-free environment, by simulating time-zone transitions and shift work or by requiring subjects to live a normal routine but on a non-24-hour day. As the data below will indicate, they all support the model previously described, namely: (*a*) REM sleep and SWS are necessary components of sleep; (*b*) the amount of SWS is determined by prior wakefulness, its distribution is generally at the beginning of sleep; (*c*) the amount of REM is not affected by prior wakefulness but its distribution is determined by a circadian oscillator.

Free-running conditions: Webb and Agnew (1974) placed subjects in a time-free environment for two weeks and allowed them to free-run as adjudged by their times of retiring and rising. They then measured the amount of REM sleep and stage 4 sleep from 22^{00} to 02^{00} on the occasions when this was in the 2nd, 4th, 6th or 8th hour of sleep and when prior wakefulness had been between 15½ and 16½ hours. A fall in the amount of stage 4 sleep and a rise in REM sleep was found as the time of recording progressed from the 2nd to the 8th hour of the sleep period. From the experimental design, this could not be attributed to the time of day or the amount of prior wakefulness and so a distribution of sleep stages in these subjects similar to that in entrained controls was indicated; that is, under free-running circumstances, the whole sleep 'infrastructure' shifted with the times of rising and retiring.

However, in detail, some changes in the distribution of REM sleep and SWS have been observed in free-running experiments. Thus previous chapters have described how, in free-running rather than entrained circumstances, endogenous rhythms become phase-advanced with respect to the rhythm of sleep and wakefulness. Such a result has been found with the rhythm of REM sleep in the free-running studies of Czeisler (1978). In free-running as compared to entrained subjects, he found a decrease in time to the first REM episode, an increased amount of REM sleep in the first 3 hours of sleep and a more rapid accumulation of REM sleep as the sleep progressed. By contrast, such changes in SWS were not observed, again suggesting that any endogenous component in this type of sleep is minimal. Nevertheless, Czeisler (1978) and Webb (1978) report that in long sleeps a SWS 'surge' is found after about 9 hours sleep; this might suggest that some (ultradian?) cyclic component is involved. The ratio of SWS to total sleep is unchanged in free-running experiments (Czeisler, 1978) and the amount of SWS is reported to be greater in subjects living a 48-hour 'day' in a cave (Chouvet et al., 1974); these results can be interpreted to reaffirm the importance of SWS and its dependence upon prior wakefulness rather than the effects of an endogenous oscillator.

Simulated time-zone transitions and shift work: A series of simulated time-zone transitions involving 8-hour advances or delays of routine has been performed with recordings of sleep stages being made during the nights before and after the shift (Webb et al., 1971; Hume, 1978). Both studies gave similar results, some of which are shown in *Fig.* 5.9, indicating that initially after the westward shift (when the subjects would be sleeping at what was 08^{00} to 16^{00}) there was an increase in REM sleep in the third of the night (08^{00} to 11^{00} pre-shift time) and a decrease in the last third of the night (13^{00} to 16^{00} pre-shift time). Similarly, after an eastward shift, on the first night there was an increase in REM sleep in

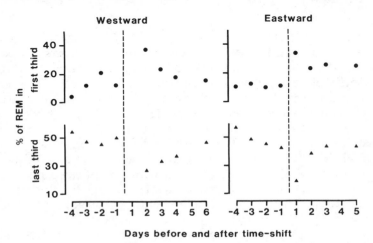

Fig. 5.9. Distribution of REM sleep in the first and last thirds of sleep before and after simulated time-shifts through 8 time zones in a westward or eastward direction. Mean of 7 subjects from westward shift; mean of 8 subjects for eastward shift. (Data from Hume, 1978.)

the first third of the night (16^{00} to 19^{00} pre-shift time) and a marked fall in the last third of the night (21^{00} to 24^{00} pre-shift time), again compared with controls. *Fig.* 5.9 indicates further that the distribution and amount of REM sleep had approached control values (that is, had adapted to the shift) after about 5 or 6 days; as a result of this adaptation, a subsequent time-zone shift resulted in very similar changes to those seen after the first shift. Finally, the amount 'of SWS seems once again to depend upon duration of prior wakefulness rather than the time (whether on a pre- or post-shift clock) at which sleep was taken.

A study by Weitzman et al. (1970) in which subjects inverted their hours of sleep (from $22^{00}-06^{00}$ to $10^{00}-18^{00}$) mimicked the effects of night work (*see also* Berger et al., 1971; Webb et al., 1971). SWS was little affected by this change but the distribution of REM sleep changed. Thus, instead of an increasing incidence as the sleep progressed, the inverse was seen, most REM sleep, like SWS, occurring early in sleep. There was evidence for some degree of adaptation to the changed routine during its second week.

Non-24-hour routines: Several non-24-hour schedules of sleep and wakefulness have been tried (for example, *see* Weitzman et al., 1974; Carskadon and Dement, 1975; Webb, 1978). The most comprehensive series has been performed by Webb and Agnew (1977) and consisted of a group of between 2 and 14 males whose sleep stages were measured while on the following sleep–wakefulness regimens (all given in hours asleep : hours awake): 3 : 6, 4 : 8, 6 : 12, 10 : 20 and 12 : 24; later (Webb, 1978) a schedule of 16 : 32 was used as well. The authors stressed the relative stability of sleep infrastructure with SWS tending to occur before REM, whatever the length of sleep. In addition, the amount of SWS depended upon the amount of prior wakefulness rather than the hour at which it was measured, the amount of REM sleep depended upon the time of measurement rather than prior wakefulness and stage 2 sleep seemed to depend upon neither factor.

Hume and Mills (1977) measured sleep stages in subjects on a 21-hour 'day' (14 hours awake, 7 hours asleep) (*see Fig.* 5.8A). The amount of REM sleep taken during each hour of the 24 showed a circadian rhythm with an acrophase at about 11^{00}, a result almost identical to that with subjects on the 'split-sleep' schedules (*see Fig.* 5.8B and C). The amount of SWS did not vary, prior wakefulness being constant throughout at 14 hours' duration. The similarity in acrophase of REM sleep in this and the 'split-sleep' studies of *Fig.* 5.8B and C might be, in fact, slightly anomalous. If one accepts that rhythms in general (and the REM rhythm in particular) cannot be entrained by a 21-hour 'day' (as discussed in earlier chapters and in Chapter 12), then one would predict free-running endogenous rhythms with a period of more than 24 hours. As a result of this, the time of peak REM sleep would become later each day and sampling from the non-stationary data on different days would not necessarily reveal the nature of that rhythm (*see* Chapter 1). The fact that, in this study, the free-running period of the endogenous components of other rhythms was so close to 24 hours (24·2 to 24·4 hours) is therefore fortunate and might explain the congruency between results on these 21-hour 'days' and the 'split-sleep' schedules. (Chapters 9 and 10 will give evidence that the 'split-sleep' schedules used here were associated with stable rhythms with an exact 24-hour period.)

Results from field situations

As has previously been stated, the cumbersome nature of the apparatus required has severely limited the number of experiments which have been performed; accounts of these studies will be given in Chapters 9 and 10.

At this stage it is sufficient to point out that the data from those field studies accord with the laboratory-based results that have just been described. Thus, the same basic sleep infrastructure exists with a cycling between REM and non-REM states. Whereas the amount of SWS seems related to the amount of prior wakefulness, that of REM sleep shows some of the properties associated with an endogenous oscillator with peak values just before noon.

4.3. The REM/non-REM cycle and some of its implications

As this account has indicated, the alternation between REM sleep and non-REM sleep has been a consistent finding in sleep studies. This has led to the view that it is the manifestation of an endogenous ultradian oscillator that is responsible for a basic 'rest–activity cycle' which can be detected throughout the nychthemeron (Kleitman, 1963). Discussion of the diurnal manifestations of such an oscillator will be deferred until Chapter 12 as will discussion of any possible link between it and the internal circadian oscillator(s). In addition, the possible relationship between sleep stages and hormone release will be considered in Chapter 7. The present account will be concerned with the evidence relating to the stability of the REM/non-REM cycle during sleep.

When the period of the REM/non-REM cycle is considered in detail, there seems to be a considerable variation both between and within individuals. Thus Webb (1974) found an average delay of 90 minutes between the first and second REM episodes of a night's sleep, but there was a large scatter about this average. Lewis also (1974) concluded that there was both a large inter- and intra-individual variation in the period of the REM/non-REM cycle as well as the delay between sleep onset and the first REM episode; and a change in the REM/non-REM cycle length in subjects with altered sleep schedules was found in the experiments of Taub and Berger (1973) (*see* s. 4.2). However, no difference from control values in the duration of the REM/non-REM cycle (about 114 minutes) was found during attempts to entrain two cave-dwellers to a 48-hour day (Jouvet et al., 1975).

Earlier in this chapter, the argument advanced by Aschoff et al. (1971) in favour of the view that sleep and wakefulness are caused by some oscillator (namely, the inverse relation that existed between sleep and prior wakefulness) was put. Similarly, Dirlich et al. (1977) investigated the duration of adjacent REM and non-REM episodes in a free-running subject; they found an inverse relationship between sleep duration and prior wakefulness but no such relationship between REM sleep duration and either preceding or succeeding non-REM sleep length. Finally, in 3 subjects who were living nychthemerally, recordings of sleep stages were made on 30 successive nights. All subjects showed the cycling of REM/non-REM sleep together with the normal tendency for the amount of REM sleep to increase as the night progressed. However, inspection of REM episodes on successive nights suggested that the exact times of REM were slightly later each night, as though determined by some mechanism that had a period slightly longer than an exact multiple of 24 hours. This result could not be attributed to the times of sleep onset which were fairly consistent between nights.

In summary therefore, the alternation between different sleep stages has been consistently found, but it has not been proved whether this process is controlled by an oscillator and, if so, it is unsure how consistent is the period of this oscillator from day to day and between individuals. In addition, it is not clear whether or not this oscillator is in phase with the circadian rhythm of sleep and wakefulness.

References

Aschoff J. (1978) Features of circadian rhythms relevant for the design of shift schedules. *Ergonomics* **21**, 739–54.

Aschoff J., Gerecke U., Kureck A. et al. (1971) Interdependent parameters of circadian activity rhythms in birds and man. In: Menaker M. (ed.) *Biochronometry*. Washington, DC, National Academy of Sciences, pp. 3–27.

Aserinsky E. (1973) Relationship of rapid eye movement density to the prior accumulation of sleep and wakefulness. *Psychophysiol.* **10**, 545–58.

Berger R. J., Walker J. M., Scott T. D. et al. (1971) Diurnal and nocturnal sleep stage patterns following sleep deprivation. *Psychon. Sci.* **23**, 273–5.

Campbell K., Weller C. and Wilkinson R. T. (1979) A portable pulse-interval modulation telemetry/multiplexing eeg recording system for use in the home. *Electroencephal. Clin. Neurophysiol.* **47**, 623–6.

Carskadon M. A. and Dement W. C. (1975) Sleep studies on a 90-min day. *Electroencephal. Clin. Neurophysiol.* **39**, 145–55.

Chouvet G., Mouret J., Coindet J. et al. (1974) Periodicité bicircadienne du cycle veille-sommeil dans des conditions hors du temps. Étude polygraphique. *Electroencephal. Clin. Neurophysiol.* **37**, 367–80.

Colin J., Timbal J., Boutelier C. et al. (1968) Rhythm of the rectal temperature during a 6-month free-running experiment. *J. Appl. Physiol.* **25**, 170–6.

Conroy R. T. W. L. and Mills J. N. (1970) *Human Circadian Rhythms*. London, Churchill.

Czeisler C. A. (1978) Human circadian physiology: internal organization of temperature, sleep–wake and neuro-endocrine rhythms monitored in an environment free of time cues. PhD Thesis, Stanford University.

Czeisler C. A. and Guilleminault C. (1980) REM sleep: a workshop on its temporal distribution. *Sleep* **2**, 285–7.

Dirlich G., Zulley J. and Schulz H. (1977) The temporal pattern of the REM sleep rhythm. Proceedings XII International Conference, International Society for Chronobiology, Washington, 1975. Milan, Il Ponte, pp. 483–93.

Fraisse P., Siffre M., Oleron G. et al. (1968) Le rythme veille-sommeil et l'estimation du temps. In: de Ajuriaguerra J. (ed.) *Cycles Biologiques et Psychiatrie*. Symposium Bel-Air III. Geneva, Georg; Paris, Masson, pp. 257–65.

Fröberg J., Karlsson C.-G., Levi L. et al. (1972) Circadian variations in performance, psychological ratings, catecholamine excretion and diuresis during prolonged sleep deprivation. *Int. J. Psychobiol.* **2**, 23–36.

Halberg F., Reinberg A., Haus E. et al. (1970) Human biological rhythms during and after several months of isolation underground in natural caves. *Bull. Natl Speleological Soc.* **32**, 89–115.

Halberg F., Reinhardt J., Bartter F. C. et al. (1969) Agreement in endpoints from circadian rhythmometry on healthy human beings living on different continents. *Experientia* **25**, 106–12.

Halberg F., Siffre M., Engeli M. et al. (1965) Étude en libre-cours des rythmes circadiens du pouls, de l'alternance veille-sommeil et de l'estimation du temps pendant les deux mois de sejour souterrain d'un homme adult jeune. *C. R. Acad. Sci. [D] Paris* **260**, 1259–62.

Hume K. I. (1978) PhD Thesis, University of Manchester.

Hume K. I. and Mills J. N. (1977) Rhythms of REM and slow-wave sleep in subjects living on abnormal time schedules. *Waking and Sleeping* **1**, 291–6.

Jouvet M. (1968) Phylogénèse et ontogénèse du sommeil paradoxal: son organisation ultradienne. In: de Ajuriaguerra J. (ed.) *Cycles Biologiques et Psychiatrie.* Symposium Bel-Air III. Geneva, Georg; Paris, Masson, pp. 185–203.

Jouvet M., Mouret J., Chouvet G. et al. (1975) Towards a 48-hour day: experimental bicircadian rhythm in man. In: Pittendrigh C. S. (ed.) *Circadian Oscillators and Organisation in Nervous Systems.* Cambridge, Mass., MIT press, pp. 491–7.

Kleitman N. (1963) *Sleep and Wakefulness*, 2nd ed. Chicago, University of Chicago Press.

Lewis S. A. (1974) The paradoxical sleep cycle revisited. In: Scheving L. E., Halberg F. and Pauly J. E. (ed.) *Chronobiology* Tokyo, Igaku Shoin, pp. 487–90.

Lewis P. R. and Lobban M. C. (1957a) The effects of prolonged periods of life on abnormal time routines upon excretory rhythms in human subjects. *Q. J. Exp. Physiol.* **42**, 356–71.

Lewis P. R. and Lobban M. C. (1957b) Dissociation of diurnal rhythms in human subjects living on abnormal time routines. *Q. J. Exp. Physiol.* **42**, 371–86.

Lund R. (1974) Personality factors and desynchronization of circadian rhythms. *Psychosom. Med.* **36**, 224–8.

Meddis R. (1968) Human circadian rhythms and the 48 hour day. *Nature* **218**, 964–5.

Mills J. N. (1964) Circadian rhythms during and after three months in solitude underground. *J. Physiol.* **174**, 217–31.

Mills J. N. (1967) Keeping in step—away from it all. *New Scientist* **33**, 350–51.

Mills J. N., Minors D. S. and Waterhouse J. M. (1974) The circadian rhythms of human subjects without timepieces or indication of the alternation of day and night. *J. Physiol.* **240**, 567–94.

Mills J. N., Minors D. S. and Waterhouse J. M. (1977) The physiological rhythms of subjects living on a day of abnormal length. *J. Physiol.* **268**, 803–26.

Mills J. N., Minors D. S. and Waterhouse J. M. (1978) The effect of sleep upon human circadian rhythms. *Chronobiologia* **5**, 14–27.

Mills J. N. and Stanbury S. W. (1952) Persistent 24-hour renal excretory rhythm on a 12-hour cycle of activity. *J. Physiol.* **117**, 22–37.

Rechtshaffen A. and Kales A. (1968) *A Manual of Standardized Terminology, Techniques, and Scoring Systems for Sleep Stages of Human Subjects.* Washington, DC, U.S. Government Printing Office.

Siffre M. (1975) Six months alone in a cave. *Nat. Geographics* **147**, 426–35.

Taub J. M. and Berger R. J. (1973) Sleep stage patterns associated with acute shifts in the sleep–wakefulness cycle. *Electroencephal. Clin. Neurophysiol.* **35**, 613–19.

Webb W. B. (1971) Sleep behaviour as a biorhythm. In: Colquhoun W. P. (ed.) *Biological Rhythms and Human Performance.* London, Academic Press, pp. 149–77.

Webb W. B. (1974) The rhythms of sleep and waking. In: Scheving L. E., Halberg F. and Pauly J. E. (ed.) *Chronobiology.* Tokyo, Igaku Shoin, pp. 482–6.

Webb W. B. (1978) The forty-eight hour day. *Sleep* **1**, 191–7.

Webb W. B. and Agnew H. W. (1971) Variables associated with split-period sleep regimens. *Aerospace Med.* **42**, 847–50.

Webb W. B. and Agnew H. W. (1974) Sleep and waking in a time-free environment. *Aerospace Med.* **45**, 617–22.

Webb W. B. and Agnew H. W. (1977) Analysis of the sleep stages in sleep–wakefulness regimens of varied length. *Psychophysiol.* **14**, 445–50.

Webb W. B., Agnew H. W. and Williams R. L. (1971) Effect on sleep of a sleep period time displacement. *Aerospace Med.* **42**, 152–5.

Weitzman E. D., Kripke D. F., Goldmacher P. et al. (1970) Acute reversal of the sleep–waking cycle in man: effect on sleep stage patterns. *Arch. Neurol.* **22**, 483–9.

Weitzman E. D., Nogeire C., Perlow M. et al. (1974) Effects of a prolonged 3-h sleep–wake cycle on sleep stages, plasma cortisol, growth hormone and body temperature in man. *J. Clin. Endocrinol. Metabol.* **38**, 1018–30.

Wever R. (1979) *The Circadian System of Man. Results of Experiments under Temporal Isolation.* Berlin, Springer-Verlag.

chapter 6 *Rhythms in Mental Performance*

An important effect of circadian rhythmicity in humans is that upon mental performance. The implications of this rhythmicity are of particular concern to people who have just undergone a flight across several time-zones or who are working at non-conventional hours; in both groups it is likely that decisions will need to be made or manipulations performed at times when the subjects are not performing at their best. Later chapters will deal with these problems but the present chapter will consider some of the laboratory findings and their possible explanation when performance is measured upon subjects living entrained to the normal day. A review of early evidence has recently appeared (Lavie, 1980).

1. Problems Associated with Psychometric Testing

Psychometric testing—the measurement of mental performance—does not produce an integrated value since the last time of sampling, as does the urine sample, but only a single datum point, as with the estimation of temperature and cardiorespiratory events. By their very nature, psychometric tests can be performed only by a waking subject and so whole stretches of data will normally be missing. The missing nocturnal data can be obtained by various changes in protocol but these often introduce further difficulties. Indeed the whole field of psychometric testing is beset with problems and it is to a brief consideration of these that we now turn.

1.1. Missing data

These data can be obtained by requiring the subjects to stay awake throughout the nychthemeron or by waking them at suitable intervals and allowing them to sleep between tests. If the subjects are kept awake for 24 hours they become fatigued (*see Fig.* 5.1). Even though fatigue is believed to exert little effect upon renal function and not much upon cardiorespiratory and temperature rhythms, there is little doubt that the general level of performance declines with sleep deprivation (Fröberg et al., 1972; Klein and Wegmann, 1979). On the other hand, if the subject is woken temporarily to perform a test, there is evidence that at least some aspects of performance *deteriorate* for some minutes after waking (Fort and Mills, 1972). This result is probably related to the more general finding that immediately after 'naps' there is a decrement of performance (Moses et al., 1978; Johnson, 1979) as though mental processes take some while to 'get going'. The loss of nocturnal data tends to diminish the usefulness of cosinor analysis, as has been stressed by Wever (1979). One possible alternative is to compare the results with some other variable (often temperature) measured simultaneously: another is to assess whether there is any variation of performance during the course of the measurements (without having to assume the time course of the variation) by an analysis of variance.

1.2. The lability of psychometric data

A finding with all psychometric tests has been their susceptibility to external factors. Thus performance can be modified by conditions of temperature, noise, lighting etc. and this has necessitated stringent control of the environment where possible. In previous chapters this result would have been attributed to a marked exogenous component and this is undoubtedly correct, but other factors are important also. One of the these is 'motivation' and it has been considered that this can prevent the appearance of rhythmicity in some tests due to 'extra effort' being made by the subject (Colquhoun, 1971; Klein and Wegmann, 1979). The factors, both exogenous and endogenous, that contribute to a subject's 'motivation' are not at all certain; what is certain is that such motivation can affect the outcome of psychometric tests far more than other variables considered in this book.

One outcome of the susceptibility of mental performance to external factors is that there has been a tendency for tests to be 'simple' (*see below*) so that they can be performed in the laboratory under conditions that are more strictly controlled. There are, of course, merits in these arguments but then the problem that arises is one of the relevance of the laboratory results, based upon artificial tests in artificial circumstances, to the situation in 'real life'. It is a problem that has often been considered (Colquhoun, 1971; Hockey and Colquhoun, 1972; Verhaegen, 1972) and will recur in this chapter and in that on shift work.

1.3. The assessment of psychometric tests

Another problem associated with psychometric tests is the 'practice effect' whereby the mean performance improves with the number of tests. The effect is marked initially but later becomes smaller (though the time course depends upon the test under consideration). The large changes can be accomplished before the experiment by means of preliminary practice tests so that during the experiment itself only smaller changes will be found. This smaller change can then be corrected by testing the subject at the same time on different days to estimate the change in performance due to practice and then correcting the other values by interpolation. Such a correction can be achieved also by fitting a compound function to the data (cosine curve plus linear trend) by computer. Alternatively, one can arrange for groups of subjects to be tested according to the same sequence of times but starting at a different point in the sequence. In this way, average values for the groups at different times of the day will have the same distribution of unpractised and practised components (see Fort and Mills, 1972).

A further problem is that most psychometric tests can be conducted and scored according to two criteria—speed and accuracy. This introduces difficulties since the two criteria need not show the same circadian variation. The instruction to subjects to perform a test 'as fast but as accurately as possible' does little to overcome this problem since the interpretation of such a command will differ between individuals. Provided that any subject interprets the criteria in the same way throughout the tests, then it is valid to assess the results for circadian rhythmicity. However, such an analysis is invalidated if the subject's strategy or interpretation of the instructions changes during the tests. Thus Colquhoun and Goldman (1972) found that subjects responded to warming (from taking exercise) by becoming more 'confident' in their decisions rather than performing them with the same degree of assurance throughout. For these reasons, there might be difficulties if results from a number of subjects are combined.

In conclusion, psychometric testing is associated with many problems in its implementation, its analysis and, as will be elaborated upon later, its interpretation. It is not surprising, therefore, that a number of authors have drawn attention to the need to standardize tests so that a comparison of results between different laboratories can be made more easily (Hamilton et al., 1972; Verhaegen, 1972).

2. Assessment of Mental Performance Rhythms

This standardization of tests has not yet taken place, as a result of which it is perhaps not unexpected to find that many types of test have been used to investigate mental performance rhythms. A description of all these would serve little purpose; some idea of the range can be gained by consulting the reviews by Kleitman (1963), Colquhoun (1971) and Hockey and Colquhoun (1972) as well as the references in the following account. This will consider some of the tests as groups; such divisions are bound to be rather arbitrary but they will attempt to distinguish subjective from objective assessments of performance and between simple and more complex tests.

2.1. Simple subjective tests

These tests rely very much upon the cooperation of the subject and there is the possibility that the answers might reflect how the subjects believe they should feel rather than their true feelings at the time. One example of a subjective test is the Activation–Deactivation checklist of Thayer (1967). The subject is required to indicate if each of a number of moods applies to him at that moment; his options are 'not at all', 'possibly', 'definitely' and 'very strongly'. The moods from the checklist can afterwards be grouped by the experimenter to measure 'General Activation' and 'Deactivation–Sleep'. There is an advantage in having groups of moods to assess these states since this gives some internal check on the consistency of the subject, a point that will be referred to again in Chapter 10.

The time taken to perform such a test, especially if the subjects are going to give equal consideration to each mood, can be considerable and there is a risk of boredom, especially if the test is to be performed frequently, and so shorter tests have been used. As a simple example, 'fatigue' can be assessed quickly by asking the subject to give a value from 1 (not at all fatigued) to 5 (extremely fatigued)—see for example Akerstedt and Fröberg (1976)—or even more simply by asking which of a group of feelings applies (indicated by a tick or a cross) at the moment of testing (Pátkai, 1971a). Alternatively the subject can place a mark somewhere along a line, the opposite ends of which indicate, for example, 'very sleepy' and 'very alert' (Pátkai, 1971a; Czeisler, 1978). The connection between low body temperature and the sensation of fatigue has been discussed already in Chapter 5. Such a relationship holds also under conditions of sleep deprivation when body temperature is lower than normal (see Chapter 9 and Colquhoun, 1971).

2.2. Simple objective tests

Many simple objective tests have been developed; these tasks can vary greatly in their complexity and the time they take to perform. One test that is relatively quick and simple is the Simple Reaction Time test (Kleitman, 1963). In this, the subject is required to push a button when a signal is heard (auditory version) or seen (visual version); it is the amount of time elapsing between the signal and response that is measured. Since neither the stimulus nor the response required varies, the test is classified as 'simple', though the observation that a delay of a few hundred milliseconds exists between stimulus and response would support the view of most neurophysiologists that, at an electrophysiological level, the processes involved were anything but simple!

As with the Simple Reaction Time test, any task will have sensory, 'central' and motor components and further, less 'simple', tests have been devised that emphasize these to different extents. Thus:

1. The sensory component can be emphasized. For example, the subject has to cancel the letter 'e' from a page of prose (Blake, 1967a) or adjacent duplicate letters in random sequences generated by computer (Fort and Mills, 1972); in these tests the motor component is still simple but the amount of 'perceptual discrimination' required by the task is increased. ('Perceptual discrimination' implies a mixture of sensory and central components; it is probable that an increase in either sensory or motor components must be associated with an increase in central activity, though not necessarily involving 'higher mental function' as defined in (3) below.)

2. The motor component can be emphasized. Examples are manipulative tests (Schubert, 1969), including the Kügel test (Bruner et al., 1960), aiming a pin-prick in the centre of a string of circles (Fort and Mills, 1972) or estimating the maximum rate of tapping with the finger (Fort et al., 1973).

3. The 'central component' can be emphasized. In practice this can be achieved in a number of ways.

First, an element of choice can be incorporated. One example of this is the Choice Reaction Time test in which the subject might be required to respond to different signals in different ways, such as pushing a button in response to a red, but not a green, light (Kleitman, 1963). Alternatively a set of cards could be sorted into various groups (Blake, 1967a).

Secondly, the ability to perform mathematical calculations could be tested (Blake, 1967a). In such a task there seems to be the need for a rapid turnover or 'throughput' of information.

In partial contrast, tests involving the use of short term memory have been devised. A straightforward example is the Digit Span Test in which the length of a random sequence of digits that can be repeated without error is found (Blake, 1967a). Short term memory is often combined with other mental facilities; for example, with the processes of logical deduction as tested by syllogisms (Fort and Mills, 1972; Folkard, 1975a) or by grammatical transformations (Folkard, 1975a); or with the process of scanning random sequences of letters, as in Memory and Search Test (MAST) of Folkard, Monk et al. (1976).

Fourthly, subjects can be asked to assess the passage of small intervals of time. This can be by: *estimation,* when the subject is asked to state the duration of an interval produced by the experimenters; *production,* when the subject is asked to produce an interval of a certain length; *reproduction,* when the subject is required to reproduce an interval given by the experimenter. (*See,* for example, Fraisse et al., 1968; Conroy and Mills, 1970.) These tests seem to have minimal sensory and motor components when compared with the central brain function required.

At this stage it must be pointed out that the extent to which these different tests of 'central' brain function overlap or assess different aspects of mental performance is quite unknown.

2.3. More complex objective tests

As will be described later, when the simpler psychometric tests are considered, there are occasions when a circadian rhythm has not been found. While this might well indicate that such a rhythm does not exist, it has been argued (Alluisi and Chiles, 1967; Hockey and Colquhoun, 1972) that it arises because the subject can correct for the effect of time of day by making an 'extra effort'. The argument continues that, when this is the case, then circumstances which 'load' the subject and push him towards the limit of his ability at performance tests are those which are most likely to result in circadian variation in performance tests. One such circumstance is increased fatigue of the subject. Another circumstance is an increase in task complexity, a further advantage of which might be that it would simulate better a field situation. However, with increasing complexity of tasks, two new problems arise. First (and this will be discussed in s.3.2) the interpretation of the results from such tasks becomes more speculative; secondly, the tasks become more difficult to standardize. These two problems apply to both field and laboratory-based tasks and so must be borne in mind in Chapter 10 also.

Some of the tests already described probably contain more than one element that is being tested but further increases in complexity have been achieved in three different ways:

1. The duration can be increased. A simple example of one way to increase the duration of a test is a modification of the Simple Reaction Test (Kleitman, 1963). In this modification, the response to the first stimulus causes presentation of the second stimulus, the response to the second stimulus causes presentation of the third stimulus and so on; the number of correct responses in 5 minutes is measured. Of course, any test can be prolonged almost indefinitely, but vigilance tests (in which subjects are required to detect 'incorrect' signals presented randomly in a continuous stream of 'correct' signals), especially when performed in simulators, are believed to mimic some field situations fairly closely (Hockey and Colquhoun, 1972; Klein and Wegmann, 1979).

2. A group of performance tests is assessed. This method has been used by, for example, Fröberg et al. (1972). A potentially important advance in this direction has been the development of a 'multiple task performance battery' (Alluisi and Chiles, 1967; Alluisi, 1972). In one example of this, the subject is required to work continuously at three vigilance-type tasks and to combine this with intermittent work at three other tasks measuring short term memory, perceptual and higher mental activity. In practice, great flexibility is possible since the concept of 'priorities' can be introduced to increase further the amounts of 'load' and 'performance stress' that are involved.

3. A single task which consists of a large number of components (often just how many being unknown) can be used. Examples would be reading

comprehension (Englund, 1979), long term memory and recall (Monk and Folkard, 1978; Folkard and Monk, 1980; Millar et al., 1980) and a 'pursuit motor task' in which the subject is required to trace the movement of a light around a square 'track' (Gibson and Allan, 1979).

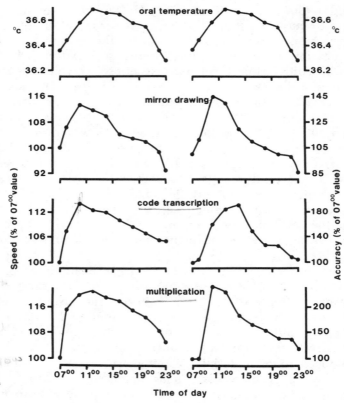

Fig. 6.1. Diurnal variations in oral temperature and the performance of various psychometric tests in a single subject (means of 20 days). For comparison, the oral temperature rhythm is plotted for both columns. Performance is expressed on the left as the reciprocal of the time taken to perform the test as a percentage of the time taken at 07^{00} and on the right as the reciprocal of the number of errors made as a percentage of the number of errors at 07^{00}. (*From* Kleitman, 1963, Fig. 16.1.)

3. Results on a Nychthemeral Routine

3.1. Simple tests

For many tests there is a remarkable amount of agreement as to the daily course of rhythms in mental performance. *Fig.* 6.1 indicates that, during the course of the day, a number of measurements of performance, whether of speed or accuracy, show increases after waking and falls before retiring.

The times of peak in this figure are fairly early (about midday) but then so is that of oral temperature; other tests by the same author showed peak values later in the day (about 18^{00}) again roughly coincident with peak temperature (Kleitman, 1963). Subsequent work has tended to confirm these findings, and the more frequent times of sampling have enabled the diurnal changes to be described in more detail. Thus Blake (1967a) divided the results from a group of tests into:

Group 1, those which rise until around evening (most simple tests, both subjective and objective, fall into this group).

Group 2, those which show an irregular time course (for example, time estimation).

Group 3, those which can reach an early peak and fall throughout the day (for example, short term memory).

In a more recent review, Hockey and Colquhoun (1972) came to very similar conclusions and stressed the difference in timing between most simple tests and short term memory.

When tests of group 1 above are performed throughout the night, a minimum performance at about 04^{00} is found. This cannot be attributed wholly to lack of sleep or to waking subjects from sleep (but *see* Fort and Mills, 1972) since performance improves after 04^{00} even if the same protocol continues. Nevertheless, fatigue does cause a general decrement in performance and so the circadian rhythm would be superimposed upon a linear trend (*see,* for example, *Fig.* 5.1). Other effects of fatigue upon circadian rhythmicity will be discussed below (s. 4.3).

Consideration will be given presently to mechanisms which are believed to play some role in the production of the observed rhythms, but it is important to consider the two exceptions to the general rule that psychometric performance improves until the evening. These two exceptions are time estimation and short term memory, groups 2 and 3 above.

Time estimation

As mentioned earlier, time estimation has small sensory and motor components and so is believed to indicate rather the working of some 'central' mechanism. However, there is not agreement as to whether or not circadian rhythms of time estimation are to be found under nychthemeral conditions (Blake, 1971; Colquhoun, 1971). In addition, the mechanisms by which circadian changes in time estimation (if they exist) are brought about are unclear. Raising or lowering body temperature artificially alters the processes of time estimation (thus, raising temperature leads to an overestimate of an estimated time interval and an underestimate of a produced time interval), but the absence of a diurnal rhythm requires the inference that the circadian changes in body temperature do not seem to be equally effective. In the free-running experiment when the subject Mairetet was living a 'circa-bi-dian' day (*see* Chapter 5), his attempts to reproduce and produce intervals gave results twice normal; that is, the passage of different intervals of time, whether of seconds or of hours, was equally affected (Fraisse et al., 1968). In addition, there was an increase in his Simple Reaction Test time (Oleron et al., 1972). However, his ability to estimate time was remarkably accurate throughout. These results are perplexing and still await a satisfactory explanation. Thus, if a single internal clock were responsible and 'ticking over' at about half speed, one would predict a halving of an estimated interval, a doubling of a produced interval and no change in a reproduced interval. One possibility is that not all assessments of the passage of time make use of the same internal clock, a view that will be considered again.

Short term memory

The evidence in favour of considering this as an exception to the general finding with performance tests is summarized by Folkard (1975a); maximum ability in short term memory tests is generally shown at or before noon with a fall throughout the rest of the day. In his own studies, Folkard has made use of syllogisms and grammatical transformation, both of which, he argues, contain short term memory components. His results indicate a peak in performance at about 14⁰⁰ with a fall thereafter (testing was performed every 3 hours, so more accurate times of peak cannot be given); this time course was considerably different from that of oral temperature which rose progressively until 20⁰⁰.

The reason why the rhythm of short term memory is phased differently is uncertain but there has been speculation that a process of 'clearing' information stored is involved (Hamilton et al., 1972). If this process showed a diurnal rhythm with the same phase as the tasks in group 1 above, then the 'store' would decay less quickly in the morning and so short term memory would be better then; conversely, group 1 tasks, which require rapid information transfer, rather than 'information storage', would be performed less well (*see* Folkard, 1975a). Data that can be used to support such an explanation of the circadian rhythm in short term memory come from the observation that this faculty improves after sleep deprivation (Hamilton et al., 1972).

3.2. Complex tests

These differences between the rhythms of short term memory, time estimation and other simpler performance tests introduce a complication when complex tasks are considered. These complex tasks will almost certainly contain different proportions of these three components all of which individually peak at different times of the day, and so the resultant time of peak for the task as a whole will be difficult to predict. By way of illustration, a test used by Folkard et al. (1976b) can be cited. This is a

Memory and Search Test (MAST) in which the subject is required to search sequentially through lines of letters for those lines which include a set of 'target' letters; the test contains short term memory and sensory elements. The memory component in this test can be increased by increasing the size of the 'target' set from two up to six letters. Under nychthemeral conditions (*see* Monk et al., 1978) the acrophase for the two-letter MAST (with a relatively high sensory component) is later than that for the six-letter MAST (with a relatively high short term memory component).

The concept has arisen (*see* Hockey and Colquhoun, 1972; Folkard, 1975a) of considering complex tasks to be a mixture of an 'information storage' or 'short term memory' component (which deteriorates as the day progresses) and an 'immediate throughput of information' or 'rapid information transfer' (which improves as the day progresses). Whatever might be the attractions of this theory, it is likely to be very difficult to test since, for any complex task, the relative importance of the two hypothetical components is not known. Further, any assessment of the anatomical and neurophysiological correlates of these two components would be pure speculation at the present time.

4. The Causal Nexus and Performance Rhythms

4.1. The role of temperature

Kleitman's results (*see Fig.* 6.1) indicate that a parallelism exists between temperature and performance rhythms and the synchrony between temperature and alertness rhythms was described in Chapter 5. Such a relationship has been found to hold in free-running subjects also (Wever, 1979), both the rhythms of fatigue and temperature becoming phase-advanced with respect to the rhythm of sleep and wakefulness (Czeisler, 1978). Earlier work by Blake (1967a) also demonstrated this parallelism but it also indicated that, in detail, differences between subjects were present. Thus, if subjects were divided into 'introverts' and 'extroverts' by means of questionnaires, then the temperature of introverts rose earlier in the morning and fell earlier in the evening when compared with that of the extroverts (Blake, 1967b) (*Fig.* 6.2). Further, with a number of psychometric tests, the introverts performed better than the extroverts in the morning but less well in the evening (Blake, 1971). Indeed, the terms 'larks' and 'owls', to describe individuals who feel less fatigued and more alert in the morning ('larks') or in the evening ('owls') is widely used. Blake (1967b) and Blake and Corcoran (1972) have argued that introverts are more highly 'aroused' than are extroverts in the morning and that the opposite holds in the evening ('aroused' will be defined in s. 4.3). More recently Colquhoun and Folkard (1978) have re-examined Blake's data and concluded that the distinction between introverts and extroverts is less clearly defined than was once thought.

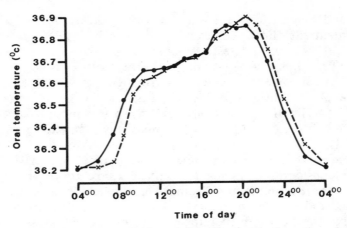

Fig. 6.2. Mean circadian rhythms of body temperature in relatively extreme introvert (●, *n* = 25) and extrovert (X, *n* = 22) groups. (Reproduced by permission *from* Blake, 1967b, *Nature* 215, 896–7. Copyright 1967 Macmillan Journals Ltd.)

In summary, these results argue for some relationship between performance and temperature rhythms, some implications of which will be considered in the chapter on shift work.

A direct test of the effect of temperature upon performance rhythms would be to change body temperature artificially. Changes within the physiological range have been achieved by immersing subjects in hot or cold water or by requiring them to take exercise or 'naps' (Fort et al., 1971, 1973). In some simple tests (for example, aiming, tapping speed, arithmetic calculation speed) the performance was better with raised temperatures and worse when body temperature had been lowered. By contrast, in other experiments, a decrement in performance in more complex tests has been observed when body temperature was raised above peak circadian values by use of a water suit (Allan et al., 1979). When the experiment was repeated with changed body temperatures falling within the physiological range, no clear changes in performance were found (Gibson and Allan, 1979). In another study, raising body temperature by exercise produced an increased number of errors in a vigilance test but no increase in the ability to detect test signals (Colquhoun and Goldman, 1972); however, as has already been mentioned, there was some evidence that the subjects' strategy in the test changed inasmuch as they felt more 'confident' that they were detecting the test signals correctly, a feeling that was, in fact, unjustified.

It is not clear whether the different results that have been obtained are due to differences in the performance tests or the means for changing body temperature that have been used. Both Allan et al. (1979) and Gibson et al. (1980) mention the unpleasantness felt by the subjects when body temperature is raised (especially to above-normal values) and it must also be remembered that all artificial changes of body temperature have overridden homeostatic thermoregulatory mechanisms; the effect upon performance tests of the stress that these factors will cause is not known with certainty, but doubtless it will be a confounding influence.

There is further evidence against the hypothesis that there is a direct link between temperature and performance rhythms.

First, Rutenfranz et al. (1972) investigated the effects of sleep deprivation. A general parallel during the course of the day between performance (assessed by measuring the reaction time) and temperature was found, but if both variables were measured at the same time on successive days, then no correlation between the two was found. Mann et al. (1972) came to the same conclusions in their study upon night workers. In a more recent study in which naval volunteers were deprived of sleep, either partially or totally, for 40 hours, again there were significant negative correlations between the diurnal courses of oral temperature, the sensation of sleepiness, errors in a vigilance test and time spent 'napping'. However, when the relationship between temperature and the other variables for the group was considered at a constant time of day, no significant correlation was found in any case (Moses et al., 1978).

Secondly, there is a transient fall in performance in the early afternoon (Colquhoun, 1971). This 'post-lunch dip' is not due to taking lunch (and dips do not occur after other meals, anyway) and its cause remains unknown. It has been speculated to be a manifestation of an ultradian oscillator that shows a minimum at this time, but data in favour of this view are lacking (see also Chapter 12). Nevertheless, temperature does not dip at this time and so the two variables can be independent.

Thirdly, 164 subjects were asked to assess what they thought their performance might be in a variety of tasks ('perceptual discrimination', 'motor skill' and 'short term memory') at different times of the day (Folkard, 1975b). These rhythms, based upon the subjective estimates, were then compared with those in the same tasks assessed objectively and with the temperature rhythm (Fig. 6.3). The parallelism between the subjective estimates for different tasks is clear, all of which incidentally show a 'post-lunch dip' (Fig. 6.3 above). However, when objective assessments of performance were made (Fig. 6.3 below) there were considerable differences between the rhythms (in accord with comments made in s. 3 above) and no rhythm exactly matched that of temperature when a detailed comparison was made. As Folkard points out, these results not only cast doubt upon the reliability of subjective estimates of performance but also indicate that temperature must be a poor predictor of performance.

The conclusion from all this is that temperature and psychometric tests are probably not as closely linked as was once thought to be the case. It is likely that temperature is a poor predictor of performance, in part

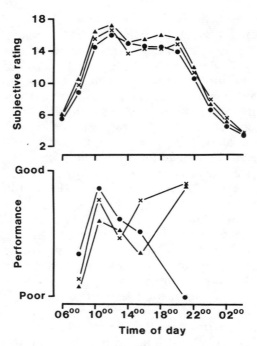

Fig. 6.3. The diurnal variation in various performance tasks. Above, subjective rating; below: objective rating. (●) Short term memory, (X) perceptual performance, (▲) motor skill. Objective ratings plotted so that amplitude of the various tasks is equal. (*From* Folkard, 1975b, Figs. 4 and 5; data of Blake, 1967a.)

because other factors can influence brain function; one of these factors might be the sympathetic nervous system, the activity of which can be inferred from catecholamine levels in the blood.

4.2. The role of the sympathetic nervous system

A group of Scandinavian workers have been the strongest proponents of the view that a connection exists between catecholamine levels, as assessed by rates of excretion in the urine, and psychometric performance. The subject has been reviewed recently by Akerstedt (1979).

A direct relationship between catecholamines and the sensation of alertness has been found in nychthemeral conditions (Akerstedt and Fröberg, 1976) and an inverse relationship between catecholamines and fatigue on a 75-hour vigil under constant conditions (Fröberg et al., 1972). In this latter study, from which *Fig.* 5.1 is taken, two further points emerged. First, there was a general increase in fatigue and fall in performance superimposed upon the circadian rhythmicity as the vigil progressed (this has already been referred to); secondly, there was also a slight progressive increase in adrenaline, again superimposed upon the circadian rhythm, and this is likely to be a response to increasing sleep deficit. Significant correlations between these variables were found, but, as has been stated on previous occasions, correlation does not prove causality.

A stronger case for a causal link can be inferred from the results of Pátkai (1971a,b) and Akerstedt and Fröberg (1976). On the basis of results from questionnaires, they classified subjects into 'morning' (lark) and 'evening' (owl) types. Pátkai found that the 'morning' type had higher levels of urinary adrenaline, felt more alert and could concentrate better at 09^{00} when compared with 'evening' types; by contrast the opposite situation held at 22^{30}. Moreover, these variables were higher at 09^{00} than 22^{30} for the 'morning' types but lower for the 'evening' group. Similar results were obtained in the study by Akerstedt and Fröberg (1976) who compared urinary catecholamines, oral temperature and self-assessments of fatigue and alertness every 3 hours throughout the daytime. Differences between the two groups were small in the evening but for all variables (except fatigue which was the inverse) values were higher in the 'morning' types from 06^{00} until at least 18^{00}.

Although there seems to be a general parallelism between catecholamine excretion and assessment of alertness, further comment is required.

As with the temperature rhythm, it has not been established that the link is causal. So far, nobody has investigated whether a correlation exists between fatigue ratings and simultaneous catecholamine levels when the time of day is kept constant. Such an investigation is more difficult in so far as the 'instantaneous' plasma catecholamine concentration rather than the 'integrated' urine level would be required (*see* Chapter 4) and vene-puncture or a venous catheter might stress the subject and complicate the picture.

It is not clear to what extent catecholamine levels are phased appro-priately for most performance tests. This problem has been raised before when cardiovascular and respiratory rhythms were considered. There is a spread of results (no doubt due to differences in type of subject and details of the protocol), some investigators placing the time of peak excretion a few hours before noon (for examples: Pátkai, 1971b; Akerstedt and Fröberg, 1976; Akerstedt, 1979), others sometime in the afternoon (for examples: Reinberg et al., 1970; Fröberg et al., 1972; Faucheux et al., 1976; Akerstedt and Levi, 1978). Whilst results which show a morning peak are acceptable for some estimates of the rhythm of blood pressure, for the rhythm in airway resistance (*see* Chapter 3) and even for the rhythm of short term memory, they seem inappropriate for other cardiovascular and respiratory rhythms (*see* Chapter 3), for many renal rhythms (*see* Chapter 4) and for most rhythms of performance, all of which peak later. Likewise, peak times in the afternoon are not accept-able for all variables. Some recent studies have complicated the issue still further. Thus two peaks of plasma adrenaline, at 14^{00} and 22^{00}, were found in a group of soldiers undergoing a regular routine of sleep, meals and light exercise (Sauerbier and von Mayersbach, 1977) and Hossmann et al. (1980) also found two peaks in plasma noradrenaline, at 09^{00} and between 15^{00} and 18^{00}, in both supine and standing subjects.

4.3. General arousal

The difficulty in wholly attributing the cause of performance rhythms to either the temperature or catecholamine rhythm has led to theories that implicate more general aspects of brain function, such as 'sympathetic nervous system performance' (Schubert, 1969; 1977) and 'arousal' (Taub and Berger, 1974). A similar line of argument, implicating the reticular formation in particular, has been developed in Chapter 3. If it is accepted that the 'arousal' of a subject might be reflected in functions such as pain and sensory thresholds (Rogers and Vilkin, 1978), somatosensory evoked potentials (Davis et al., 1978), auditory and visual evoked potentials (Kerkhof et al., 1980), as well as withstanding pain (Folkard, Glynn et al., 1976; Glynn et al., 1976), then in all cases a higher degree of 'arousal' in the evening than morning has been found. However, the subjects of Rogers and Vilkin (1978) also measured six moods and no correlation was found between the moods and the pain or sensory thresholds. This result, coupled with that of Strempel (1977), who found different phases for the rhythms for protopathic and epicritic pain thresholds, renders it unlikely that all rhythms in brain function can be controlled by a single factor, even one as general as 'arousal'. This conclusion accords with one reached earlier and has been discussed by Folkard and Monk (1981).

Another approach, but one that need in no way be incompatible, has been developed by psychologists. It is based upon the 'inverted U curve' and has been discussed by Colquhoun (1971) and Blake and Corcoran (1972) (*Fig.* 6.4). This curve indicates that performance efficiency increases with increasing level of arousal up to a peak value, after which efficiency falls off with further increases in arousal, a state of 'hyper-arousal' being reached. (The relationship between 'arousal', as used here and as used previously—to indicate one of the manifestations of reticular formation activity—is not known, but this does not affect the usefulness of the model.) As a result, changes in arousal level will affect performance efficiency far more when the overall level of arousal is low than when it is higher.

It is beyond the scope of the present text to consider all implications of this curve; for example, it has been considered by Blake (1971) and Blake and Corcoran (1972) as a partial explanation of some of the differences between introverts and extroverts. An implication that is more important for present purposes is that it also offers a possible explanation for the effects of stress, fatigue and motivation upon circadian rhythms of performance, especially in complex tests. Thus, it is generally accepted that fatigue will lower and stress will raise the general level of arousal (Colquhoun, 1971; Klein and Wegmann, 1979) and such views, coupled with use of the 'inverted U curve' (*Fig.* 6.4) would explain the following results:

1. As has already been mentioned, 'motivation', by increasing arousal, could reduce or remove the circadian variation in performance efficiency by moving the subject to the top of the curve.

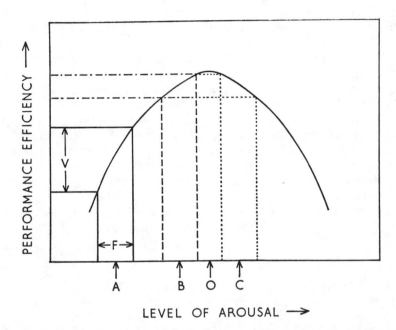

Fig. 6.4. The 'inverted U' relationship between the level of performance and arousal. A given fluctuation (F) in arousal level results in a greater variation (V) in performance level when the overall level of arousal is low (A) than when it is relatively high (B or C) and near the optimal point (O) for the performance task. (With permission from Colquhoun, 1971. Copyright by Academic Press Inc. (London) Ltd.)

2. Fatigue would produce a large amplitude of oscillation in a rhythm or result in a rhythm appearing where previously none had existed (Colquhoun, 1971; Bugge et al., 1979).

3. Stress, whether produced by overloading the subject ('performance stress') or by administering tests designed for such a purpose (for example, the Stroop test described in Pátkai, 1971b) causes rhythmicity to appear or be accentuated, the subjects exhibiting a 'hyper-aroused' state (Alluisi and Chiles, 1967; Alluisi, 1972).

5. Summary

This account has tried to draw attention to some of the complexities that face experimenters investigating rhythms of human performance. These complexities arise at all stages of testing, such as devising the test, administering and scoring it and interpreting the results that are obtained; complexities also arise when the subject is considered, since external disturbances, fatigue, motivation and the psychological make-up of the subject all exert an effect.

Considerable advances seem to have been made where the circadian rhythms of individual tests are concerned. However, two major problems persist. The first is the endogenous cause of the rhythms in mental performance. Attempts to explain the results of psychometric tests in terms of the circadian rhythm of temperature or sympathetic nervous system activity, or even something as general as 'arousal', have met with only limited success. It begins to seem as though the concept of a single cause, no matter how general, is too simple. Instead, the endogenous 'component' might be derived from a number of strands of the causal nexus (*see* Chapter 12); the relative strength of these strands might depend upon the mental processes involved and so differ between individual psychometric tests. Clearly a vast amount of further research needs doing here.

The second problem involves the relationship between the laboratory-based test and the field conditions; it will be discussed in Chapter 10.

References

Akerstedt T. (1979) Altered sleep/wake patterns and circadian rhythms. Laboratory and field studies of sympathoadrenomedullary and related variables. *Acta Physiol. Scand.* Suppl. 469, pp. 1–48.

Akerstedt T. and Fröberg J. E. (1976) Interindividual differences in circadian patterns of catecholamine excretion, body temperature, performance and subjective arousal. *Biol. Psychol.* **4**, 277–92.

Akerstedt T. and Levi L. (1978) Circadian rhythms in the secretion of cortisol, adrenaline and noradrenaline. *Eur. J. Clin. Invest.* **8**, 57–8.

Allan J. R., Gibson T. M. and Green R. G. (1979) Effect of induced cyclic changes of deep body temperature on task performances. *Aviat. Space. Environ. Med.* **50**, 585–9.

Alluisi E. A. (1972) Influence of work–rest scheduling and sleep loss on sustained performance In: Colquhoun W. P. (ed.) *Aspects of Human Efficiency. Diurnal Rhythm and Loss of Sleep.* London, English Universities Press, pp. 199–215.

Alluisi E. A. and Chiles W. D. (1967) Sustained performance, work–rest scheduling, and diurnal rhythms in man. *Acta Psychologica* **27**, 436–42.

Blake M. J. F. (1967a) Time of day effects on performance in a range of tasks. *Psychon. Sci.* **9**, 349–50.

Blake M. J. F. (1967b) Relationship between circadian rhythm of body temperature and introversion-extraversion. *Nature* **215**, 896–7.

Blake M. J. F. (1971) Temperature and time of day. In: Colquhoun W. P. (ed.) *Biological Rhythms and Human Performance.* New York, Academic Press, pp. 109–48.

Blake M. J. F. and Corcoran D. W. J. (1972) Introversion-extroversion and circadian rhythms. In: Colquhoun W. P. (ed.) *Aspects of Human Efficiency. Diurnal Rhythm and Loss of Sleep.* London, English Universities Press, pp. 261–72.

Bruner H., Jovy D. and Klein K. E. (1960) Ein objectives Messverfahren zur Feststellung der psychomotorischen Leisungsbereitschaft. *Int. Z. Angew. Physiol. Einschl. Arbeitsphysiol.* **18**, 306–18.

Bugge J. F., Opstad P. K. and Magnus P. M. (1979) Changes in the circadian rhythm of performance and mood in healthy young men exposed to prolonged, heavy physical work, sleep deprivation and caloric deficit. *Aviat. Space Environ. Med.* **50**, 663–8.

Colquhoun W. P. (1971) Circadian variations in mental efficiency. In: Colquhoun W. P. (ed.) *Biological Rhythms and Human Performance.* London, Academic Press, pp. 39–107.

Colquhoun W. P. and Folkard S. (1978) Personality differences in body temperature rhythm, and their relation to its adjustment to night work. *Ergonomics* **21**, 811–17.

Colquhoun W. P. and Goldman R. F. (1972) Vigilance under induced hyperthermia. *Ergonomics* **15**, 621–32.

Conroy R. T. W. L. and Mills J. N. (1970) *Human Circadian Rhythms.* London, Churchill.

Czeisler C. A. (1978) Human circadian physiology: internal organization of temperature, sleep–wake and neuroendocrine rhythms monitored in an environment free of time cues. PhD Thesis, Stanford University.

Davis G. D., Buchsbaum M. S. and Bunney W. E. (1978) Naloxone decreases diurnal variation in pain sensitivity and somatosensory evoked potentials. *Life Sci.* **23**, 1449–60.

Englund C. E. (1979) The diurnal function of reading rate, comprehension and efficiency. *Chronobiologia* **6**, 96.

Faucheux B., Kuchel O., Cuche J. L. et al. (1976) Circadian variations of the urinary excretion of catecholamines and electrolytes. *Endocrinol. Res. Commun.* **3**, 257–72.

Folkard S. (1975a) Diurnal variation in logical reasoning. *Br. J. Psychol.* **66**, 1–8.

Folkard S. (1975b) The nature of diurnal variations in performance and their implications for shift work studies. In: Colquhoun P., Folkard S., Knauth P. et al. (ed.) *Experimental Studies of Shift Work.* Oplanden, Westdeutscher Verlag GmbH, pp. 113–22.

Folkard S., Glynn C. J. and Lloyd J. W. (1976a) Diurnal variation and individual differences in the perception of intractable pain. *J. Psychosom. Res.* **20**, 289–301.

Folkard S. and Monk T. H. (1980) Circadian rhythms in human memory. *Br. J. Psychol.* **71**, 295–307.

Folkard S. and Monk T. H. (1981) Circadian rhythms in performance—one or more oscillators? In Sinz R. (ed.) *Proceedings of Psychophysiology.* Berlin, VEB Gustav Fischer (in the press).

Folkard S., Monk T. H., Knauth P. et al. (1976b) The effect of memory load on the circadian variation in performance efficiency under a rapidly rotating shift system. *Ergonomics* **19**, 479–88.

Fort A., Gabbay J. A., Jackett R. et al. (1971) The relationship between deep body temperature and performance on psychometric tests. *J. Physiol.* **219**, 17P–18P.

Fort A., Harrison M. T. and Mills J. N. (1973) Psychometric performance: Circadian rhythms and effect on raising body temperature. *J. Physiol.* **231**, 114P.

Fort A. and Mills J. N. (1972) Influence of sleep, lack of sleep and circadian rhythms on short psychometric tests. In: Colquhoun W. P. (ed.) *Aspects of Human Efficiency. Diurnal Rhythm and Loss of Sleep.* London, English Universities Press, pp. 115–27.

Fraisse P., Siffre M., Oleron G. et al. (1968) Le rythme veille-sommeil et l'estimation du temps. In: de Ajuriaguerra J. (ed.) *Cycles Biologiques et Psychiatrie.* Geneva, Georg; Paris, Masson, pp. 257–65.

Fröberg J., Karlsson C.-G., Levi L. et al. (1972) Circadian variations in performance, psychological ratings, catecholamine excretion, and diuresis during prolonged sleep deprivation. *Int. J. Psychobiol.* **2**, 23–36.
Gibson T. M. and Allan J. R. (1979) Effect on performance of cycling deep body temperature between 37·0 and 37·6 °C. *Aviat. Space Environ. Med.* **50**, 935–8.
Gibson T. M., Allan J. R., Lawson C. J. et al. (1980) Effect of induced cyclic changes of deep body temperature on performance in a flight simulator. *Aviat. Space Environ. Med.* **51**, 356–60.
Glynn C. J., Lloyd J. W. and Folkard S. (1976) The diurnal variation in perception of pain. *Proc. R. Soc. Med.* **69**, 369–372.
Hamilton P., Wilkinson R. T. and Edwards R. S. (1972) A study of four days partial sleep deprivation. In: Colquhoun P. W. (ed.) *Aspects of Human Efficiency. Diurnal Rhythm and Loss of Sleep.* London, English Universities Press, pp. 101–13.
Hockey G. R. J. and Colquhoun W. P. (1972) Diurnal variation in human performance: a review. In: Colquhoun W. P. (ed.) *Aspects of Human Efficiency. Diurnal Rhythm and Loss of Sleep.* London, English Universities Press, pp. 1–23.
Hossmann V., Fitzgerald G. A. and Dollery C. T. (1980) Circadian rhythms of baroreflex reactivity and adrenergic vascular response. *Cardiovasc. Res.* **14**, 125–9.
Johnson L. C. (1979) Sleep disturbance and performance. In: *Sleep, Wakefulness and Circadian Rhythm.* AGARD Lecture Series No. 105, AGARD, ch. 8.
Johnson L., Naitoh P., Lubin A. et al. (1972) Sleep stage and performance. In: Colquhoun W. P. (ed.) *Aspects of Human Efficiency. Diurnal Rhythm and Loss of Sleep.* London, English Universities Press, pp. 81–100.
Kerkhof G. A., Korving H. J., Willemse-v.d. Geest H. M. M. et al. (1980) Diurnal differences between morning-type and evening-type subjects in self-rated alertness, body temperature and the visual and auditory evoked potential. *Neurosci. Letters* **16**, 11–15.
Klein K. E. and Wegmann H.-M. (1979) Circadian rhythms of human performance and resistance: operational aspects. In: *Sleep, Wakefulness and Circadian Rhythms.* AGARD Lecture Series No. 105, AGARD, ch. 2.
Kleitman N. (1963) *Sleep and Wakefulness*, 2nd ed. Chicago, University of Chicago Press.
Lavie P. (1980) The search for cycles in mental performance from Lombard to Kleitman. *Chronobiologia* **7**, 247–56.
Mann H., Pöppel E. and Rutenfranz J. (1972) Untersuchungen zur Tagesperiodik der Reaktionszeit bei Nachtarbeit. III. Wechselbeziehungen zwischen Körpertemperatur und Reaktionszeit. *Int. Arch. Arbeitsmed.* **29**, 269–84.
Millar K., Styles B. C. and Wastell D. G. (1980) Time of day and retrieval from long-term memory. *Br. J. Psychol.* **71**, 407–14.
Monk T. H. and Folkard S. (1978) Concealed inefficiency of late-night study. *Nature* **273**, 296–7.
Monk T. H., Knauth P., Folkard S. et al. (1978) Memory based performance measures in studies of shiftwork. *Ergonomics* **21**, 819–26.
Moses J., Lubin A., Naitoh P. et al. (1978) Circadian variation in performance, subjective sleepiness, sleep and oral temperature during an altered sleep–wake schedule. *Biol. Psychol.* **6**, 301–8.
Oleron G., Fraisse P., Zuili N. et al. (1972) The effects of variations in the sleep–wakefulness cycle during a 'Time-Isolation' experiment on reaction time and spontaneous tempo. In: Colquhoun W. P. (ed.) *Aspects of Human Efficiency. Diurnal Rhythm and Loss of Sleep.* London, English Universities Press, pp. 171–6.
Pátkai P. (1971a) The diurnal rhythm of adrenaline secretion in subjects with different working habits. *Acta Physiol. Scand.* **81**, 30–4.
Pátkai P. (1971b) Interindividual differences in diurnal variations in alertness, performance and adrenaline excretion. *Acta Physiol. Scand.* **81**, 35–46.

Reinberg A., Ghata J., Halberg F. et al. (1970) Rythmes circadiens du pouls, de la pression artérielle, des excrétions urinaires en 17-hydroxycorticostéroides catécholamines et potassium chez l'homme adulte sain, actif et au repos. *Ann. Endocrinol. Paris* **31**, 277–87.

Rogers E. J. and Vilkin B. (1978) Diurnal variation in sensory and pain thresholds correlated with mood states. *J. Clin. Psychiatry* **39**, 431–2.

Rutenfranz J., Aschoff J. and Mann H. (1972) The effects of a cumulative sleep deficit, duration of preceding sleep period and body-temperature on multiple choice reaction time. In: Colquhoun W. P. (ed.) *Aspects of Human Efficiency. Diurnal Rhythm and Loss of Sleep.* London, English Universities Press, pp. 217–29.

Sauerbier I. and von Mayersbach H. (1977) Circadian variations of catecholamines in human blood. *Horm. Metab. Res.* **9**, 529–30.

Schubert D. S. P. (1969) Simple task rate as a direct function of diurnal sympathetic nervous system predominance: a law of performance. *J. Comp. Physiol. Psychol.* **68**, 434–6.

Schubert D. S. P. (1977) Alertness and clear thinking as characteristics of high naturally occurring autonomic nervous system arousal. *J. Gen. Psychol.* **97**, 179–84.

Strempel H. (1977) Circadian cycles of epicritic and protopathic pain threshold. *J. Interdiscipl. Cycle Res.* **8**, 276–80.

Taub J. M. and Berger R. J. (1974) Diurnal variations in mood as asserted by self-support and verbal content analysis. *J. Psychiat. Res.* **10**, 83–8.

Thayer R. E. (1967) Measurement of activation through self-report. *Psychol. Rep.* **20**, 663–78.

Verhaegen P. (1972) General discussion. In: Colquhoun W. P. (ed.) *Aspects of Human Efficiency. Diurnal Rhythm and Loss of Sleep.* London, English Universities Press, pp. 241–5.

Wever R. A. (1979) *The Circadian System of Man. Results of Experiments under Temporal Isolation.* Berlin, Springer-Verlag.

chapter 7 *The Endocrine System*

Hormones that affect metabolism (Chapter 3), or the kidney (Chapter 4), or which are believed to be manifestations of the state of 'arousal' in general (Chapter 6) have already been considered. The present chapter will deal with the other members of the endocrine system. Most consideration will be given to the pituitary-adreno-cortical axis. This is not only because rhythmicity in this system has been studied more than in the other endocrines but also because the system acts as a model for other systems. As will be seen, the properties of the rhythm of growth hormone are in many respects the opposite to those of cortisol; this hormone will be considered second and the other hormones, which have been studied far less fully, will be seen to fit somewhere within a framework formed by these two extremes. Two problems recur— the extent to which nychthemeral changes in hormone levels can be considered as part of a 'homeostatic' system, and the relationship between hormone secretion and the stages of sleep— and these will be considered at the end of the chapter. Changed hormone rhythms in disease states will be referred to in Chapter 11.

1. Problems

A major difficulty associated with studies of circadian rhythmicity in the endocrine system arises because the concentrations of free hormone in the plasma are so low. As a result of this, the volume of blood that was required for chemical or biological assay (which anyway were not always very specific) was large. This limited the number of blood samples that could be taken and hence the accuracy with which a rhythm could be assessed. An additional problem was that the technique of repeated venepuncture could be stressful. Therefore, it is not surprising that, as an alternative, the rhythm of some excretory metabolite in the urine has been measured. This method has the advantage that the metabolite is present in a higher concentration and that larger volumes of urine can be sampled. The disadvantages are that the relationship between the rhythm of the hormone in the plasma and that of its metabolite in the urine might render the interpretation of the data more complex, that there might be other sources of the metabolite being measured and that the hormone might be degraded to more than one product, not all of which are being measured.

Just as the study of the circadian rhythms of cardiovascular variables has been transformed by the development of continuous ambulatory monitoring devices (Chapter 3), so too has the study of circadian hormone rhythms by the development of radioimmunoassay systems. These have a greater specificity than many chemical or bioassay systems and require far smaller quantities of plasma so that a higher frequency of sampling becomes possible. Coupled with these improvements has been the more frequent use of an indwelling catheter by which samples can be taken whenever required; ingenious devices exist whereby samples can be continuously taken (for example, Kowarski et al., 1971). This technique produces an integrated value since the last sampling time in a similar way to that when the bladder is catheterized or micturition takes place. Such sampling techniques are suitable only for laboratory environments in which a restriction of mobility matters less and the subject can be supervised. Accordingly, in field situations, studies continue to be performed in conditions in which the difficulties involved have reduced the frequency of sampling blood or urine.

2. Cortisol, Adrenocorticotrophic Hormone and Corticotrophin Releasing Factor

2.1. Nychthemeral studies

The results of earlier studies are reviewed by Daly and Evans (1974). Similar conclusions were reached whether plasma concentrations of 11-hydroxycorticosteroids (11-OHCS) or urinary excretion of 17-hydroxycorticosteroids (17-OHCS) were considered, namely values were highest at about waking, declined throughout the daytime until retiring and rose again during sleep. It was also a general finding that the pattern was remarkably reproducible from day to day and from subject to subject (Daly and Evans, 1974; Krieger, 1975; 1978). Therefore pooling data from different days or subjects is an acceptable procedure resulting in smooth circadian curves with maximum variances at times when averages are highest, that is, at about the time of waking.

This position was radically altered by studies in which blood samples were taken every 20 (Weitzman, Fukushima et al., 1974; Weitzman, 1976) or 30 (Krieger et al., 1971) minutes for extended periods of time. The plasma level of cortisol varies in a pulsatile manner suggesting a process of 'episodic' secretion by the adrenal gland (*Fig.* 7.1). There were between seven and thirteen secretory episodes per day with about one-quarter of the total time spent in secretory activity. These episodes were not distributed uniformly throughout the nychthemeron but divided instead into four phases (Weitzman, Fukushima et al., 1974). These were: 6 hours 'minimal secretory activity' (4 hours before, 2 hours after lights out); 3 hours 'preliminary nocturnal secretory phase' (3rd to 5th hour of sleep); 4 hours 'main secretory phase' (6th to 8th hour of sleep and 1st hour after

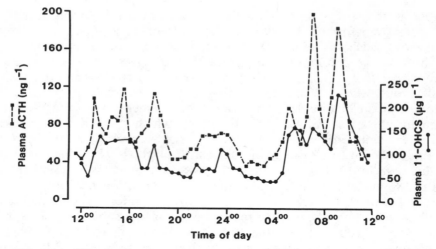

Fig. 7.1. Circadian variations of plasma 11-hydrocorticosteroid (OHCS) and ACTH concentrations in a female subject, determined by half-hourly sampling over a 24-hour period. (*From* Krieger et al., 1971, Fig. 2.)

waking); 11 hours 'intermittent waking secretory activity' (rest of nychthemeron). Since the exact time of the secretory bursts was not constant in different subjects or on different days, the consequences of pooling would have been to smooth out the episodic nature of the secretion and to produce a curve with a highest mean value on waking and with highest variance at those times when there were most frequent bursts, that is during the 'main secretory phase' at about the end of sleep.

2.2. Cause of the Cortisol Rhythm

In the above account it has been assumed that the episodic variations in plasma cortisol concentrations are due to episodic secretory activity by the adrenal cortex. This assumption is warranted since the rhythmic changes in the plasma level of free cortisol have been shown not to result from variation in the rate of hormone removal (DeLacerda et al., 1973), in the amount of hormone binding protein (DeMoor et al., 1962) or the metabolic pool of the hormone (Weitzman et al., 1971). In turn, these episodes result from the periodic secretion of adrenocorticotrophic hormone (ACTH). Close parallels have been found in studies in which ACTH and cortisol have been simultaneously measured in plasma samples removed each hour (Vague et al., 1974) or half-hour (Krieger et al., 1971) from healthy subjects on a normal diurnal routine. *Fig.* 7.1 shows the relationship that exists between plasma ACTH and cortisol when samples were taken every half-hour. A similar parallelism was found when sampling every 5 minutes (Gallagher et al., 1973) but occasionally two close episodes of ACTH secretion produced only a single burst of cortisol. It is believed that changes in the half-life of ACTH cannot account for the circadian variation that has been observed (Krieger, 1979) but, since its half-life is about 8 minutes (Krieger, 1975), a high sampling frequency will be required to fully substantiate this claim.

The two hormones ACTH and cortisol are known to be involved in a feedback loop whereby cortisol inhibits ACTH secretion at the level of the pituitary. However, this situation is complex since there is evidence that the responsiveness of the adrenal glands to ACTH and of the pituitary gland to cortisol both show circadian variation. Thus the adrenal response to an ACTH injection in the evening is less than that to the same dose in the morning (Krieger, 1979). There is evidence to indicate that this adrenal responsiveness is proportional to the concentration of plasma ACTH. Thus the adrenal response to an ACTH injection in the evening was greater if ACTH levels were already high due to a 'priming' injection of ACTH (Perkoff et al., 1959) and constant plasma levels of ACTH produced by continuous infusion were associated with loss of circadian rhythm in plasma cortisol (Nugent et al., 1960). Also, the suppression of ACTH release by the pituitary gland following glucocorticoid administration is greater in the evening than in the morning (Nichols et al., 1963), the clinical significance of which will be discussed in Chapter 11. However, more recent work upon rats has shown that the rhythm of adrenal responsiveness persists after hypophysectomy and after suppression of endogenous ACTH release by dexamethasone (Dallman et al., 1976) and so the position remains unclear.

In spite of all these complexities and their potential for inducing rhythmicity in a feedback system, the rhythmic changes in plasma ACTH are believed not to depend upon feedback control from cortisol since they continue (but at a higher mean value) after adrenalectomy (Krieger and Gewirtz, 1974). Instead, the rhythm is believed to depend upon CRF release from the hypothalamus which in turn results from rhythmic changes in transmitter release in the hypothalamus. Evidence for these statements has been found in some animal species but not yet in man (*see* Krieger (1979) for details). This is an aspect of circadian rhythms in which man is clearly an unsuitable species for study; nevertheless, it is an important area and will be mentioned again in Chapter 12 when neurotransmitters in general are considered. At this stage it would be sufficient to note that a vagolytic drug (Ferrari et al., 1977) and a β-adrenergic blocker (Cugini et al., 1975) have both been observed to be at least partially effective in preventing the morning rise of cortisol secretion.

2.3. The endogenous component

There is general agreement that the rhythm of cortisol, like that of deep body temperature (Chapter 2), has a small exogenous component. The main approaches by which the large endogenous component has been found will be described briefly below. It should be borne in mind that, in most studies, the sampling frequency has not been high and so little information is available from these studies about the presence or otherwise of an 'episodic' secretory pattern; further, with infrequent sampling, only approximate estimates of rates and amounts of adaptation can be made.

Nychthemeral conditions

The general nature of the nychthemeral rhythm has already been described (*see Fig.* 7.1) and its high degree of reproducibility has already been remarked upon. This can be confirmed by comparing results from transverse studies of different individuals (Aschoff, 1978; 1979) or from longitudinal studies (Czeisler, 1978). This day-to-day reproducibility is clearly shown in Czeisler's study in which the average rhythm is 'educed' (*see* Chapter 1); such analysis, like other methods of averaging data, smooths out the episodic components, but the relevant finding is that the variance about the educed curve is small. Reinberg et al. (1970) compared the urinary 17-OHCS rhythms in subjects under two routines. In both they were in the dark from 23^{00} to 08^{00}, but, in the first, normal activity and meals were allowed whereas, in the second, 4-hourly isocaloric meals were taken and the subjects were confined to bedrest for the whole sampling period. The authors found that the rhythms of urinary steroid excretion were 'almost indistinguishable' on the two routines. A similar conclusion was reached by Krieger et al. (1971) as a result of studies upon subjects who underwent different feeding schedules; thus the circadian plasma cortisol rhythms were similar whether subjects ate normal meals, identical snacks throughout the day or received intravenous glucose infusions. However, the published data do suggest that the episodic secretory pattern was more marked when normal meal times were taken, a result that has been found also in a study by Quigley and Yen (1979).

Free-running conditions

Under free-running conditions, the cortisol rhythm continues with only a slightly reduced amplitude though its phase is advanced with respect to the sleep—wakefulness cycle when compared with entrained conditions. As a result, the minimum plasma concentration is now some hours before retiring and a slight rise before sleep is often seen; the 'main secretory phase' begins earlier in sleep but continues until waking, a result that indicates an increase in duration of this phase (Weitzman et al., 1979). There have been similar findings when the rhythm of deep body temperature under the same conditions has been considered (*see* Chapter 2). This similarity exists too in subjects showing internal desynchronization (Czeisler, 1978; Wever, 1979), the main component of both rhythms showing a period of about 25 hours even when the sleep—wakefulness period is greater than 30 hours.

However, this stability of the cortisol rhythm in the face of irregularities of sleep and wakefulness has not invariably been found. Thus when Lafferty was tested after his 127-day isolation underground by sampling plasma every 4 hours for 72 hours, five peaks in plasma 11-OHCS concentration were found (Mills et al., 1974). The extent to which this result indicates a 'disintegration of circadian rhythmicity' (*see* Chapter 5) or rather is spurious, resulting from a combination of episodic secretion and the frequency of sampling, is not known.

Real and simulated time-zone transitions and shift work

As has already been stated, frequent blood sampling is not practical outside the laboratory. As a result, relatively few field studies have considered steroids and the exceptions have generally used urine samples collected infrequently enough for only gross changes to be measured. Nevertheless, those data that have been gathered all suggest that the rhythm adapts only slowly to real or simulated time-zone shifts (Aschoff et al., 1975). The rate of adaptation is not known with any degree of confidence but it seems to be slower than that of many other urinary constituents. Equally, adaptation to nightwork was found only in permanent night workers; the criteria for adaptation were a high plasma 11-OHCS on waking and a decline throughout the subject's 'day' to low values on retiring. Using these criteria, Conroy et al. (1970) concluded that adaptation of the steroid rhythm to nightwork took place very slowly and was complete only when nightwork had been 'universal, accepted and lifelong'. Weitzman et al. (1968) studied sleep stages, urine volume and urinary 17-OHCS excretion in 5 adults who underwent sleep reversal, now sleeping from 10^{00} to 18^{00}. Urine was collected by catheter every 30 minutes on alternate 'nights' for 2 weeks after reversing habits. The 'nocturnal' rhythms of urine volume rapidly adapted to the changed routine but those of temperature and 17-OHCS did not reverse until the second week. This is a particularly useful study since the laboratory environment permitted relatively frequent sampling and accurate assessment of rhythms. Another study, performed in the far more difficult conditions of the Arctic summer, also indicated a slow rate of adaptation of urinary steroids to a reversal of habits (Sharp et al., 1961); more details of this group of experiments have been given in Chapter 4.

Non-24-hour days

The important experiments in which subjects lived 21- or 27-hour 'days' have been mentioned in earlier chapters. In one part of these studies the subjects gave regular urine samples and the concentration of 17-OHCS in them was determined. In the original mathematical analysis (Simpson and Lobban, 1967), it was shown that the 17-OHCS rhythm did not disappear, nor did it adapt to the 21-hour day; instead an 'unadapted', 24-hour rhythm persisted. Later analyses of these data (Simpson et al., 1970; Simpson, 1974, 1977) indicated that the 'unadapted' rhythm had a period of 24·2 hours, that is, it was free-running. Moreover, when a 'circadian amplitude ratio' was calculated (*see* Chapter 1) its value was higher for the 17-OHCS rhythm than for potassium and the other urinary constituents. In other words, the endogenous component of the urinary 17-OHCS rhythm was very strong. Such a conclusion has been reached also by Weitzman, Nogeire et al. (1974) using a 3-hour sleep–wakefulness schedule and measuring plasma cortisol and by Wever (1979) using subjects on a 28-hour day and assaying urine. In this last study, the amplitude of the 28-hour component of the 17-OHCS rhythm was only one-third of that of the free-running, 24·8 hour component.

2.4. The exogenous component

In summary, therefore, the cortisol rhythm, like that of deep body temperature, has a large endogenous component and some implications of this will be considered in Chapter 12. However, there is some exogenous influence, and it is this that must be considered next.

The effect of the exogenous component is seen most easily in the non-24-hour studies that have just been described. Although the endogenous component was strong, it was not the only one present; thus the 'circadian amplitude ratio' was not infinitely high (Simpson et al., 1970). A 3-hour component was present in the study of Weitzman, Nogeire et al. (1974) and a 28-hour component was present in the experiments of Wever (1979). Further, in an earlier series of experiments, subjects on a metabolic ward were placed on 12-, 19- or 33-hour days, in each case one-third of the time being spent asleep (Orth et al., 1967). Plasma cortisol did not seem to adapt either immediately or fully to any of the abnormal schedules, but an influence of the sleep–wakefulness cycle was present since plasma levels fell on waking and rose during sleep whatever the length of 'day' (*see also* Liddle, 1965). A direct effect of sleep has been claimed as a result of free-running studies of Czeisler (1978). As described earlier, the phase-advance in plasma 11-OHCS observed in these conditions meant that the minimum level was some hours before sleep and that steroid levels were rising when the subject retired. It was observed that sleep was associated with a small fall of hormone concentrations. This was a result that was seen more clearly when steroid levels were expressed relative to retiring times (*Fig.* 7.2B). In addition, on waking a further rise of plasma steroid was observed, again seen more clearly when the educed wave had been obtained from steroid levels expressed relative to waking times. Presumably, under entrained nychthemeral conditions, the sleep-

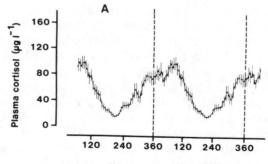

Time (degrees: 360 degrees=24.7 h)

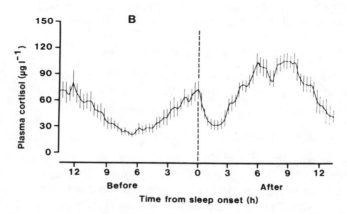

Time from sleep onset (h)

Fig. 7.2. A, Plasma cortisol rhythm in a subject under free-running conditions. Data represent means ± s.e. from 13 cycles for a cycle length of 24·7 h. B, Data derived from (A) but plotted relative to time of sleep onset (means from 14 sleep periods). (Data of Czeisler, 1978.)

induced fall is missed (as the hormone concentration is already close to zero) and the waking-induced increase accentuates the endogenous rhythm. An effect of the pattern of sleep and wakefulness upon the steroid rhythm is illustrated also by a comparison between British and Amerindian subjects (Halberg and Simpson, 1967) based upon 2-hourly urine samples. A statistically significant difference in acrophase was found, but the sleep–wakefulness patterns of the two populations were different. When the steroid data were expressed with reference to midsleep, no difference existed.

The argument for referring rhythms of hormones in general to midsleep has been summarized by Aschoff (1979). It must be remembered that such considerations do not enable a distinction to be made between the effects of sleep and wakefulness *per se* and the associated rhythms of posture, of light and dark or of eating and fasting. Such a problem has been raised before and will be referred to again in the last chapter when zeitgeber are considered; in the present context, in which exogenous influences are small, the effects of light (reviewed by Daly and Evans, 1974) and food (*see* s. 2.3) are slight.

The result of *Fig.* 7.2A and B illustrate also some of the difficulties that exist in interpreting data when they are pooled inappropriately: pooling successive 24 hours of free-running data would result in a loss of rhythm (unless the period were exactly 24 hours); pooling successive periods of free-running data would indicate the general nature of the rhythm but irregularities of sleep and wakefulness on successive days would distort the shape of the pooled rhythm if an exogenous effect of sleep were present. It will be noted that a similar problem was encountered in Chapter 3 when it was considered whether or not a rise in blood pressure before waking took place (*see also Fig.* 3.2A and B).

3. Growth Hormone

As with cortisol, frequent sampling and the development of a radio-immunoassay for growth hormone (GH) have led to considerable interest in the field recently.

3.1. The nychthemeral rhythm

Fig. 7.3 shows data from 30 young men undergoing a conventional routine with respect to sleep, activity, meals and lighting. There is a marked peak in hormone secretion associated with the first part of sleep, a finding that has been consistently reported (Daly and Evans, 1974; Weitzman, 1976; Quabbe, 1977; Parker et al., 1979; Weitzman, 1979). Smaller episodes of secretion can be seen later in sleep and some bursts are found diurnally, plasma levels otherwise dropping to negligible values. The diurnal peaks of GH have been related to times of feeding, since pre-prandial hypoglycaemia would act as a stimulus of GH release. In support of this is the study of Parker et al. (1979) in which GH was measured when meal times were advanced by 2½ hours; the diurnal secretory bursts too were advanced by about this time. However, in an earlier study by Parker et al. (1972) a comparison was made between sleep-related GH release in subjects on an 80-hour fast and on a glucose-controlled diet for 6 days. Sleep and GH patterns were normal, although the absolute level of GH was higher during the fast. Glucose infusions have been found to be poor suppressors of episodes of GH secretion whether during daytime 'naps' (Parker et al., 1979) or a normal nocturnal sleep (Daly and Evans, 1974). These results suggest that, whereas the basal level of GH can be influenced by plasma glucose levels, the episodes of secretion (which give rise to the circadian rhythm) are far less easily suppressed whether they occur diurnally or nocturnally.

The daily amounts of GH secreted vary with the stage of maturity of the individual (Weitzman, 1979). Amounts are low before puberty and rise towards this event, during which episodes of secretion are spread throughout the nychthemeron. As subjects continue past adolescence the total secretion falls progressively; it is worth pointing out that most subjects for the study of circadian rhythms are young adults (for example, *Fig.* 7.3).

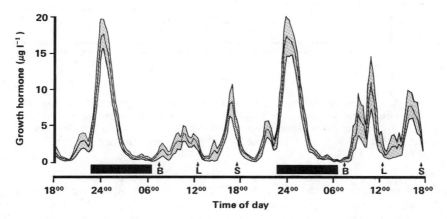

Fig. 7.3. Mean growth hormone (± s.e.) concentrations at 20-minute intervals from 30 studies over 36 hours of normal young men. B, L and S indicate the times of breakfast, lunch and supper, respectively. Black bars indicate time of sleep. (*From* Parker et al., 1979, Fig. 1. Reproduced from Krieger D. T. (ed.) *Endocrine Rhythms,* 1979, by kind permission of Raven Press, New York.)

3.2. The exogenous component of the growth hormone rhythm

Sleep is a potent influence upon GH secretion, and changes in the pattern of sleep and wakefulness indicate that this has a dominant exogenous effect upon the hormone secretion. Some of the experimental approaches are: sleep can be delayed or advanced (Takahashi et al., 1968; Honda et al., 1969); the sleep–wakefulness cycle can be reversed (Sassin et al., 1969; Parker et al., 1979); 'naps' can be taken during the daytime (Weitzman, 1975). In all cases, the changed time of sleep has been associated with an immediate change in GH release and there has been no evidence of a continuation of GH secretion during the 'old' sleep time. This argument is illustrated in *Fig.* 7.4 in which a comparison with cortisol—a rhythm with a high endogenous component—can also be made.

Confirmation of the view that there is a large exogenous component in the rhythm comes from two further types of experiment. In the free-running experiments of Czeisler (1978), no dissociation between sleep

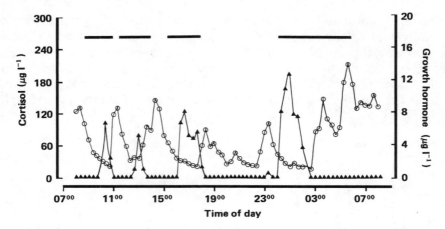

Fig. 7.4. Circadian changes in plasma cortisol (○) and growth hormone (▲) concentrations measured by sampling blood every 20 minutes. Horizontal bars indicate periods when sleep was taken. Note that each sleep period is associated with an episode of secretion of growth hormone. (*From* Weitzman, 1975, Fig. 1.)

and GH was found even when an internal dissociation between the sleep–wakefulness and temperature rhythms had taken place. Moreover, when dissociation of the rhythms had not occurred, release of the hormone did not advance relative to sleep, as was found for temperature and cortisol rhythms. This observation suggests that the GH rhythm was produced directly by sleep rather than by an independent endogenous oscillator.

In the second type of experiment, Weitzman and his colleagues (Weitzman, Nogeire et al., 1974) have placed subjects on a 3-hour 'day', i.e. 2 hours awake and 1 hour asleep. Growth hormone showed an ultradian period of 3 hours but no circadian component; this was in contrast to cortisol which, although it showed an ultradian rhythm (and therefore confirmed that the rhythm of sleep–wakefulness affects this hormone) continued to show a marked circadian component. Parker et al. (1979) studied 4 subjects on a 30-hour day (21 hours awake, 9 hours asleep) and gave them iso-caloric meals at 2-hourly intervals throughout the daytime. Even though the subjects were aware of the conventional 24-hour routine of others, the secretion of GH showed a marked association with sleep.

3.3. The endogenous component

In summary, therefore, the GH rhythm seems to be dominated by the rhythm of sleep and wakefulness. Is there evidence for *any* endogenous component to the rhythm? One possibility comes from the study by Parker et al. (1979) in which subjects lived a 38-hour day. Analysis of the secretion of GH by autocorrelation indicated a dominant 30-hour (exogenous) period, but a much smaller one at about 24 hours was found in addition. Whether this should be regarded as an endogenous component, or rather a response to any persistent nychthemeral cues, is not known.

4. Other Hormones

4.1. Prolactin

The recent development of a radioimmunoassay system has led to a detailed description of the nychthemeral rhythm (for review, *see* Frantz, 1979). Sassin et al. (1972) studied 6 subjects living on a normal routine

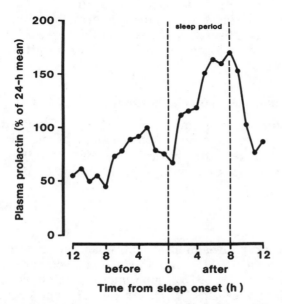

Fig. 7.5. Plasma prolactin levels in 6 normal subjects measured at 20-minute intervals. The hourly values derived by averaging 3 20-minute values are plotted as means for all 6 subjects. (*From* Sassin et al., 1972, Fig. 2. Copyright 1972 by the American Association for the Advancement of Science.)

who were sampled every 20 minutes over the course of 24 hours. The average plasma concentration of the group is shown in *Fig.* 7.5. Maximum secretion took place during the night with a series of increasing peaks as the sleep progressed, the last and highest being at, or just after, waking. Thereafter the level fell rapidly, was low for much of the daytime and began to rise before sleep. When each subject was considered separately, the same general pattern was found though individual secretory episodes (smoothed out by the averaging process that was used to produce *Fig.* 7.5) were evident. A similar nocturnal peak has been found in a large transverse study of two groups of women from Japan and America (Haus et al., 1980); the former group showed the more marked peak, but the possible explanation (dietary, climatic, genetic) is unknown. A recent study (Beck and Wuttke, 1980) showed a nocturnal peak in plasma prolactin in pubertal and pre-pubertal children of both sexes.

The importance of sleep has been shown in studies in which sleep was delayed by 3 or 6 hours or a reversal of sleeping habits took place, subjects sleeping from 11^{00} to 19^{00} (Sassin et al., 1973). In all cases changes in the timing of prolactin secretion were immediate indicating a strong exogenous component. In agreement with this is the observation that daytime 'naps' were associated with rises in prolactin concentration (Parker et al., 1973). In spite of the importance of sleep in determining prolactin levels, the smaller rise before retiring must have another cause. Whether this is a manifestation of an endogenous component or an effect of feeding has not been determined.

4.2. Gonadotrophins and gonadal hormones

The situation for these hormones is complex since rhythmicity differs between the sexes and depends upon the stage of the menstrual cycle in women. In both sexes, the rhythm depends upon the state of sexual maturity (as was the case with growth hormone but in contrast to prolactin).

Females

In pre-pubertal girls, low-amplitude circadian rhythms in both follicular stimulating hormone (FSH) and luteinizing hormone (LH) have been described (Kulin et al., 1976), superimposed upon which are episodic components (Penny et al., 1977). As puberty approaches, there is an increase in size of the secretory episodes and raised concentrations of both gonadotrophins are observed during sleep (*Fig.* 7.6) (Beck and Wuttke, 1980); as puberty progresses, episodic activity increases during the waking period also, so that the circadian rhythm is only transitional (Rebar and Yen, 1979). The burst of LH is associated with sleep rather than an endogenous rhythm since changes in sleep times produce immediate changes in the time of the LH burst (Weitzman, 1976). In mature women, episodic secretion of the gonadotrophins is marked and its frequency and magnitude depends upon the stage of the menstrual cycle. Thus the episodes of LH secretion are greatest about ovulation and their frequency least during the late luteal phase. For FSH, these components are less marked but most noticeable in the early follicular phase (Yen et al., 1973). Sleep influences LH levels as at puberty but now the result is a *fall* during the early stages of sleep onset (Kapen et al., 1973) and one that immediately follows changes in the time of sleep (Weitzman, 1976). In post-menopausal women there are no circadian rhythms in gonadotrophins but the absolute levels are high and episodic secretion is marked for both LH and FSH.

When steroid rhythms are considered, at mid-puberty, oestradiol levels show a circadian rhythm with a minimum during sleep and a maximum about 14^{00} (*Fig.* 7.6) (Boyar et al., 1976). This 'inappropriate' timing is believed to reflect a delay in girls between a gonadotrophic stimulus and the ovarian response (Rebar and Yen, 1979). Steroid rhythms in mature

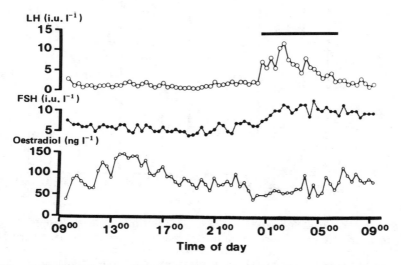

Fig. 7.6. Circadian variations in plasma concentrations of luteinizing hormone (LH), follicle stimulating hormone (FSH) and oestradiol in a pubertal girl. Blood sampled every 20 minutes. Horizontal bar indicates time of sleep. (*From* Boyar et al., 1976, Fig. 1.)

women are more difficult to assess because these substances originate from both the ovaries and the adrenal glands. The observation that many steroids are secreted in increasing amounts as sleep progresses (Rebar and Yen, 1979) is appropriate for a role of ACTH but the relative importance of this and of the gonadotrophins in the species man has not yet been established.

The extent to which the secretions of steroid, gonadotrophin and hypothalamic releasing factor are causally linked is not as clear in women as men (*see below*). A study upon 2 women from whom blood was taken every 5 minutes for periods of about an hour indicated only an approximate relationship between luteinizing hormone releasing factor (LRF) (assessed by rat bioassay) and LH (assessed by radioimmunoassay) as far as plasma concentration and secretory episodes were concerned (Seyler and Reichlin, 1974).

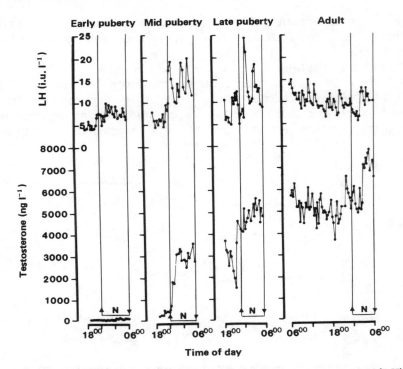

Fig. 7.7. Changes in plasma concentrations of luteinizing hormone (LH) and testosterone in pubertal and adult males. N, represents sleep period. Note that early puberty is characterized by small nocturnal increases of both hormones, these nocturnal increases become more prominent in mid- and late puberty. During late puberty and adulthood diurnal increases are also seen. (*From* Judd, 1979, Fig. 3. Reproduced from Krieger D. T. (ed.) *Endocrine Rhythms,* 1979, by kind permission of Raven Press, New York.)

Males

There is disagreement as to whether or not there are changes in LH levels in the pre-pubertal boys. This disagreement might result in part from differences in the age at which the subjects are studied (Judd, 1979). Thus, he has distinguished between young and older pre-pubertal boys (10·5 and 13·7 years of age, respectively). In the former there were no episodic components or circadian rhythms; in the latter there was evidence for a rise of both LH and testosterone during sleep, especially of testosterone. This observation awaits confirmation by studies upon larger groups of subjects. Small nocturnal increases in both FSH and testosterone have been found and a causal link between LH and testosterone bursts is possible (Parker et al., 1975).

As in girls, at puberty there are prominent episodes of LH and testosterone secretion at night (Beck and Wuttke, 1980) and, as puberty progresses, pulses begin to occur more in the daytime also so that a marked circadian rhythm is observed only transiently (*Fig. 7.7*). Parker et al. (1975) have found also similar, but smaller, changes in FSH secretion during the various stages of puberty. Again they note a temporal relationship between LH and testosterone levels which supports the view that there is a causal link between them. There is evidence for an endogenous component to these rhythms in addition to a marked influence of sleep. Thus, even though Kapen et al. (1974) found substantial changes in gonadotrophin release following inversion of the sleep—wakefulness pattern, these took 3 days before they were complete; and nocturnal wakefulness was associated with higher LH levels than were found during diurnal wakefulness in the controls.

In mature males, episodic secretion of LH and FSH is found throughout the nychthemeron. Once again the changes in concentration of FSH are smaller and this can be attributed to its longer half-life when compared with LH (4 hours rather than 30 minutes) and to its relative insensitivity to LRF (*see* Judd, 1979). Testosterone levels too show rapid and irregular fluctuations and claims have been made for some, but not perfect, coincidence between LH and testosterone pulses, especially if a delay of 1—3 hours is allowed between LH and testosterone (Rowe et al., 1975).

Low-amplitude circadian changes in testosterone levels have been widely reported, generally peaking during or at the end of sleep (*see* review by Aschoff, 1979). The phasing of this rhythm, together with the competence of the adrenal gland to secrete sex steroids, has led to speculation as to the role that ACTH might play. However, unlike that of cortisol, this rhythm is not abolished by dexamethasone and so is attributed to the gonads rather than the adrenal glands (Judd et al., 1974). This group has shown also that during sleep there is some coincidence between testosterone levels and prior bursts of LH, but again this parallelism is by no means exact; accordingly, suggestions have been made that the testes show a circadian rhythm of receptivity to circulating LH levels, either through blood flow or prolactin, but evidence so far is sparse and contradictory (*see* Judd, 1979). However, on the basis of sleep-reversal studies, a claim has recently been made that the normal nocturnal rise in plasma testosterone levels is not due to sleep or to LH secretion (Miyatake et al., 1980).

4.3. Thyrotrophin, tri-iodothyronine and thyroxine

Many studies have indicated that there is a circadian rhythm of thyrotrophin (TSH) concentration (*see* Aschoff, 1979 for a review) with a maximum at or about the time of sleep. In a study by Parker et al. (1976), in which blood samples were taken from 10 males every 20 minutes over the course of 24—48 hours, times of maximum coincided with retiring, nocturnal values showing a fall throughout the sleep period and then during the first part of the daytime to reach a minimum about 15^{00}. When the rhythm was considered in more detail, there was evidence for episodic fluctuations as with other hormones (Alford et al., 1973).

Rhythmicity in tri-iodothyronine (T_3) and thyroxine (T_4) also has been sought, as has any relationship that might exist between all three hormones. In one study (Azukizawa et al., 1976), T_3 and T_4 secretions were found to be episodic, but circadian components were low or absent. There was a small positive correlation between T_4 and T_3 levels, but no relation was found between either of these and TSH levels simultaneously measured. In another study (Weeke and Gundersen, 1978), in which samples were taken from 5 males every 30 minutes during the course of 24 hours, synchronous rhythms of T_3 and TSH were found with highest values at night; T_4 levels also were highest at this time, but insignificantly so (*Fig.* 7.8). In further experiments upon another 5 males sampled every 5 minutes for about 7 hours, results indicated that, superimposed upon the diurnal changes, all three hormones showed an ultradian rhythm of low amplitude.

The rhythm of TSH seems to possess endogenous and exogenous components, the latter being associated with sleep. Thus, in one study (Vanhaelst et al., 1972), the same rhythm in TSH was found whether the subjects lived nychthemerally or were recumbent throughout the 24 hours (but sleeping in the dark and taking meals at normal times): this indicates that the daytime fall in TSH was not mediated by activity. A report (Patel et al., 1972) that 'naps' taken in the daytime raised TSH levels (as they did also for GH) indicates an exogenous influence of sleep, but the view more frequently expressed is that sleep depresses TSH. By way of example, the study by Parker et al. (1976) already referred to can be cited. Further experiments by this group showed that if the time of sleep was advanced, then the fall of TSH started earlier; as a result, the maximum plasma concentration was decreased. If sleep was delayed, the maximum was increased, as TSH levels climbed for longer. These results indicate that sleep onset acts exogenously and initiates the fall in TSH levels. However, if sleep reversal took place, that is the subjects stayed up all night and slept from 11^{00} to 19^{00} the next day, then the rise continued to a peak at about 04^{00} on the first sleepless night and fell thereafter; on the second sleepless night a smaller peak at 04^{00} was observed. These results were interpreted by the authors to indicate that there is an endogenous component in the rhythm of TSH that, when not influenced by sleep, peaks at about 04^{00}. A problem is that this peak was far less on the second night of sleep reversal and persistence of rhythmic component in the absence of external time cues is an important criterion of endogeneity (*see* Chapter 1). In this study, the daytime sleep was not associated with marked changes in TSH, but levels were low anyway and so a fall would not be easy to measure (compare with the effect of nocturnal sleep upon cortisol rhythms described above). It is not known whether a pre-sleep rise and

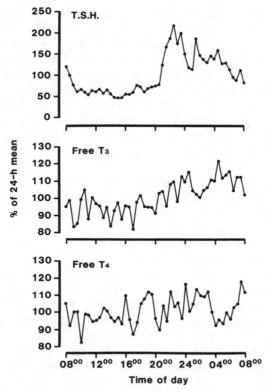

Fig. 7.8. Circadian variations of serum thyrotrophin (TSH), free tri-iodothyronine (T_3) and free thyroxine (T_4) concentrations measured at 30-minute intervals. Data represent means from 5 male subjects. (*From* Weeke and Gundersen, 1978, Fig. 5.)

sleep-induced fall of TSH would occur with prolonged time spent on this schedule.

4.4. Melatonin

Under nychthemeral circumstances there is general agreement that the levels are higher when subjects are asleep in the dark than when awake in the light. This result has been found in studies using both radioimmuno-assay and frog melanophore bioassay techniques and with both plasma and urine sampling (*see* Weitzman et al., 1978). The melatonin rhythm is remarkable for its amplitude, nocturnal values being 30–70 times diurnal concentrations. Daytime values are not negligible, however, and they are higher when subjects are active rather than taking bedrest (Weitzman et al., 1978; Weinberg et al., 1979). As with other hormones, frequent sampling demonstrates that there is epidosic secretion (Vaughan, Allen et al., 1979; Weinberg et al., 1979).

If results in chickens and rats apply to man (Brownstein, 1975; Binkley, 1976; Klein, 1979; Zimmerman and Menaker, 1979) a major influence upon this rhythm will be an inhibitory effect of light acting via a sympathetic input to the gland. Such details in man are not yet available, though the observation that hormone levels are not much affected by waking subjects and exposing them to light for one hour during the course of a night's sleep argues against light exerting a strong inhibitory influence in man (Vaughan, Bell et al., 1979). Nevertheless, the further observation that hormone secretion adjusts rapidly to changes in sleep—wakefulness schedules indicates that a strong exogenous influence does exist (Weitzman et al., 1978).

There are enough pieces of evidence to suggest that a substantial endogenous component exists as well (*see* review by Klein, 1979). In a recent study by Weitzman's group (Weitzman et al., 1978), 3 subjects were studied during free-running conditions after some control days on a nychthemeral routine. One subject showed no relationship between the sleep—wakefulness and plasma melatonin rhythms. The other two showed increased secretion nocturnally during the control days. Under free-running conditions one subject developed a period of about 25 hours for both rhythms, that of metatonin now being phase-advanced with respect to the sleep—wakefulness cycle. The other subject exhibited internal desynchronization, his sleep—wakefulness rhythm having a period of 38·5 hours while his melatonin rhythm showed 9 peaks in 11 days. In another study by Akerstedt et al. (1979), subjects were sleep-deprived for 64 hours and placed on a regular regimen. The urinary melatonin rhythm paralleled that of subjective fatigue, that is it showed a circadian period superimposed upon a rising trend. The results from both of these studies are most easily interpreted on the assumption that an endogenous component is present in the melatonin rhythm, but they do not permit inferences to be made as to the causal links with other rhythms, for example that of temperature, that might exist.

5. Some General Problems

This account of circadian rhythms in the endocrine system of humans has indicated that differences between the hormones exist when the relative size of their endogenous and exogenous components is considered. On the other hand, some problems are common to the different hormones and they must now be commented upon. These problems are the relationship between ultradian rhythms, episodic secretion and homeostatic mechanisms and the relationship between hormone secretion and sleep stages.

5.1. Episodic secretion, ultradian rhythms and homeostasis

The results of frequent blood sampling indicate that the level of hormones does not remain constant or change smoothly throughout the nychthemeron. The process of 'eduction', used by Czeisler when studying the rhythms of cortisol, is a way of extracting underlying trends from complex data and it has been used with success to detect circadian rhythmicity (*see above*). However, superimposed upon these circadian rhythms are erratic fluctuations of concentration of the blood hormone (assessed both by size and frequency of occurrence). The term 'ultradian' would correctly describe the time course of the changes (*see* Chapter 1) but 'pulsatile' and 'episodic' are more commonly used (*see,* for example, Krieger et al., 1972). These latter terms are used to imply that the fluctuations are due to erratic bursts of secretory activity by the glands rather than the manifestation of some oscillator. In agreement with this distinction are the observations that the rate of decline of hormone after a burst accords with what is known of the rate of removal of the hormone from the plasma rather than with the view that it is the controlled descending phase of an oscillatory process. If this view is accepted, then the circadian rhythmicity can be considered to result from some, as yet unknown, process whereby the frequency and/or size of these bursts varies (a process which is strongly influenced by sleep in the case of most hormones).

Furthermore, and this point has been stressed often by Weitzman (1976, 1979), the episodic secretory processes are not easily understood in terms of conventional homeostatic mechanisms. Instead, he argues, they represent a 'programmed' sequence of episodes that is entrained to a nychthemeral existence by zeitgebers, generally by some component of sleep. The position is presumed to be different from, for example, the circadian changes in the 'set-point' of deep body temperature. Thus, whereas attempts to change the deep body temperature initiate thermoregulatory reflexes that attempt to maintain the temperature at some value appropriate to the time of the nychthemeron, this is often not the case when the episodic secretion of hormones is considered. Examples of this inability to change the episodic secretion of hormones by feedback control have been given earlier in the chapter. However, the nature of the 'programmes' and the means by which they are implemented are quite unknown.

Nevertheless there is a vast amount of evidence that hormone levels can be influenced by feedback control. Thus, results of the administration of cortisol antagonists (Daly and Evans, 1974), measurement of gonadotrophin levels in women after menopause (Reber and Yen, 1979) and a whole wealth of clinical disorders (*see,* for examples, Daly and Evans, 1974; Weitzman, 1979; Chapter 11) indicate that feedback mechanisms operate to control the *mean* level of hormone concentration seemingly independently of circadian changes. A distinction can be drawn, therefore, between the longer term control of hormone levels by negative feedback and the shorter term 'programmed' changes accounting for episodic release and circadian rhythmicity. It has been suggested (Daly and Evans, 1974) that the hypothalamic sites controlling these mechanisms are different. In addition, it is known that 'stress' changes the secretion of many hormones. In a study performed upon patients just before severe elective surgery, Czeisler et al. (1976) found that their circadian rhythms of cortisol were indistinguishable from those of control patients except for a large burst when preoperative preparations began. This result suggests that a further mechanism influencing hormone release—acute changes due to stress—might operate independent of episodic, circadian and homeo-static mechanisms. (*But see also* Follenius and Brandenberger (1980).)

5.2. The relationship between hormone release and sleep stages

The present account has stressed the important role of sleep where circadian rhythms in hormones are concerned. This association applies also to the hormones that affect the kidney (Chapter 4) but the cate-cholamines (Chapter 6) seem to be the exception. Since sleep can be divided into a number of stages, the question whether these and the epi-sodes of hormone secretion are related has often been considered (Daly and Evans, 1974; Weitzman, Fukushima et al., 1974; Krieger, 1975; Weitzman, 1976; Quabbe, 1977; Aschoff, 1978; Oswald, 1978; Rubin et al., 1978; Aschoff, 1979; Besset et al., 1979; Tolis et al., 1979; Weitzman, 1979; Mullen et al., 1980). As judged from the times of peak and trough of plasma levels and simultaneous measurement of sleep stages, the following relationships are the main ones that have been suggested:
1. Cortisol secretion is stimulated during REM sleep.
2. GH secretion is stimulated during SWS.
3. Prolactin secretion is inhibited during REM sleep.
Such relationships might be fortuitous rather than causal, of course, and considerable amounts of data both against and in favour of a causal link have been adduced. Detailed evidence can be found in the reviews cited above, but the principal approaches, illustrated mainly by reference to the rhythm of GH which possesses a large, sleep-related exogenous component, are as follows:
1. Detailed studies of the relationship between hormone levels and sleep stages. The relationship can be investigated in more detail as the frequency with which blood samples are taken rises. Such studies have not found the correlation between them to be marked. Illustrative examples are the findings that GH concentrations sometimes rise before the first SWS episode and that the later SWS episodes during sleep often are not associ-ated with GH peaks.
2. Investigations of the effect of changed schedules. If 'naps' are taken during the daytime, or the time of sleep is changed in some other way, hormone secretion and sleep stages can be compared. Unfortunately, simultaneous EEG recording and frequent blood sampling are comparatively rare but it has been found by this approach that REM and cortisol secretion can be dissociated but that, as the amount of prior wakefulness increases

(and hence of SWS, *see* Chapter 5), there is an increase in the amount of GH released when the delayed sleep is taken.

3. Use of drugs or other procedures that alter sleep profiles. Two examples can be given. First, exercise increases the amount of both SWS and GH secretion during the following night and, second, benzodiazepines block SWS but not GH release.

4. Use of clinical and blind subjects. In narcoleptic patients (in whom sleep times and profiles are abnormal) no correlation between these sleep stages and hormone release has been found. Similarly, obese subjects show less nocturnal GH secretion but show a normal distribution of SWS. On the other hand, blind subjects with both sleep-stage profiles and GH secretion that are abnormal have been reported.

In summary, therefore, the correspondence between hormone secretion and sleep stages seems not to be exact or causal; possibly it is analogous to the relationship discussed earlier between the rhythms of temperature, catecholamines and mental performance (Chapter 6). Perhaps this is not surprising: correspondence between electrical events in the cerebral cortex and humoral events in the hypothalamus is being sought. Both could be manifestations of deeper processes so that no direct link between the two would be expected. Further, any link that might exist would undoubtedly be complex and enable all kinds of disturbing influence to intrude. Nevertheless, sleep as a whole does exert a powerful effect upon hormone secretion, a secretion that is affected also by homeostatic mechanisms, 'programmed' sequences and stress; the complex interaction between all these factors remains to be elucidated.

References

Akerstedt T., Fröberg J. E., Friberg Y. et al. (1979) Melatonin excretion, body temperature and subjective arousal during 64 hours of sleep deprivation. *Psychoneuroendocrinology* **4**, 219–25.

Alford F. P., Baker H. W. G., Burger H. G. et al. (1973) Temporal patterns of integrated plasma hormone levels during sleep and wakefulness. I. Thyroid-stimulating hormone and cortisol. *J. Clin. Endocrinol. Metab.* **37**, 841–7.

Aschoff J. (1978) Circadiane Rhythmen im endocrinen System. *Klin. Wochenschr.* **56**, 425–35.

Aschoff J. (1979) Circadian rhythms: general features and endocrinological aspects. In: Krieger D. T. (ed.) *Endocrine Rhythms.* New York, Raven Press, pp. 1–61.

Aschoff J., Hoffmann K., Pohl H. et al. (1975) Re-entrainment of circadian rhythms after phase-shifts of the zeitgeber. *Chronobiologia* **2**, 23–78.

Azukizawa M., Pekary A. E., Hershman J. M. et al. (1976) Plasma thyrotrophin, thyroxine, and tri-iodothyronine relationships in man. *J. Clin. Endocrinol. Metab.* **43**, 533–42.

Beck W. and Wuttke W. (1980) Diurnal variations of plasma luteinizing hormone, follicle-stimulating hormone and prolactin in boys and girls from birth to puberty. *J. Clin. Endocrinol. Metab.* **50**, 635–9.

Besset A., Bonardet A., Billiard M. et al. (1979) Circadian patterns of growth hormone and cortisol secretions in narcoleptic patients. *Chronobiologia* **6**, 19–31.

Binkley S. (1976) Pineal gland biorhythms: N-acetyltransferase in chickens and rats. *Fed. Proc.* **35**, 2347–52.

Boyar R. M., Wu R. H. K., Roffwarg H. et al. (1976) Human puberty: 24-hour estradiol patterns in pubertal girls. *J. Clin. Endocrinol. Metab.* **43**, 1418–21.

Brownstein M. J. (1975) Mini-review. The pineal gland. *Life Sci.* **16**, 1363–74.

Conroy R. T. W. L., Elliott A. L. and Mills J. N. (1970) Circadian rhythms in plasma concentration of 11-hydroxycorticosteroids in men working on night shift and in permanent night workers. *Br. J. Industr. Med.* **27**, 170–4.

Cugini P., Giovannini C., Rossi G. et al. (1975) Influenza del propranololo sulle variazioni diarie della cortisolemia nell'uomo. *Boll. Soc. Ital. Biol. Sper.* **51**, 442–47.

Czeisler C. A. (1978) Human circadian physiology: internal organization of temperature, sleep–wake and neuroendocrine rhythms monitored in an environment free of time cues. PhD Thesis, Stanford University.

Czeisler C. A., Moore-Ede M. C., Regestein Q. R. et al. (1976) Episodic 24-hour cortisol secretory patterns in patients awaiting elective cardiac surgery. *J. Clin. Endocrinol. Metab.* **42**, 273–83.

Dallman M. F., Engeland W. C. and Shinsako J. (1976) Circadian changes in adrenocortical responses to ACTH. *Proc. 58th Meeting Endocrinol. Soc.* **58**, Abstr. 4.

Daly J. R. and Evans J. I. (1974) Daily rhythms of steroid and associated pituitary hormones in man and their relationship to sleep. *Adv. Steroid Biochem. Pharmacol.* **4**, 61–110.

DeLacerda L., Kowarski A. and Migeon C. J. (1973) Diurnal variation of the metabolic clearance rate of cortisol. Effect on measurement of cortisol production rate. *J. Clin. Endocrinol. Metab.* **36**, 1043–9.

DeMoor P., Heirwegh K., Hermans J. K. et al. (1962) Protein binding of corticoids studied by gel filtration. *J. Clin. Invest.* **41**, 816–27.

Ferrari E., Bossolo P. A., Vailati A. et al. (1977) Variations circadiennes des effets d'une substance vagolytique sur le système ACTH-secrétant chez l'homme. *Ann. Endocrinol. (Paris)* **38**, 203–13.

Follenius M. and Brandenberger G. (1980) Evidence of a delayed feedback effect on the mid-day plasma cortisol peak in man. *Horm. Metab. Res.* **12**, 638–9.

Frantz A. G. (1979) Rhythms in prolactin secretion. In: Krieger D. T. (ed.) *Endocrine Rhythms.* New York, Raven Press, pp. 175–86.

Gallagher T. F., Yoshida K., Roffwarg H. D. et al. (1973) ACTH and cortisol secretory patterns in man. *J. Clin. Endocrinol. Metab.* **36**, 1058–73.

Halberg F. and Simpson H. (1967) Circadian acrophases of human 17-hydroxycorticosteroid excretion referred to midsleep rather than midnight. *Human Biology* **39**, 405–13.

Haus E., Lakatua D. J., Halberg F. et al. (1980) Chronobiological studies of plasma prolactin in women in Kyushu, Japan, and Minnesota, USA. *J. Clin. Endocrinol. Metab.* **51**, 632–40.

Honda Y., Takahashi I., Takahashi S. et al. (1969) Growth hormone secretion during nocturnal sleep in normal subjects. *J. Clin. Endocrinol. Metab.* **29**, 20–9.

Judd H. L. (1979) Biorhythms of gonadotropins and testicular hormone secretion. In: Krieger D. T. (ed.) *Endocrine Rhythms.* New York, Raven Press, pp. 299–324.

Judd H. L., Parker D. C., Rakoff J. S. et al. (1974) Elucidation of mechanism(s) of the nocturnal rise of testosterone in men. *J. Clin. Endocrinol. Metab.* **38**, 134–41.

Kapen S., Boyar R. M., Finkelstein J. W. et al. (1974) Effect of sleep–wake cycle reversal on luteinizing hormone secretory pattern in puberty. *J. Clin. Endocrinol. Metab.* **39**, 294–99.

Kapen S., Boyar R., Perlow M. et al. (1973) Luteinizing hormone: changes in secretory pattern during sleep in adult women. *Life Sci.* **13**, 693–701.

Klein D. C. (1979) Circadian rhythms in the pineal gland. In: Krieger D. T. (ed.) *Endocrine Rhythms.* New York, Raven Press, pp. 203–23.

Kowarski A., Thompson R. G., Migeon C. J. et al. (1971) Determination of integrated plasma concentrations and true secretion rates of human growth hormone. *J. Clin. Endocrinol.* **32**, 356–60.

Krieger D. T. (1975) Circadian pituitary adrenal rhythms. *Adv. Exp. Med. Biol.* **54**, 169–89.

Krieger D. T. (1978) Factors influencing the circadian periodicity of ACTH and corticosteroids. *Med. Clin. North Am.* **62**, 251–9.

Krieger D. T. (1979) Rhythms in CRF, ACTH, and corticosteroids. In: Krieger D. T. (ed.) *Endocrine Rhythms.* New York, Raven Press, pp. 123–42.

Krieger D. T., Allen W., Rizzo F. et al. (1971) Characterization of the normal pattern of plasma corticosteroid levels. *J. Clin. Endocrinol. Metab.* **32**, 266–84.

Krieger D. T. and Gewirtz G. P. (1974) The nature of the circadian periodicity and suppressibility of immunoreactive ACTH level in Addison's disease. *J. Clin. Endocrinol. Metab.* **39**, 46–52.

Krieger D. T., Ossowski R., Fogel M. et al. (1972) Lack of circadian periodicity of human serum FSH and LH levels. *J. Clin. Endocrinol. Metab.* **35**, 619–23.

Kulin H. E., Moore R. G. and Santner S. J. (1976) Circadian rhythms in gonadotropin excretion in pre-pubertal and pubertal children. *J. Clin. Endocrinol. Metab.* **42**, 770–3.

Liddle G. W. (1965) An analysis of circadian rhythms in human adrenocortical secretory activity. *Am. Clin. Climatol. Assoc.* **77**, 151–60.

Mills J. N., Minors D. S. and Waterhouse J. M. (1974) The circadian rhythms of human subjects without timepieces or indication of the alternation of day and night. *J. Physiol.* **240**, 567–94.

Miyatake A., Morimoto Y., Oishi T. et al. (1980) Circadian rhythm of serum testosterone and its relation to sleep: comparison with the variation in serum luteinizing hormone, prolactin and cortisol in normal men. *J. Clin. Endocrinol. Metab.* **51**, 1365–71.

Mullen P. E., James V. H. T., Lightman S. L. et al. (1980) A relationship between plasma renin activity and the rapid eye movement phase of sleep in man. *J. Clin. Endocrinol. Metab.* **50**, 466–9.

Nichols T., Nugent C. A. and Tyler F. H. (1963) Diurnal variation in suppression of adrenal function by glucocorticoids. *J. Clin. Endocrinol. Metab.* **25**, 343–9.

Nugent C. A., Eik-Nes K., Kent H. S. et al. (1960) A possible explanation for Cushing's syndrome associated with adrenal hyperplasia. *J. Clin. Endocrinol. Metab.* **20**, 1259–68.

Orth D. N., Island D. P. and Liddle G. W. (1967) Experimental alteration of the circadian rhythm in plasma cortisol (17-OHCS) concentration in man. *J. Clin. Endocrinol. Metab.* **27**, 549–55.

Oswald I. (1978) Editorial. Sleep and hormones. *Eur. J. Clin. Invest.* **8**, 55–6.

Parker D. C., Judd H. L., Rossman L. G. et al. (1975) Pubertal sleep–wake patterns of episodic LH, FSH and testosterone release in twin boys. *J. Clin. Endocrinol. Metab.* **40**, 1099–1109.

Parker D. C., Pekary A. E. and Hershman J. M. (1976) Effect of normal and reversed sleep–wake cycles upon nyctohemeral rhythmicity of plasma thyrotropin: evidence suggestive of an inhibitory influence in sleep. *J. Clin. Endocrinol. Metab.* **43**, 318–29.

Parker D. C., Rossman L. G., Kripke D. P. et al. (1979) Rhythmicities in human growth hormone concentrations in plasma. In: Krieger D. T. (ed.) *Endocrine Rhythms.* New York, Raven Press, pp. 143–73.

Parker D. C., Rossman L. G. and Vanderlaan E. F. (1972) Persistence of rhythmic human growth hormone release during sleep in fasted and nonisocalorically fed normal subjects. *Metabolism* **21**, 241–52.

Parker D. C., Rossman L. G. and Vanderlaan E. F. (1973) Sleep-related, nyctohemeral and briefly episodic variation in human plasma prolactin concentrations. *J. Clin. Endocrinol. Metab.* **36**, 1119–24.

Patel Y. C., Alford F. P. and Burger H. G. (1972) The 24-hour plasma thyrotrophin profile. *Clin. Sci.* **43**, 71–7.

Penny R., Olambiwonnu N. O. and Frasier S. D. (1977) Episodic fluctuations of serum gonadotropins in pre- and post-pubertal girls and boys. *J. Clin. Endocrinol. Metab.* **45**, 307–11.

Perkoff G. T., Eik-Nes K., Nugent C. A. et al. (1959) Studies of the diurnal variation of plasma 17-hydroxycorticosteroids. *J. Clin. Endocrinol. Metab.* **16**, 432–43.

Quabbe H.-J. (1977) Chronobiology of growth hormone secretion. *Chronobiologia* **4**, 217–46.

Quigley M. E. and Yen S. S. C. (1979) A mid-day surge in cortisol levels. *J. Clin. Endocrinol. Metab.* **49**, 945–7.

Rebar R. W. and Yen S. C. C. (1979) Endocrine rhythms in gonadotropins and ovarian steroids with reference to reproductive processes. In: Krieger D. T. (ed.) *Endocrine Rhythms.* New York, Raven Press, pp. 259–98.

Reinberg A., Ghata J., Halberg F. et al. (1970) Rythmes circadiens du pouls, de la pression artérielle, des excrétions urinaires en 17-hydroxycorticosteroïdes catécholamines et potassium chez l'homme adulte sain, actif et au repos. *Ann. Endocrinol. (Paris)* **31**, 277–87.

Rowe P. H., Racey P. A., Lincoln G. A. et al. (1975) The temporal relationship between the secretion of luteinizing hormone and testosterone in man. *J. Endocrinol.* **64**, 17–26.

Rubin R. T., Poland R. E., Gouin P. R. et al. (1978) Secretion of hormones influencing water and electrolyte balance (antidiuretic hormone, aldosterone, prolactin) during sleep in normal adult man. *Psychosom. Med.* **40**, 44–59.

Sassin J. F., Frantz A. G., Kapen S. et al. (1973) The nocturnal rise of human prolactin is dependent on sleep. *J. Clin. Endocrinol. Metab.* **37**, 436–40.

Sassin J. F., Frantz A. G., Weitzman E. D. et al. (1972) Human prolactin: 24-hour pattern with increased release during sleep. *Science* **177**, 1205–7.

Sassin J. F., Parker D. C., Mace J. W. et al. (1969) Human growth hormone release: relation to slow wave sleep and sleep–waking cycles. *Science* **165**, 513–15.

Seyler L. E. and Reichlin S. (1974) Episodic secretion of luteinizing hormone-releasing factor (LRF) in the human. *J. Clin. Endocrinol. Metab.* **39**, 471–8.

Sharp G. W. G., Slorach S. A. and Vipond H. J. (1961) Diurnal rhythms of keto- and ketogenic steroid excretion and the adaptation to changes of the activity–sleep routine. *J. Endocrinol.* **22**, 377–85.

Simpson H. W. (1974) The human circadian system and aerospace travel. In: Scheving L. E., Halberg F. and Pauly J. E. (ed.) *Chronobiology.* Tokyo, Igaku Shoin, pp. 448–50.

Simpson H. W. (1977) Human 21-h day studies in the high Arctic, a review of analyses carried out on the 1960 Spitsbergen and 1969 Devon Island studies. *Nova Acta Leopoldina* **225**, 407–29.

Simpson H. W. and Lobban M. C. (1967) Effect of a 21-hour day on the human circadian excretory rhythms of 17-hydroxycorticosteroids and electrolytes. *Aerospace Med.* **38**, 1205–13.

Simpson H. W., Lobban M. C. and Halberg F. (1970) Arctic chronobiology. Urinary near-24-hour rhythms in subjects living on a 21-hour routine in the Arctic. *Arctic Anthropology* **7**, 144–64.

Takahashi Y., Kipnis D. M. and Daughaday W. H. (1968) Growth hormone secretion during sleep. *J. Clin. Invest.* **47**, 2079–90.

Tolis G., Banovac K., McKenzie J. M. et al. (1979) Circadian rhythms of anterior pituitary hormone secretion: effects of dexamethasone. *J. Endocrinol. Invest.* **2**, 433–6.

Vague Ph., Oliver Ch. and Bourgoin J. Y. (1974) Circadian rhythms in plasma ACTH in healthy adults. In: Sheving L. E., Halberg F. and Pauly J. E. (ed.) *Chronobiology.* Tokyo, Igaku Shoin, pp. 112–4.

Vanhaelst L., Van Cauter E., Degante J. P. et al. (1972) Circadian variations of serum thyrotropin levels in man. *J. Clin. Endocrinol. Metab.* **35**, 479–82.

Vaughan G. M., Allen J. P. and de la Pena A. (1979) Rapid melatonin transients. *Waking and Sleeping* **3**, 169–73.

Vaughan G. M., Bell R. and de la Pena A. (1979) Nocturnal plasma melatonin in humans: episodic pattern and influence of light. *Neurosci. Letters* **14**, 81–4.

Weeke J. and Gundersen H. J. G. (1978) Circadian and 30 minutes variations in serum TSH and thyroid hormones in normal subjects. *Acta Endocrinol.* **89**, 659–72.

Weinberg U., D'Eletto R. D., Weitzman E. D. et al. (1979) Circulating melatonin in man: episodic secretion throughout the light-dark cycle. *J. Clin. Endocrinol. Metab.* **48**, 114–18.

Weitzman E. D. (1975) Neuroendocrine pattern of secretion during the sleep–wake cycle of man. *Prog. Brain Res.* **42**, 93–102.

Weitzman E. D. (1976) Circadian rhythms and episodic hormone secretion in man. *Ann. Rev. Med.* **27**, 225–43.

Weitzman E. D. (1979) Sleep stage organization: neuro-endocrine relations. In: *Sleep, Wakefulness and Circadian Rhythm.* AGARD Lecture Series No. 105, AGARD, Ch. 3.

Weitzman E. D., Czeisler C. A. and Moore-Ede M. C. (1979) Biological rhythms of man living in isolation from time cues. In: *Sleep, Wakefulness and Circadian Rhythms.* AGARD Lecture Series No. 105, AGARD, ch. 7.

Weitzman E. D., Fukushima D. and Nogeire C. (1971) Twenty-four hour pattern of the episodic secretion of cortisol in normal subjects. *J. Clin. Endocrinol. Metab.* **33**, 14–22.

Weitzman E. D., Fukushima D., Nogeire C. et al. (1974) Studies on ultradian rhythmicity in human sleep and associated neuroendocrine rhythms. In: Scheving L. E., Halberg F. and Pauly J. E. (ed.) *Chronobiology.* Tokyo, Igaku Shoin, pp. 503–5.

Weitzman E. D., Goldmacher D., Kripke D. et al. (1968) Reversal of sleep–waking cycle: effect on sleep stage pattern and certain neuroendocrine rhythms. *Trans. Am. Neurol. Assoc.* **93**, 153–7.

Weitzman E. D., Nogeire C., Perlow M. et al. (1974) Effects of a prolonged 3-hour sleep–wake cycle on sleep stages, plasma cortisol, growth hormone and body temperature in man. *J. Clin. Endocrinol. Metab.* **38**, 1018–30.

Weitzman E. D., Weinberg U., D'Eletto R. et al. (1978) Studies of the 24 hour rhythm of melatonin in man. *J. Neural Transmission* Suppl. 13, 325–37.

Wever R. A. (1979) *The Circadian System of Man. Results of Experiments under Temporal Isolation.* Berlin, Springer-Verlag.

Yen S. S. C., Rebar R., VandenBerg G. et al. (1973) Pituitary gonadotropin responsiveness to synthetic LRF in subjects with normal and abnormal hypothalamic-pituitary-gonadal axis. *J. Reprod. Fertil.* **20**, (Suppl. 1), 137–61.

Zimmerman N. H. and Menaker M. (1979) The pineal gland: a pacemaker with the circadian system of the house sparrow. *Proc. Natl Acad. Sci. USA* **76**, 999–1003.

chapter 8 *Rhythms in the Infant and the Aged*

The data so far described have usually been obtained from the adult. A consideration of circadian rhythms in infants and aged subjects is useful, not only for the sake of completeness, but also because it might be possible to gain some insight into the mechanisms by which rhythms operate and interact if they are studied during times of development and change.

Special problems are associated with experimentation upon such groups. Obviously it is not possible to gain informed consent from an infant and procedures such as urine collection become far more difficult than in the adult. Frequent plasma sampling from so small an individual is not possible and there are ethical considerations which must severely limit the kind of experiment that would be scientifically useful and permissible upon a volunteer adult. As a result of these factors, there are even less data from the infant than the adult and any statements made must be regarded with more caution. The only exception to this is the readily available supply of prepuces upon which mitotic counts can be made.

In the case of the aged too, it is important that unnecessary trauma be avoided in a subject who is becoming frail. The ease of performance of studies upon geriatric subjects is often decreased by the failing mental powers of the subjects or the difficulty they can find in cooperation, for example by having to micturate at set times.

Nevertheless, some data from infants and the aged have been obtained and an outline of these and some of their implications will be given. A fuller review has appeared recently (Minors and Waterhouse, 1981).

I. THE INFANT
I.1. Results
I.1.1. Sleep and wakefulness

The assessment of whether an infant is awake or asleep can be made by either an observer or machine. The observer requires little more than patience, noting unobtrusively the time when the child wakes and goes back to sleep. Devices that measure the amount of movement of the infant require far less time of the researcher, but the assumption that a stationary infant is sleeping is not always warranted.

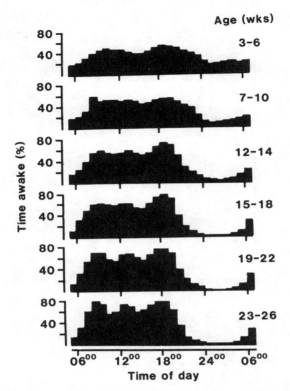

Fig. 8.1. The percentage time spent awake in each hour in six successive 4-week periods of observations of a group of infants. (*From* Kleitman and Engelmann, 1953, Fig. 1.)

The total amount of time spent asleep per day is about 16 hours in the neonate, decreasing to 14 by 3 weeks and more slowly thereafter to about 13 hours at 1 year (Webb, 1974). There then follows a progressive decline to about 8 hours in the second decade of life, a value fairly constant until old age.

The distribution of sleep throughout the nychthemeron alters dramatically. During its first week, the neonate spends equal fractions of the day and night asleep but by the third week a slightly greater proportion of time spent awake in the daylight becomes apparent (*Fig.* 8.1). This imbalance in sleeping habits between night and day becomes progressively more marked with age until, by about 6 months, the infant is awake during most of the daytime and most of its sleep is concentrated during the night (Hellbrügge, 1960). By the age of 4 years, daytime 'naps' and a broken night's sleep become comparatively rare, though in this as in many other aspects of the development of circadian rhythms, one's subjective experience as a parent might suggest that a wide variation exists in the age at which these changes take place.

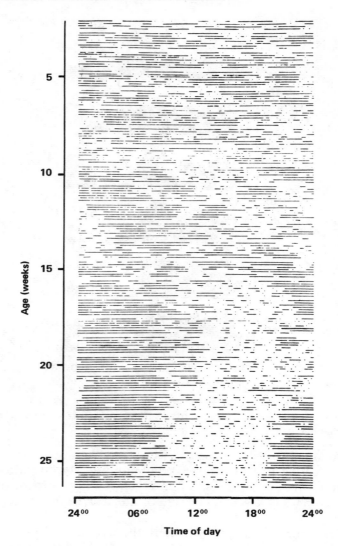

Fig. 8.2. The distribution of sleep (black bars) and wakefulness in an infant from the 4th to 182nd day of life. Dots indicate feeds. (*From* Kleitman and Engelmann, 1953, Fig. 3.)

Fig. 8.2 shows the development during the first 6 months of the pattern of sleep and wakefulness in a single child that was demand-fed and subjected to the alternation of light and dark (Kleitman and Engelmann, 1953). It indicates that the sleeps were frequent, short and irregularly placed during the first 4 weeks. A pattern approximating to that in the adult with most sleep at night and most wakefulness during the day did not develop until about week 16; between weeks 4 and 15, the

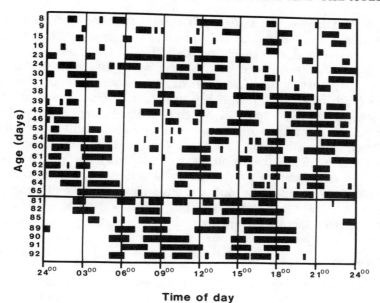

Time of day

Fig. 8.3. Distribution of sleep (white) and wakefulness (black) in an infant who, from the eighth day of life, was kept in isolation in uniform illumination and fed on demand. After the eightieth day of life the infant was exposed to the normal alternation of light and dark. Note that the ordinate does not represent successive days. (Data of Martin-du-Pan, 1970.)

periods of sleep gradually merged into a single session with what appears to be a 'free-running' rhythm during weeks 9 to 15. Although the pattern of a number of short 'naps' merging into a single longer sleep is normally seen in infants, the observation in the present case of the 'free-running' rhythm between weeks 9 and 15 is less common; it can be interpreted to indicate that the circadian clock has developed before the ability of the child to respond to external synchronizers, and further comment about this will be made later.

It seems likely that the darkness and comparative inattention that the infant would experience at night coupled with light, noise and attention during the daytime would be potent forces in developing a circadian rhythmicity and entraining it to the nychthemeral environment. There is some evidence that a regular 4-hourly routine facilitates the development of a 24-hour rhythm (Sander et al., 1970). Thus, a group of 9 babies that had been nursed communally on a 4-hour feeding schedule had a more marked rhythm of sleep and wakefulness at an age of 10 days than had 9 other babies of the same age who had previously been demand-fed in individual rooms. Even so, it is not certain that other nychthemeral cues were identical in the two groups. Of particular interest is a study upon two infants who were kept in constant light and demand-fed from being 8 days old (Martin-du-Pan, 1970, 1974). Data from one of them, *Fig.* 8.3, indicate that there were irregularly placed short sleeps at first but that, by the seventh week, sleep periods became longer at some times of day than others. By days 60–65, this distribution was quite marked but, at this stage, activity was concentrated during the night-time with most sleep between 06^{00} and 15^{00}; presumably this resulted from a lack of cues as to the alternation of day and night. When the infants were aged 80 days they were exposed to a normal light and dark cycle, the light being turned out at night, and it can be seen that the activity pattern rapidly adjusted to a normal pattern of greater amounts of nocturnal sleep and diurnal wakefulness. The importance of this study is not only that it suggests that a rhythmic environment is not a prerequisite for the development of circadian rhythmicity in the infant, but also that it emphasizes the importance of external factors in synchronizing an infant to an adult pattern.

I.1.2. Sleep stages

As adjudged by conventional polygraphic recording methods, the distinction between the different stages of sleep is not as clear in the neonate as in the adult and the terms 'active REM', 'quiet sleep' and 'indeterminate sleep' have been used instead (Emde and Walker, 1976). Whichever system of nomenclature is used, it is agreed that the neonate spends about 50 per cent of its sleep time in REM sleep and the amount decreases with age to reach the adult value of 20 per cent at about 2 years (Webb, 1974). As in the adult (Chapter 5), there is a circadian variation in the tendency for REM to occur, it being claimed (Sostek et al., 1976) that in infants the lowest amount occurs during the afternoon.

By contrast, it has been reported that the neonate does not manifest slow wave sleep (SWS) until about 1–3 months of age, a result that is attributed to the immaturity at birth of the thalamo-cortical system which is believed to be responsible for this type of sleep (Webb, 1974); in the infant, as in the adult, there does not seem to be a circadian rhythm in SWS. The association between the nocturnal secretion of some hormones and SWS that has been claimed in the adult (*see* Chapter 7) would receive some 'negative' support from the observation that neither SWS nor a circadian rhythm of growth hormone secretion is present in the neonate (*see below*).

I.1.3. Temperature

Because of the relative ease of measurement of rectal temperature in infants, the development of the temperature rhythm has been studied more extensively than many others. The general finding is that a rhythm is absent in the first weeks after birth and that thereafter the amplitude of the rhythm increases progressively until the adult value is reached by

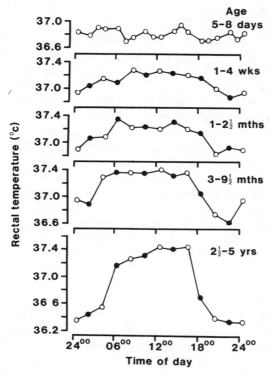

Fig. 8.4. The development of a circadian rhythm of body temperature in relation of age. Measurements made 4-hourly except in the youngest group. The different symbols indicate different groups, each of 3–18 infants, each observed over 2–11 days. (Data of Jundell, 1904.)

about the fifth year (Jundell, 1904; Hellbrügge, 1960; Hellbrügge et al., 1964; Hellbrügge, 1968). These points are illustrated in *Fig.* 8.4. Two more recent studies have confirmed these findings. In the first (Abe et al., 1978) a large number of subjects, ranging in age from newborn to adult, were studied transversely, temperature being recorded every 4 hours over a single nychthemeron. In the second (Abe and Fukui, 1979), 3 females and 1 male were studied longitudinally, the temperature rhythm being determined every 3 months from 3 until 18 months of age. The results from both studies were very similar; thus a circadian temperature rhythm was absent in neonates and of low amplitude and inappropriately phased (that is, at an earlier time than in the adult) until 6 months of age. The phasing became 'normal' by about the end of the first year. In infants older than 6 months, the amplitude increased and from 9 months until at least 2 years was significantly greater than that in adults. This increased amplitude is due to lower nocturnal values (diurnal values are the same) and might well result from the higher surface area/volume ratio of the infant.

In the adult, the rhythm of body temperature is produced by changes in heat loss and vasomotor tone which in turn are controlled by the hypothalamus (*see* Chapter 2). Even though studies to investigate the mechanisms of heat loss and gain by the infant have been performed (Hill and Rahimtulla, 1965; Hey and Katz, 1969), the relationship between these mechanisms and the time of day has not been considered. Indirect evidence that the rate of heat loss is important in the generation of the body temperature rhythm in the infant also has been obtained by measuring the fall in galvanic skin resistance (associated with sweating) at different times of the day (Hellbrügge, 1960). This rhythm is slight during the first week of life but increases progressively during the first 9 months. The time-course of this increase in amplitude is appropriate for that of the temperature rhythm; further, since the lowest resistance (most sweating) is found in the evening and the highest during the forenoon period, the phasing of this rhythm also is appropriate for a role in body temperature control.

I.1.4. Urinary rhythms

A reduced production of urine at night together with sufficient bladder control to preclude nocturnal enuresis are events that parents must look forward to! Studies of the development of urinary rhythms in infants have been severely handicapped by the difficulty in obtaining urine samples, especially from girls. The difficulties have resulted in the number of samples that have been collected during the course of 24 hours being small, and this has made the demonstration of rhythmicity, even more its exact timing, all the more difficult.

The general finding for urine flow is again one of a rhythm with an amplitude that is low in the first week of extra-uterine life and then increases until adult values are reached by about 6 months (Hellbrügge, 1960; Hellbrügge et al., 1964). If this is generally true then the widespread finding that nocturnal enuresis continues beyond this age indicates some aspects of bladder function—for example, the process by which a partially distended bladder is not reflexly emptied—have not matured. Even though it is natural to consider that the rhythm in flow is secondary to one of fluid intake, there is evidence against this being the sole cause. Thus the rhythm began to appear in infants who were still receiving regular 4-hourly feeds and, when the night feed was omitted, urine flow rate was falling before the last evening feed and rising before the first morning one (Hellbrügge, 1960).

There is some evidence, that will be commented upon again later, that the rhythm of urine flow is at first exogenous in origin and only becomes endogenous at about 1 year. Thus, a child of 6 weeks who showed rhythmicity when living with others in a nursery, became arrhythmic when studied in continuous darkness; on the other hand, a similar experiment performed upon a 12-month-old infant indicated no loss of rhythm in the dark (Beyer and Kayser, 1949).

In the adult, the rhythm of urine flow is at least in part an osmotic consequence of the rhythm in excretion of total electrolytes (*see* Chapter 4). Some evidence for an association between flow and osmotic content in children is the observation that a 7-year-old who suffered from nocturnal enuresis had inverted rhythms of sodium, chloride and water and an arrhythmic potassium loss (Lewis et al., 1970).

The time courses of the development of rhythms in the rates of urinary excretion of individual substances have not yet been defined satisfactorily; by far the largest assemblage of data (Hellbrügge et al., 1964; Hellbrügge, 1968) was analysed before cosine curve fitting techniques were routinely used. Further, the pooling of data from infants of a rather wide age span in these studies has meant that the time course of development cannot be defined except in the broadest outline. In brief, therefore, the claims made by these authors that rhythms of chloride, phosphate and creatinine excretion do not develop until the second year of life, together with the implication that these claims have, must be treated with caution. No evidence exists that clearly separates the development of rhythms of flow and of individual electrolytes.

I.1.5. The endocrine system

Studies of the development of rhythms in the endocrine system in infants have been severely limited by the restricted number of plasma samples that can be taken from any individual during the course of a single day; the problem has been exacerbated by the demonstration, at least in adults, that superimposed upon the circadian trend there is episodic secretion (*see* Chapter 7), definition of which requires a high rate of sampling.

Most investigations have been performed upon the adrenal glucocorticoids. Since the adult pattern shows a maximum at about the time of waking and a fall throughout the rest of the day, there have been attempts (Franks, 1967) to define the development of the rhythm by taking only two samples at 08⁰⁰ and 20⁰⁰ (*Fig.* 8.5). *Fig.* 8.5 indicates that, in premature babies, no significant difference between the average values for the group at these two times were found. The variation between the individuals, however, was large and it must be stressed that, if each infant had possessed a circadian rhythm with peaks at times different from each other or from those in the adult, the sampling times chosen in this study might not have demonstrated these rhythms. *Fig.* 8.5 also shows that in full-term infants the evening values were generally lower than those in the morning by the tenth month and reliably so by 3 years. In apparent contradiction to this finding of a slow development of the steroid rhythm is the result of a study upon 2 infants aged 1 and 2 months, in both of whom the 09⁰⁰

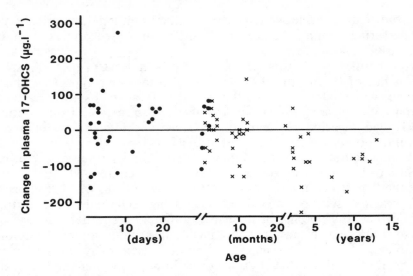

Fig. 8.5. Changes in the plasma concentrations of 17-hydroxycortico-
steroids from 08⁰⁰ to 20⁰⁰ in children of different ages. ● indicates
children born prematurely; X indicates children born at full term. Negative
change indicates 20⁰⁰ less than 08⁰⁰ value. (Data of Franks, 1967.)

plasma concentration was much higher than that at 21⁰⁰ (Martin-du-Pan
and Vollenweider, 1967).

A larger study has been reported more recently (Zurbrügg, 1976) in
which 4-hourly plasma samples have been taken from 08⁰⁰ until 12⁰⁰ the
next day. Samples were taken from 8 neonates (3–9 days), 5 infants (1–6
months), 11 children (1½–11½ years) and a control group of adults. The
average concentration for each group (measured over 24 hours) was
similar, but increasing age was associated with an increasing amplitude of
the rhythm together with an increase in its period from about 12 hours in
the neonate to 24 hours in the adult group. Further, two neonates were
sampled hourly and rhythms with periods of 6 and 12 hours were claimed.
However, in no case were the estimates of amplitude or period made by
formal statistical analysis.

Such a sampling frequency in infants as in the experiments just described
is extremely rare and other ways to obtain more than two samples per day
have been tried. One is to collect urine samples and analyse these for
17-hydroxycorticosteroids (17-OHCS). Such a study has been performed
upon male subjects, 5–105 days old, from whom urine samples were
collected every 3 hours (Martin-du-Pan and Vollenweider, 1967). The
daytime (06⁰⁰ to 18⁰⁰) excretion of 17-OHCS exceeded that at night by
about the fourth week, an age that also showed urine flow, potassium
excretion and wakefulness to be higher over this time span. When the
urine excretion of steroids of each child was considered separately, there

were at least two excretory maxima in the 24 hours, another example of ultradian components in the infant. Another method of obtaining a comparatively large number of points during the 24 hours has been used upon 37 neonates divided into three groups of 9 and one of 10 (Sisson et al., 1974). Each group was sampled every 8 hours for 48 hours beginning about 2 days after birth. By staggering the timing of the first sample the authors obtained 9 or 10 samples at each even hour throughout the 48 hours. The samples in this study were assayed for growth hormone and showed three evenly spaced peaks in the 24 hours, the pattern repeating itself in the second 24 hours. Since the nocturnal peak was the highest of the three, the authors concluded that both circadian and ultradian components were present. The fact that the frequency of the ultradian rhythm equalled that with which samples were taken from each group of infants raises the possibility that this component was spurious and arose from the sampling procedure.

I.2. Inferences and Speculations

I.2.1. The causal nexus

Having outlined the development of different rhythms in the infant, can any information relating to the causal nexus between them be deduced?

The methods by which the causal nexus has been investigated in adults—namely, living in isolation without any cues as to real time; living a routine determined by a non-24-hour clock; undergoing real or simulated time-zone transitions; undergoing a reversal of the sleep–wakefulness cycle—have not been used with infants. Accordingly one can do no more than compare the ages at which different rhythms are first observed and infer the causal nexus from this. Even the value of this approach is slight since rarely are data full enough to define closely such time courses of development (Mills, 1975; Minors and Waterhouse, 1981).

Nevertheless, in addition to the relationships already commented upon in this chapter, some further speculations are possible and are enumerated below:

1. The development of rhythms in blood pressure and heart rate is likely to be produced by the combined effects of the temperature and activity rhythms and that of aldosterone by rhythmic changes in posture and activity. Rhythmic changes in respiration are also undoubtedly influenced by the developing sleep–wakefulness pattern, but there is some evidence (Hoppenbrouwers et al., 1979) that the depth of sleep is an additional influence (compare with Bülow (1963) in Chapter 3).

2. The regular ultradian period of feeding and maternal attention is believed to be at least partly responsible for the ultradian rhythms in the cardiovascular system in the infant (Hellbrügge, 1960).

3. The whole causal nexus involving the kidney (the sympathetic nervous system, plasma concentration, glomerular filtration rate, hormones

and tubular function) is impossible to deduce, the available data being totally inadequate for a task of such complexity. Nevertheless, Krauer (1970) has shown that the nychthemeral variation in the rate of urinary elimination of some drugs is less in infants and she has attributed this to the poorly developed rhythm of urinary acid secretion that exists at this age (*see* Chapter 11).

4. The mitotic rhythm in human prepuces is well developed at birth (Cooper, 1939) and later analysis of the data by cosinor analysis indicates an acrophase at about 19^{00}, a result similar to that from skin biopsy in the adult (Fisher, 1968). Further, since each neonate contributes only one datum point, the observed rhythmicity indicates also that the phasing of the rhythm in different subjects must be similar. There has been some discussion of the possibility that a link exists between these rhythms and that of cortisol (Mills, 1973). However, the mitotic rhythm is present at birth and the steroid rhythm is one of the latest to develop (*see Fig.* 8.5).

I.2.2. Is the development of circadian rhythms genetically or environmentally determined?

In answer to this question, at one extreme is the belief that, since most rhythms develop over the course of some months in a nychthemeral environment, some components of this regular external rhythmicity induce rhythmicity in the animal. Ultimately the rhythmicity inside the animal can continue even in the absence of the external influence, that is, the animal has acquired an internal clock. Evidence that can be adduced in favour of this theory comes from three sources. (This evidence will be considered more fully in Chapter 12 when zeitgeber are studied.)

1. The rhythms of communities born and living near to the poles (where external rhythmicity is not at all marked for some seasons) have low amplitudes (Lobban, 1967).

2. Blind subjects often have poorly developed circadian rhythms (Tokura and Takagi, 1974; Pauly et al., 1977).

3. As mentioned earlier, infants exposed to a 4-hourly, regular routine developed a rhythm of sleep and wakefulness more rapidly than infants reared alone and fed on demand (Sander et al., 1970).

However, the poorer development of rhythms seen in these studies might result in part from a decreased exogenous component of the circadian rhythm rather than a decreased endogenous component even though this internal component, according to this hypothesis, is 'induced' by external rhythmicity. A distinction between these possibilities could be made by measuring the amplitude of the rhythms when the subjects were in a constant environment, but this has not, in general, been performed upon infants. One exception is the observations upon urine flow already described in which two infants that showed circadian rhythmicity were then studied in darkness (Beyer and Kayser, 1949). Rhythmicity disappeared in the 6-week-old infant, but not in the other infant aged 1 year. The original presence of a urine flow rhythm in the younger child would seem therefore to have been caused by the rhythmic environment (exogenous) rather than an internal clock (endogenous). A further difficulty is that a model attributing the development of an internal clock to the rhythmicity of the external environment needs to explain why the internal clock in humans has been observed almost invariably to run slow in the absence of nychthemeral cues (Wever, 1979); a scatter of free-running periods centred around 24 hours would seem to be a more obvious prediction if this model were true.

At the other extreme is the view that the internal clock is genetically determined and develops independently of environmental periodicity. Evidence in favour of this possibility (which will be considered again in Chapter 12) consists of the following observations:

1. In those two cases in which a child has been kept in constant light and been given no clue by its attendants as to real time, a rhythm in sleep and wakefulness was observed to develop (Martin-du-Pan, 1970, 1974) (*Fig.* 8.3).

2. Animal and plant (but not human) experiments have produced a number of mutants that show no rhythmicity or an abnormal period. Such differences appear to be inherited in a manner in accord with Mendelian laws of inheritance (Konopka and Benzer, 1971).

3. The rhythm of mitosis is developed at birth. This has been discussed earlier and it has been inferred that a rhythmic environment is not needed for the development of an internal clock (but *see below*).

In this 'genetic' theory, the external environment is important only in so far as it synchronizes the internal clock. The observation that in one individual the sleep—wakefulness rhythm appeared to free-run before it was synchronized to the external environment (*see Fig.* 8.2) supports such a view; so too do the reports of blind subjects who showed free-running rhythms whilst living in normal society (Miles et al., 1979; Orth et al., 1979; Chapter 12).

Other evidence suggests that the distinction that has been drawn between the 'environmental' and 'genetic' theories to account for the development of rhythmicity in infants is not always clear-cut. Thus:

1. It has been found that the circadian rhythms of temperature and sleep—wakefulness develop more slowly in premature infants (Hellbrügge, 1968, 1974); this can be interpreted either as a later maturation of the ability of the infant to respond to environmental rhythmicity or as a later maturation of some inherent mechanism.

2. It has often been assumed that the human is not exposed to a rhythmic environment until he is born, but the fetus is exposed to changes resulting from alterations in maternal posture (derived in turn from the mother's nychthemeral sleep—wakefulness pattern) and to rhythmic changes in maternal nutrients, metabolites and hormones. Thus there is

evidence that fetal movement shows circadian variation, with peak values at about the time of retiring, whether this movement is assessed by the mother herself (Minors and Waterhouse, 1979; Birkenfield et al., 1980), by observers or by abdominal strain gauges (Wood et al., 1977) or by ultrasonic scanning techniques (Roberts et al., 1978). Additionally, measurements of maternal plasma oestriol (a steroid produced by the placenta from fetal adrenal precursors) by some (for examples, Reck et al., 1978; Patrick et al., 1979), but not all (for examples, Compton et al., 1979; Wisser et al., 1979), groups indicate a circadian rhythm of small amplitude with a phasing that is opposite to that of plasma cortisol. The relevance of these observations is that the presence of a mitotic rhythm at birth can no longer be interpreted unambiguously as evidence that rhythms develop independent of 'external' rhythmicity. However, the observed rhythmicity of the fetus need not show that it possesses a clock (that it 'loses' temporarily after birth) but rather that it can respond to its rhythmic, exogenous, uterine environment. Accordingly, the lack of circadian rhythmicity in the neonate might indicate no more than that it is unable (through immaturity or some other reason) to respond to the new external rhythms (Minors and Waterhouse, 1979).

In summary, it would seem that the easiest compromise is to speculate that an interaction between an inherited component and external rhythmicity takes place during the normal development of circadian rhythms. Such a conclusion is in accord with that from studies that have been performed upon other aspects of brain development, for example the visual cortex of kittens (Blakemore and Cooper, 1970).

I.2.3. The relationship between circadian and ultradian rhythms

If the neonate has poorly developed circadian rhythms, that does not mean that it is arrhythmic; mention has already been made on more than one occasion of the presence of ultradian rhythms in the young infant before circadian rhythms had developed. In addition, Prechtl (1974) studied the neonate infant and found ultradian components (with a period of 30—90 minutes) in the EEG, respiratory and heart rates and the alternation between quiet and active sleep. Ultradian rhythmicity is present in the fetus also. Thus, both subjective means of assessment and the use of ultrasonic scanning devices have indicated the presence of ultradian rhythms of movement (Sterman, 1972), respiration (Patrick et al., 1978; Roberts et al., 1978) and urine production (Campbell et al., 1973). As has already been described, fetal circadian rhythms might well be responses to circadian changes in the mother; the external cause of the ultradian rhythms is not known. However, the observation that the ultradian components do not disappear at birth has led to the inference that they are endogenous; even so, the possible exogenous effects of regular feeding and maternal attention should not be overlooked (*see above*).

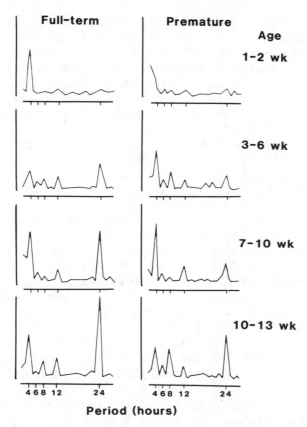

Fig. 8.6. Variance spectra of the sleep—wakefulness rhythm of full-term and premature infants of different ages. Abscissa linear in period. Note the circadian (24-hour) component develops sooner in the full-term infants. (*From* Hellbrügge, 1974, Fig. 1.)

The ultradian rhythms found in the human infant have been summarized by Hellbrügge (1973); in addition, he has speculated on the relationship between ultradian and circadian rhythms. He has suggested that the circadian rhythms 'grow out of' the ultradian ones, a result that reflects some maturation process within the infant (*see also* Halberg, 1963). As additional support for this hypothesis, he described some results (Hellbrügge, 1974) from a comparison between full-term and premature infants (*Fig.* 8.6). The times of sleep and wakefulness were measured and the data analysed by spectral analysis to estimate the important frequency components. The results support his view that a circadian component grows out of an ultradian one; furthermore, the sequence was delayed in the premature children by an amount similar to their prematurity. In a study upon two premature babies (Ullrier, 1974), REM activity was measured continuously from the 38th week of gestation. Spectral analysis of the data indicated as before a declining ultradian and an increasing circadian influence with age of the infant.

To investigate whether Hellbrügge's hypothesis holds for hormonal and urinary rhythms would require more frequent sampling than is acceptable. Unfortunately, since respiratory and cardiovascular rhythms have such a small endogenous component, frequent sampling of these functions, which would be far easier, would be of little value. However, body temperature would seem a fruitful area for study, frequent measurement being an acceptable technique and the rhythm having a suitably large endogenous component (at least in the adult).

Even though Hellbrügge's hypothesis seems to describe the present body of data, two types of reservation exist.

First, the mechanism whereby the ultradian components 'add' to form a circadian rhythm is unknown. One possibility—that the period gradually increases from an ultradian to a circadian value—receives some support from the studies upon the length of the REM/non-REM cycle (Hartman, 1968) and upon plasma cortisol rhythms (Zurbrügg, 1976).

An alternative mechanism can be inferred from data on feeding times in an infant that was demand-fed but kept in a normal nychthemeral environment (Morath, 1974). At first there were 6 feeds per day about 4 hours apart. From the age of 11 days, the child took only 5 feeds, that at 04^{00} being omitted (even though the child was restless at the time of this missed feed). At 76 days, the midnight feed too was missed. In other words the transition from an ultradian to a substantially diurnal pattern of feeding was not by a lengthening of period but rather by what appeared to be the missing out of cycles. In addition, in this study, if the size or timing of a meal was altered, the infant woke for the next feed 'on time' and corrected for the size of the earlier meal if necessary. This last observation favours the existence of an endogenous ultradian rhythm controlling waking or feeding rather than the view that the infant wakes because of a requirement for calory intake. From investigations of sleep and wakefulness upon a single infant who was demand-fed in a nychthemeral environment (Meier-Koll et al., 1978; Meier-Koll, 1979), essentially similar conclusions—that ultradian cycles 'drop out' to produce a circadian rhythm—have been drawn. Thus, for the first 4 weeks there were 3 periods of wakefulness diurnally and 3 nocturnally; from weeks 5 to 11, there were 3 periods diurnally and 2 (and then 1) periods nocturnally: and from week 11 onwards, there was little nocturnal waking.

The second reservation relates to the observation that, in adults, ultradian as well as circadian components are present. This has been mentioned earlier (Chapter 5) in connection with the REM/non-REM cycle as has the concept of a basic 'rest–activity' cycle of Kleitman (1963). The relationship between this concept and that of Hellbrügge requires further consideration and this will be given in Chapter 12.

II. THE AGED

Some of the problems associated with studies upon aged subjects have already been referred to. Nevertheless, the volume of literature pertaining to circadian rhythms in the aged is so small that it is difficult to avoid the conclusion that this branch of the field has suffered neglect.

Most of the studies have been performed upon subjects in institutions (and so subjected to closely controlled routines) and all, it would seem, in nychthemeral conditions. For example, in a study in which the sleep–wakefulness cycle of aged subjects living in a welfare institution was investigated, analysis of the amount of time spent sleeping at each hour showed a strong 24-hour rhythm with a nocturnal maximum (Wessler et al., 1976). No study seems to have been performed that investigates only the internal rhythms; instead it is the combined effect of the exogenous and endogenous components that has been investigated. Therefore, any differences from younger adults cannot be unambiguously attributed to changes in an internal clock, for they may instead reflect a modified effect upon the body of exogenous influences.

II.1.1. Results

The total amount of time spent asleep decreases beyond the sixtieth year. Other changes in the sleep pattern are a greater number of occasions when the subject wakes at night and an increase in the number of diurnal 'naps'. Polygraphic recording of sleep stages has shown a fall in the amount of SWS that is taken (Webb, 1974). This last finding accords with another study in which blood samples were taken by venous catheter during sleep (Murri et al., 1977; 1980). Results showed that there was a normal adult pattern of prolactin secretion but that growth hormone (GH) did not show a nocturnal peak. This offers some evidence for the hypothesis that GH secretion is associated with SWS (*see* Chapter 7). A decreased nocturnal secretion of GH in aged subjects has also been found by D'Agata et al. (1974).

Another hormone to have been studied is plasma cortisol. Two investigations upon subjects aged 70 years and over have both found an increased variability between individuals in timing of the peak when compared with younger controls (Serio et al., 1970; Milcu et al., 1978). In the study of Milcu et al., cosinor analysis indicated also that both the amplitude and mesor of the best-fitting curve were decreased.

These three findings—a fall in the amplitude and mean value of a rhythm together with a rise in variability of its phasing—apply to many of the variables that have been studied. Examples are: plasma adrenaline (Descovich et al., 1974); oral temperature, peak expiratory flow rate, grip strength, ability to estimate time intervals, blood pressure and heart rate Scheving et al., 1974) and urine flow and urinary excretion of various constituents (Elithorn et al., 1966; Scheving et al., 1974). However, not all variables show a fall in average value; thus the mean daily concentration of a variety of plasma metabolites such as glucose, lactate, pyruvate and glycerol (Alberti et al., 1975), serum magnesium concentration (Touitou et al., 1978) and peak expiratory flow rate (Reinberg et al., 1979) all showed an increased mean value.

II.1.2. Some implications and inferences

One of the most distressing results of the changes of rhythms associated with old age is the increased number of occasions on which sleep is broken, often by the need to micturate. Such a result might derive from changes in the sleep process or in some aspect of urine formation or voiding. Evidence of some changes that might affect sleep comes from the altered timing of the cortisol rhythm. Thus the earlier acrophases that were found in one study were surmised to cause early waking (Milcu et al., 1978); by contrast, in another study there was reported to be no relationship between the cortisol acrophase and whether or not the subject believed himself to be an 'insomniac' (Serio et al., 1970). However, for both groups, the variation in cortisol acrophases between individuals was so great that only very marked differences would have been statistically significant. Nevertheless, as has already been stated, there is a decreased amount of SWS and of GH release and the hormone data do suggest that, in some unknown way, the sleep process is modified in old age.

Ageing is associated also with changes in bladder and kidney function. It has been argued that one of the consequences of ageing is an 'uninhibited neurogenic bladder' (Brocklehurst et al., 1971). This condition results not only in a difficulty with the act of micturition and in problems associated with 'stress incontinence' but also (and this is most relevant in the present context) in difficulty in holding large volumes of urine. In addition, the fall in amplitude of the circadian rhythm of urine flow will increase the amount of urine produced during the night and so the subject will be more likely to wake. The causes for this are speculative; a likely contributory factor is the decreased amplitude of electrolyte rhythms and the osmotic effect this will produce (*see* Chapter 4), but the cause of the electrolyte changes is unknown. Another possibility, based upon the observation of a decreased amplitude of the creatinine rhythm, is that age is associated with an increased GFR response to changes in posture; that is, by comparison with younger controls, GFR is lower when standing and higher when lying and this will act to oppose that part of the circadian rhythm of GFR that does not depend upon posture (Chapter 4) (Lindeman, 1975).

II.1.3. Possible causes of the changed rhythms in the aged

Although the amount of data available is small, it is remarkable for its uniformity; this suggests that the changes are common to most variables rather than different in each case.

One change—the greater variation between individuals in the timing of their rhythms—suggests that in any or all changes in routine, the responsiveness of the individual to exogenous influences or the endogenous clock itself could be involved. Some workers have rejected the first possibility since the studies have been performed in situations in which the environment was highly controlled (Scheving et al., 1974). Nevertheless, the circadian rhythmicity of sleep decreases and, to the extent that sleep acts as an exogenous, masking influence (*see* Chapter 5), this would tend to decrease the nocturnal–diurnal differences in other variables. Offering some support for this view is the finding (Lobban and Tredre, 1967) that the changes in amplitude and phase of urinary potassium excretion (with a large endogenous component) were less than those in other urinary variables (with smaller endogenous components). Whether the internal clock and/or the influence of other exogenous factors is changed with age is quite unknown.

The other general finding—the decline in amplitude and mean of many rhythms—emphasizes again the relationship between these two variables that has already been discussed in Chapter 4. (Of course the change with age in the average value of a rhythm remains unexplained.) In part it may be spurious because all estimates of amplitude have been derived from grouped data. These group estimates will be decreased if, as is observed, the times of peak of the different subjects vary more in aged subjects than in younger controls.

It has been argued that one consequence of ageing is a deterioration of homeostatic mechanisms in general (Alberti et al., 1975; Comfort, 1979). It is tempting to apply such an argument to circadian rhythms but such a view is purely speculative and, as has been discussed in Chapter 7 and will be considered again in Chapter 12, the relationship between feedback mechanisms and circadian rhythms is not at all clear.

References

Abe K. and Fukui S. (1979) The individual development of circadian temperature rhythm in infants. *J. Interdiscipl. Cycle Res.* **10**, 227–32.

Abe K., Sasaki H., Takebayashi K. et al. (1978) The development of circadian rhythm of human body temperature. *J. Interdiscipl. Cycle Res.* **9**, 211–16.

Alberti K. G. M. M., Durnharst A. and Rowe A. S. (1975) Metabolic rhythms in old age. *Biochem. Soc. Trans.* **3**, 132–3.

Beyer P. and Kayser C. (1949) Établissement du rythme nycthéméral de la sécrétion urinaire chez le nourrisson. *C. R. Séance. Soc. Biol. (Paris)* **143**, 1231–3.

Birkenfield A., Laufer N. and Sadovsky E. (1980) Diurnal variation of fetal activity. *Obstet. Gynecol.* **55**, 417–19.

Blakemore C. and Cooper G. F. (1970) Development of the brain depends on the visual environment. *Nature* **228**, 477–8.

Brocklehurst J. C., Fry J., Griffiths L. L. et al. (1971) Dysuria in old age. *J. Am. Geriat. Soc.* **19**, 582–92.

Bülow K. (1963) Respiration and wakefulness in man. *Acta Physiol. Scand.* Suppl. **209**, 1–110.

Campbell S., Wladimiroff J. W. and Dewhurst C. J. (1973) The antenatal measurement of fetal urine production. *J. Obstet. Gynaecol. Br. Commonw.* **80**, 680–6.

Comfort A. (1979) Physiology, homeostasis and aging. In: *The Biology of Senescence*, 3rd ed. Edinburgh, Churchill Livingstone, pp. 251–9.

Compton A. A., Kirkish L. S., Parra J. et al. (1979) Diurnal variations in unconjugated and total plasma estriol levels in late normal pregnancy. *Obstet. Gynecol.* 53, 623–6.

Cooper Z. K. (1939) Mitotic rhythm in human epidermis. *J. Invest. Derm.* 2, 289–300.

D'Agata R., Vigneri R. and Polosa P. (1974) Chronobiological study on growth hormone secretion in man: Its relation to sleep–wake cycles and to increasing age. In: Scheving L. E., Halberg F. and Pauly J. E. (ed.) *Chronobiology*. Tokyo, Igaku Shoin, pp. 81–7.

Descovich G. C., Kuhl J. F. W., Halberg F. et al. (1974) Age and catecholamine rhythms. *Chronobiologia* 1, 163–71.

Elithorn A., Bridges P. K., Lobban M. C. et al. (1966) Observations on some diurnal rhythms in depressive illness. *Br. Med. J.* 2, 1620–3.

Emde R. N. and Walker S. (1976) Longitudinal study of infant sleep: results of 14 subjects studied at monthly intervals. *Psychophysiol.* 13, 456–61.

Fisher L. B. (1968) The diurnal mitotic rhythm in the human epidermis. *Br. J. Dermatol.* 80, 75–80.

Franks R. C. (1967) Diurnal variation of plasma 17-hydroxycorticosteroids in children. *J. Clin. Endocrinol. Metab.* 27, 75–8.

Halberg F. (1963) Periodicity analysis. A potential tool for biometeorologists. *Int. J. Biometeor.* 7, 167–91.

Hartmann E. (1968) The 90-minute sleep-dream cycle. *Arch. Gen. Psychiat.* 18, 280–6.

Hellbrügge T. (1960) The development of circadian rhythms in infants. *Cold Spring. Harbor. Symp. Quant. Biol.* 25, 311–23.

Hellbrügge T. (1968) Ontogénèse des rythmes circadiaires chez l'enfant. In: Ajuriaguerra J. (ed.) *Cycles Biologiques et Psychiatrie*. Symposium Bel-Air III, Geneva, September, 1967. Geneva, Georg; Paris, Masson, pp. 159–83.

Hellbrügge T. (1973) Ultradian rhythms in childhood. *Int. J. Chronobiol.* 1, 331.

Hellbrügge T. (1974) The development of circadian and ultradian rhythms of premature and full-term infants. In: Scheving L. E., Halberg R. and Pauly J. E. (ed.) *Chronobiology*. Tokyo, Igaku Shoin, pp. 339–41.

Hellbrügge T., Lang J. E., Rutenfranz J. et al. (1964) Circadian periodicity of physiological functions in different stages of infancy and childhood. *Ann. NY Acad. Sci.* 117, 361–73.

Hey E. N. and Katz G. (1969) Evaporative water loss in the new-born baby. *J. Physiol.* 200, 605–19.

Hill J. R. and Rahimtulla K. A. (1965) Heat balance and the metabolic rate of newborn babies in relation to environmental temperature; and the effect of age and weight on basal metabolic rate. *J. Physiol.* 180, 239–65.

Hoppenbrouwers T., Jensen D., Hodgman J. et al. (1979) Respiration during the first six months of life in normal infants. II. The emergence of a circadian pattern. *Neuropädiatrie* 10, 264–80.

Jundell J. (1904) Über die nykthemeralen Temperaturs Schwankungen im I. Lebensjahre des Menschen. *Jb. Kinderheil.* 59, 521–619.

Kleitman N. (1963) *Sleep and Wakefulness*, 2nd ed. Chicago, University of Chicago Press.

Kleitman N. and Engelmann T. G. (1953) Sleep characteristics of infants. *J. Appl. Physiol.* 6, 269–82.

Konopka R. J. and Benzer S. (1971) Clock mutants of *D. melanogaster*. *Proc. Natl Acad. Sci. USA* 68, 2112–16.

Krauer B. (1970) Intraindividuelle Variabilität der Arneimittelelimination. *Z. Kinderheilk.* 108, 231–7.

Lewis H. E., Lobban M. C. and Tredre B. E. (1970) Daily rhythms of renal excretion in a child with nocturnal enuresis. *J. Physiol.* **210**, 42–3.

Lindeman R. D. (1975) Age changes in renal function. In: Goldman R. and Rockstein M. (ed.) *The Physiology and Pathology of Human Ageing.* New York, Academic Press, pp. 19–38.

Lobban M. C. (1967) Daily rhythms of renal excretion in arctic-dwelling Indians and Eskimos. *Q. J. Exp. Physiol.* **52**, 401–10.

Lobban M. C. and Tredre B. E. (1967) Diurnal rhythms of renal excretion and of body temperature in aged subjects. *J. Physiol.* **188**, 48P–49P.

Martin-du-Pan R. (1970) Le rôle du rythme circadian dans l'alimentation du nourrisson. *La Femme l'Enf.* **4**, 23–30.

Martin-du-Pan R. (1974) Some clinical applications of our knowledge of the evolution of the circadian rhythm in infants. In: Scheving L. E., Halberg F. and Pauly J. E. (ed.) *Chronobiology.* Tokyo, Igaku Shoin, pp. 342–7.

Martin-du-Pan R. and Vollenweider L. (1967) L'apparition du rythme circadien des 17-hydroxy-stéroides chez le nourrisson. Sa modification sous l'effet de la consommation de corticostéroides. *Praxis* **56**, 138–44.

Meier-Koll A. (1979) Interactions of endogenous rhythms during postnatal development. Observations of behaviour and polygraphic studies in one normal infant. *Int. J. Chronobiol.* **6**, 179–89.

Meier-Koll A., Hall U., Hellwig U. et al. (1978) A biological oscillator system and the development of sleep-waking behavior during early infancy. *Chronobiologia* **5**, 425–40.

Milcu S. M., Bogdan C., Nicolau G. Y. et al. (1978) Cortisol circadian rhythm in 70–100-year-old subjects. *Rev. Roum. Méd.-Endocrinol.* **16**, 29–39.

Miles L. E. M., Raynal D. M. and Wilson M. A. (1977) Blind man living in normal society has circadian rhythms of 24·9 hours. *Science* **198**, 421–3.

Mills J. N. (1973) Transmission processes between clock and manifestations. In: Mills J. N. (ed.) *Biological Aspects of Circadian Rhythms.* London, Plenum Press, pp. 27–84.

Mills J. N. (1975) Development of circadian rhythms in infancy. *Chronobiologia* **2**, 363–71.

Minors D. S. and Waterhouse J. M. (1979) The effect of maternal posture, meals and time of day on fetal movements. *Br. J. Obstet. Gynaecol.* **86**, 717–23.

Minors D. S. and Waterhouse J. M. (1981) Development of circadian rhythms in infancy. In: Davis J. A. and Dobbing J. (ed.) *Scientific Foundations of Paediatrics,* 2nd ed. London, Heinemann (in press).

Morath M. (1974) The four-hour feeding rhythm of the baby as a free-running endogenously regulated rhythm. *Int. J. Chronobiol.* **2**, 39–45.

Murri L., Barreca T., Gallamini A. et al. (1977) Prolactin and somatotropin levels during sleep in the aged. *Chronobiologia* **4**, 135.

Murri L., Barreca T., Cerone G. et al. (1980) The 24-h pattern of human prolactin and growth hormone in healthy elderly subjects. *Chronobiologia* **7**, 87–92.

Orth D. N., Besser G. M., King P. H. et al. (1979) Free running circadian plasma cortisol rhythm in a blind human subject. *Clin. Endocrinol.* **10**, 603–17.

Patrick J., Challis J., Natale R. et al. (1979) Circadian rhythms in maternal plasma cortisol, esterone, estradiol, and estriol at 34 to 35 weeks gestation. *Am. J. Obstet. Gynecol.* **135**, 791–8.

Patrick J., Natale R. and Richardson B. (1978) Patterns of human fetal breathing activity at 34 to 35 weeks' gestational age. *Am. J. Obstet. Gynecol.* **132**, 507–13.

Pauly J. E., Scheving L. E., Burns E. R. et al. (1977) Studies of the circadian system in blind human beings. Proceedings of XII International Conference of the International Society for Chronobiology, Washington, 1975. Milan, Il Ponte, pp. 19–28.

Prechtl H. F. R. (1974) The behavioural states of the newborn infant. *Brain Res.* **76**, 185–212.

Reck G., Nowostowskyj H. and Breckwoldt M. (1978) Effect of ACTH and dexamethasone on the diurnal rhythm of unconjugated oestriol in pregnancy. *Acta Endocrinol.* **87**, 820–7.

Reinberg A., Vieux N., Ghata J. et al. (1979) Consideration of the circadian amplitude in relation to the ability to phase shift circadian rhythms of shift workers. *Chronobiologia* Suppl. 1, 57–63.

Roberts A. B., Little D. and Campbell S. (1978) 24 hour studies of fetal respiratory movements and fetal body movements in normal and abnormal pregnancies. In: Beard R. W. and Campbell S. (ed.), *Current Status of Fetal Heart Rate Monitoring and Ultrasound in Obstetrics.* London, Royal College of Obstetricians and Gynaecologists, pp. 209–20.

Sander L. W., Stechler G., Burns P. et al. (1970) Early mother–infant interaction and 24-hour patterns of activity and sleep. *J. Am. Acad. Child Psychiatry* **9**, 103–23.

Scheving L. E., Roig C., Halberg F. et al. (1974) Circadian variations in residents of a 'senior citizens' home. In: Scheving L. E., Halberg F. and Pauly J. E. (ed.) *Chronobiology.* Tokyo, Igaku Shoin, pp. 353–7.

Serio M., Piolanti P., Romano S. et al. (1970) The circadian rhythm of plasma cortisol in subjects over 70 years of age. *J. Gerontol.* **25**, 95–7.

Sisson T. R. C., Root A. W., Kendall N. et al. (1974) Biologic rhythm of plasma human growth hormone in newborns of low birth weight. In: Scheving L. E., Halberg F. and Pauly, J. E. (ed.) *Chronobiology.* Tokyo, Igaku Shoin, pp. 348–52.

Sostek A. M., Anders T. F. and Sostek A. J. (1976) Diurnal rhythms in 2- and 8-week old infants: sleep-waking state organization as a function of age and stress. *Psychosom. Med.* **38**, 250–6.

Sterman M. B. (1972) The basic rest–activity cycle and sleep: developmental considerations in man and cats. In: Clemente C. D., Purpura D. P. and Mayer F. E. (ed.) *Sleep and the Nervous System.* London, Academic Press, pp. 175–97.

Tokura H. and Takagi K. (1974) Comparison of circadian oral temperature rhythms between blind and normal subjects. *J. Physiol. Soc. Japan* **36**, 255–6.

Touitou Y., Touitou C., Bogdan A. et al. (1978) Serum magnesium circadian rhythm in human adults with respect to age, sex and mental status. *Clin. Chim. Acta* **87**, 35–41.

Ullrier R. E. (1974) On the development of ultradian rhythms: the rapid eye movement activity in premature children. In: Scheving, L. E., Halberg F. and Pauly J. E. (ed.) *Chronobiology.* Tokyo, Igaku Shoin, pp. 478–81.

Webb W. (1974) The rhythms of sleep and waking. In: Scheving L. E., Halberg F. and Pauly J. E. (ed.) *Chronobiology.* Tokyo, Igaku Shoin, pp. 482–6.

Wessler R., Rubin M. and Sollberger A. (1976) Circadian rhythm of activity and sleep–wakefulness in elderly institutionalized persons. *J. Interdiscipl. Cycle Res.* **7**, 333–48.

Wever R. A. (1979) *The Circadian System of Man. Results of Experiments under Temporal Isolation.* Berlin, Springer-Verlag.

Wisser H., Knoll E., Hartmann G. et al. (1979) Diurnal variations of estriol and cortisol in serum and urine in late pregnancy with and without fenoterol-therapy. *Clin. Chim. Acta* **92**, 221–8.

Wood C., Walters W. A. W. and Trigg P. (1977) Methods of recording foetal movement. *Br. J. Obstet. Gynaecol.* **84**, 561–7.

Zurbrügg R. P. (1976) Hypothalamic-pituitary-adrenocortical regulation: a contribution to its assessment, development and disorders in infancy and childhood with special reference to plasma circadian rhythm. In: *Monographs in Paediatrics,* 7. Basle, Karger Press.

chapter 9   *Time-Zone Transitions*

If short time before comp, maybe maintain rhythm

1. Introduction

A traveller journeying in an eastward direction will have to advance his watch continually as he crosses time zones: by contrast, a traveller from England to the United States has to delay his watch in order to be in accord with his new surroundings. When in these new surroundings all time cues are changed by the same amount (light, social influences, meal times etc.) and it would be pointless as well as extremely inconvenient to retain the timing of one's original time zone unless an imminent return to it was planned. When intercontinental trips were made by train or boat, the rate of movement through time zones was slow enough for the endogenous components of rhythms to adjust (Sasaki, 1964), that is the period of the zeitgeber was within the range of entrainment of the internal 'clock' (*see* Chapter 1). With rapid time-zone transitions, dissociation occurs between endogenous and exogenous components with the endogenous rhythms taking some time to adjust to the new time zone. (*See* previous chapters in which evidence for an internal component was being sought.) Whilst this adjustment is taking place, the subject is said to show 'desynchronosis' (Strughold, 1971) or 'transmeridian dyschronism' (Halberg and Lee, 1974) and often suffers from a feeling of general discomfort popularly known as 'jet-lag'.

Studies of time-zone transitions can be divided between those involving real flights and those involving simulated time-zone transitions, the results from both of which are very similar. The relative merits of each type of experiment are similar to those commented upon in Chapter 6 and to be discussed further in Chapter 10 when the differences between laboratory-based and field tests of mental performance are considered. In brief, advantages of simulated time changes performed in an isolation unit are that the experiment is cheaper and the clock adjustment can be accomplished immediately and will cause less apprehension on the part of the subjects than will a real flight. However, stress, as well as the time-zone change itself, might be a component of real flights and it is possible that subjects will see the isolation unit experiment as 'unreal'. In addition, they will have to be isolated from the rest of the community in an artificial way that might modify the process of adaptation. These points will all be considered during this chapter.

2. Results

2.1. Sleep and fatigue

Whether it results from unfamiliarity with new surroundings or (in the case of aircrew) from the exigencies of duty, flight is often associated with stress and loss of sleep. Indeed it has been argued (Siegel et al., 1969) that there are many stressful factors associated with a trip but also both before and after the flight itself. Subjectively, most subjects feel tired, 'out-of-sorts' and suffer from gastrointestinal disorders and loss of appetite (Conroy, 1971; Endo et al., 1978; Richards and Jacobson, 1978). Even though their total sleep time may increase (Klein and Wegmann, 1979), they complain of poor and broken sleep at night time and yet feel tired at other times when they are supposed to be working. The time of day when these symptoms are most noticeable depends upon the flight direction and can be predicted if the assumption is made that the symptoms are a manifestation of rhythmic changes in fatigue (presumably connected with temperature changes, *see* Chapters 5 and 6) still phased more appropriately to the old time. The problem is not only that the traveller will wish to sleep at times inconvenient to his hosts but also that he will have difficulty in sleeping during their night time. Thus, after a long flight to the west (where new time is behind the original) the traveller feels tired in the late afternoon and begins to wake up as midnight approaches (Evans et al., 1972). After an eastward flight (where the new time zone is ahead of the original) the subject will be more tired in the morning of his new time zone and yet feel alert when his hosts prepare for bed.

A more objective assessment of sleep can be made by electroencephalographic (EEG) recording. This is rare after real flights due to difficulties in obtaining an adequate recording together with problems of interpretation that may arise when sleep in a strange environment is considered. Nevertheless, those results which have been obtained agree closely with laboratory-based experiments (*see* Chapter 5). For example, *Fig.* 9.1 shows the pre- and post-flight EEG of a female who underwent a westward time-zone transition (Klein et al., 1976). It shows clearly that the sleep infrastructure during the control pre-flight days showed a normal increase in REM throughout the night; by contrast, the infrastructure was changed immediately after the flight, with most REM now appearing earlier during the sleep at a time that coincided with late sleep on 'old' time, a time associated with peak REM activity. *Fig.* 9.1 also shows that adaptation to the time-zone shift appeared to be fairly rapid since a normal infrastructure, adapted to the new time zone, was present by the third night. Other data from real flights (for example, Moiseeva et al., 1976) have shown an increase in total REM after a westward flight (when sleep was delayed) and a decrease after an eastward flight (when sleep was taken earlier than on 'old' time).

The most comprehensive series of experiments has been summarized in a recent paper by Endo and his colleagues (Endo et al., 1978). Measurements of sleep stages were made upon a number of subjects before and during recovery from flights in both eastward and westward directions. An increase in the amount of REM sleep and a decrease in its latency after the westward flight were especially marked during the first half of sleep; the

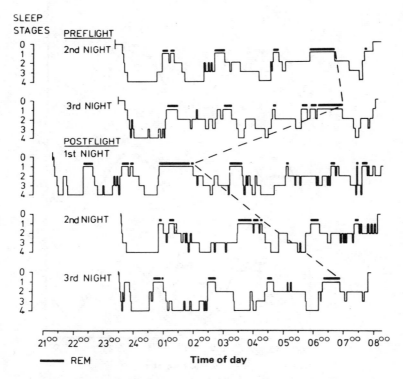

Fig. 9.1. Distribution of sleep stages before and after a flight from east to west through 6 time zones. The dashed line indicates the shift in the time of maximum REM sleep portions. (*From* Klein et al., 1976, Fig. 7.)

opposite changes were seen after a flight to the east. Further, experiments in which flights to the north and south were taken (as a result of which no time shift was produced) resulted in much less marked and only transient changes that the authors attributed to fatigue. In a final experiment, the rates of adaptation of sleep stages in a single subject to flights in both directions were assessed. The REM sleep infrastructure took about 8 days to adjust to the new time zone and possibly less to readjust back to home time. On the other hand, SWS was in all cases associated more with the amount of prior wakefulness whether or not a time-zone transition was involved.

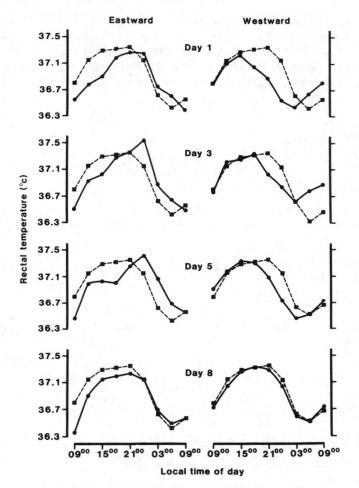

Fig. 9.2. The effect of transmeridian flight through 6 time zones in an eastward or westward direction upon the rhythm of rectal temperature. ■--■, preflight data, represented for clarity on each postflight day. ●—● postflight data. Data represent the means from 8 subjects. (*From* Klein, Wegmann et al., 1972, Fig. 2.)

2.2. Other physiological and psychological variables

Fig. 9.2 shows the circadian changes in rectal temperature in a group of 8 subjects undergoing first an eastward shift through six time zones and then, 18 days later, a westward return journey (Klein, Wegmann et al., 1972). The initial effect after both journeys was that the rhythm was inappropriately phased (being too late after the eastbound and too early after the westbound flight). This effect was transient and adaptation seems to have been complete by about day 8.

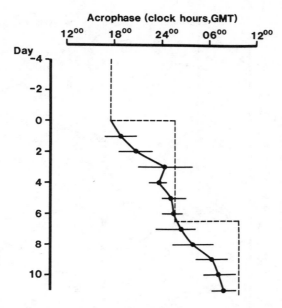

Fig. 9.3. Acrophases of the rectal temperature rhythm of a single subject before and after two simulated flights across 8 time zones in a westward direction. Acrophases after time shift (●) derived from cosine curves fitted to single days. Horizontal bars represent 95 per cent confidence limits of acrophase derived from cosinor analysis. Dashed line indicates the course the acrophase would take if adaptation to each time-shift were immediate and complete. Acrophase during control period derived from cosinor analysis of all control days data. (Data of Minors, 1975.)

The difficulty with this type of analysis, where the raw or average data are presented, is that stages intermediate between 'completely unadapted' and 'completely adapted' are quantitatively difficult to define. A number of ways of overcoming this problem have been used. In one of them, cosinor analysis can be performed and the acrophases, sometimes with their confidence intervals (*see* Appendix) can be included (for example, Levine et al., 1974) *Fig.* 9.3 illustrates this approach for a single subject undergoing two simulated westbound flights through 8 time zones. One advantage of this means of assessment is that the rate of adaptation can be studied (the implications of which will be considered in a later section). However, this approach is not without its criticisms. Thus, sometimes the peak and trough of the rhythm become closer together (Klein and Wegmann, 1979); sometimes the rhythmicity disappears, irregular changes being seen until an appropriately phased rhythm 'grows' out of the irregularity (*see,* for example, Aschoff et al., 1975, Fig. 40); sometimes the rhythm becomes biphasic, the peak at the 'old' time declining while that at the 'new' increases (Elliott et al., 1972). In all

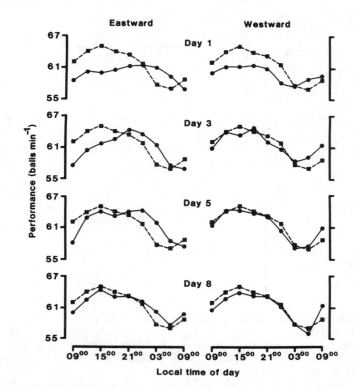

Fig. 9.4. The effect of transmeridian flight through 6 time zones in an eastward or westward direction on psychomotor performance (the sorting of differently sized steel balls and insertion into the corresponding size hold of a rapidly rotating cylinder). (Other details as *Fig.* 9.2.) (*From* Klein, Wegmann et al., 1972, Fig. 3.)

cases, changes in the acrophase might produce spurious estimates of the process of adaptation. A similar reservation about cosinor analysis was made before (*see* Chapter 4 and Appendix); the method of eduction that was used when endocrine rhythms were studied is not applicable to the present situation in which the data are not stationary and the rhythm changes day by day. Another possibility is the use of maximum and minimum values (*see* Wever, 1979). This requires neither stationary nor sinusoidal data, but, as with any analysis based upon a single datum point, it would seem to be more susceptible to aberrant readings. A more elaborate analysis involving cross-correlation techniques can be used under certain circumstances and this will be described later.

Fig. 9.4 shows the results from a test of psychomotor performance (the 'Kügel' test, *see* Brüner et al., 1960) performed in the same experiment as that from which *Fig.* 9.2 was taken (Klein, Wegmann et al., 1972). *Fig.* 9.4 shows: (*a*) the relationship between temperature (*Fig.* 9.2) and psychomotor performance still holds to some extent (but *see* s. 3.2); (*b*) not only is there a temporary change in time of peak as assessed by local time but also the curve becomes 'flatter' immediately after the shift. This last observation suggests some interaction between competing influences, probably internal and external, and is a problem that will recur, especially when shift work is considered in Chapter 10.

These two findings, that there are transient changes in the phase of rhythms (initially too early after westward and too late after eastward flights) and that there is a temporary flattening of the rhythms, have been reported in many studies upon real and simulated time-zone shifts, in males and females and in different species. These findings apply also to a whole range of endocrine, renal, respiratory, cardiovascular and psychometric rhythms as well as to that of deep body temperature. To detail these results would be unnecessary in the present work (even though some examples have been given in previous chapters), but the details are to be found in many reviews, some of the more recent being Klein, Brüner et al., 1972), Sasaki (1972), Mills (1973a), Aschoff et al. (1975) and Klein and Wegmann (1979).

2.3. Evidence that the effects on rhythms relate to changes in time zone rather than fatigue

It has already been stated that, after a time-zone shift, subjects feel tired and that they have been stressed. To what extent will the flight itself cause the symptoms and the changes in rhythms that have just been described? Clearly the finding that similar results are found in isolation unit experiments (Elliott et al., 1972; Aschoff et al., 1975) argues against this view, though a sojourn in an isolation unit might produce its own kind of stress. Two more direct approaches to this problem have been made.

Hauty and Adams (1966a,b,c.) performed a classic set of experiments. They compared the response of subjects to one of three protocols: a flight to the east, a flight through a similar number of time zones to the west; and a flight for a similar distance and duration but southwards (so that no time shift was produced). (It will be noted that a similar protocol had been used by Endo et al. (1978) in their study of sleep stages, s. 2.1.) All flights produced an increase in fatigue, a decrement in performance in a number of psychometric tests and changes in temperature, respiratory and cardiovascular rhythms during the first day after the flights; however, normal rhythms were recovered by the second post-flight day after the southward flight. By contrast, after the other two flights, the rhythms remained phased more appropriately to pre-shift than post-shift time for from 4 to 8 days. More recently, Gerritzen and Strengers (1974) have obtained similar results; since their studies were performed under conditions in which posture and food and fluid intakes were controlled, this argues that an endogenous rhythm is involved, even though the alternations of light and dark and social rhythms were still present.

Lafontaine et al. (1967) flew subjects through 11 time zones from Paris to Anchorage and then returned them to Paris either immediately or after 5 days in Anchorage. Regular samples of urine were taken and analysed for potassium and 17-hydroxycorticosteroids. The results clearly showed that the rhythms did not adapt immediately to the time-zone shift but that this process took place in subjects who remained in Anchorage for 5 days. After the return flight, the rhythms were immediately appropriate to Paris time in the case of subjects who returned straightaway whereas, in subjects who stayed for 5 days in Anchorage, rhythms were now appropriately phased to that location rather than to Paris. In other words, brief stays meant that a subject's rhythms were unadapted to the new, but were appropriate to the old, time zone, whereas, with longer stays, the rhythms slowly adapted; in this latter case, return to the original time zone required a further period of readaptation.

Conclusions similar to those described in both approaches above have more recently been reached upon a single subject by Sekiguchi et al. (1976).

In summary, therefore, a large amount of data indicates that adaptation of rhythms to time-zone shifts is not immediate; this has important implications for the transmeridian traveller and, in some unknown way, seems to be associated with the 'jet-lag' syndrome from which he suffers.

3. Differences Produced by Time-zone Transitions

So far the similarities in effects of time-zone transitions upon different circadian rhythms have been considered, but there is much evidence to indicate that there are differences according to the direction of flight, physiological variable, subject and detailed circumstances under which the transition is made. Far from only confusing the issue, these differences hint at possible mechanisms involved in the changes accompanying time-zone shifts and have practical implications when the question of ameliorating the problems of 'jet-lag' is considered.

3.1. Differences due to direction of flight

Fig. 9.5 summarizes results from a number of experiments with different subjects and upon different variables. These results all indicate that the rate of adaptation to an eastward shift (when the new time is ahead of the old) is slower than that to a shift in a westward direction. With one exception (*see* below), this result has been found by studies both in the field (reviewed by Klein and Wegmann, 1979) and in an isolation unit (Elliott et al., 1972).

Why should this difference exist? Earlier suggestions had been that the

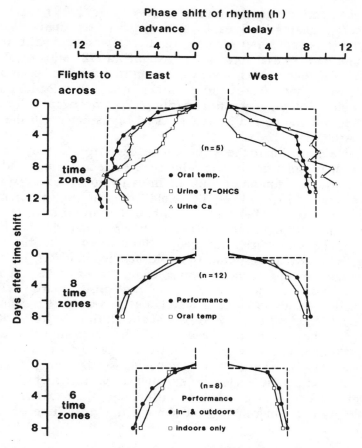

Fig. 9.5. Changes in the phases of circadian rhythms after transmeridian flights in a eastward or westward direction. Dashed lines indicate the course the phase would take if adaptation were immediate and complete. (*From* Aschoff 1978, Fig. 10; data of Halberg et al., 1970, 1971; Klein et al., 1970; Klein and Wegmann, 1974.)

difference depended upon whether the subject was flying to or from his 'home' time zone and whether he was flying at night (normal for long eastbound flights) or during the day (normal for long westbound flights). However, it has been shown that directional asymmetry exists, that is, adaptation to a delay in time (westward) is easier than to an advance, whatever time the flight is made and whether it is to or from home (*see* Klein and Wegmann, 1979).

It is now generally accepted that the 'asymmetry effect' is some manifestation of the manner in which the human oscillator system responds to advances and delays of the zeitgeber. Since, in most humans, the free-running period is greater than 24 hours (Wever, 1979), it is argued that the process of adaptation is easier to accomplish if the phase of the rhythms is required to become later each day (as after a westward shift) rather than to become earlier (eastward shift) (Aschoff, 1978; Mills et al., 1978a). Evidence in favour of such a view has been found in studies upon animals in which those with free-running periods of less than 24 hours adapt to eastward shifts of zeitgeber more easily than to westward shifts and those with a free-running period greater than 24 hours behave in the same way as man (Aschoff et al., 1975).

A notable exception to these findings has been from a study by Aschoff's group in their isolation unit in which six subjects were found to adjust more rapidly to a simulated shift in an eastward direction (*see* Aschoff, 1978; Wever, 1979, 1980). The explanation of this difference is not known. Elliott et al. (1972)—who obtained 'conventional' results—told their subjects to adjust to the shifted lighting regimen whereas the German group changed the lighting regimen without forewarning the subjects or advising them how to respond. In addition, their subjects had the use of dim auxiliary lighting at 'night' and so could show a certain independence of the imposed routine. The more 'natural' protocol would be that of Elliott and co-workers since a traveller is certain to be aware' that his schedule has changed and try to be wakeful during the hours of light and sleep during the dark hours. Wever (1979) has suggested that the lack of stress and fatigue in their experiments were possible causes of the observed differences. If this is so, then the mechanism by which the Germans' result is altered by the exigencies experienced in real flights is unknown. Whatever the explanation of these differences, their existence highlights the difficulty in comparing real and simulated flights (Siegel et al., 1969).

3.2. Differences between different variables

Fig. 9.5 indicates also that the rate of adaptation to time-zone shift is not the same for all variables. This has been found in a number of studies that between them cover many variables (*see*, for examples, Elliott et al., 1972; Aschoff et al., 1975; Halberg and Nelson, 1978, Fig. 25; Klein and Wegmann, 1979). Although, in detail, different results have been obtained from different studies (no doubt a reflection of differences in protocol, etc.) in general it is possible to group variables into those which adapt rapidly (for example, noradrenaline, mental performance tests, urine flow and calcium excretion) and others which adapt far more slowly (for example, deep body temperature, urinary 17-OHCS and potassium excretion). One explanation of this division assumes that the rate observed nychthemerally is determined by the relative importance of endogenous and exogenous components. Thus, the endogenous component adapts slowly to a time-zone shift so that those variables which have a large endogenous component will appear to adapt more slowly than those with a large exogenous component, this latter changing immediately on arrival in the new time zone (Klein and Wegmann, 1979). In support of this is the finding that the slowly adapting variables tend to have high endogenous components as assessed by the criteria put forward in Chapter 1 and the data discussed in earlier chapters.

The different rates at which variables adapt to time-zone shifts can be used also to infer something about the causal nexus that exists between them (*see* Chapter 12 and Mills, 1973b). Thus, if two rhythms adapt at different rates, then neither can be the sole cause of the other. For example, as shown earlier in *Fig.* 4.5, whereas the rhythm of sodium appears fully adapted on the first day after a simulated eastward shift of 8 hours, that of potassium still adheres to the old time and so the two rhythms are in antiphase. Therefore, the two rhythms of urinary excretion are not strongly causally related (in this subject), as has already been inferred in the chapter on renal rhythms. The observation that the deep body temperature rhythm adjusts more slowly than mental performance tests argues against a strong causal link here also (*see* Chapter 6).

The effect that this internal dissociation between different variables will have upon general health, together with the question of whether this dissociation accounts for the symptoms of 'jet-lag' syndrome, are two problems to which the answer is not known, but they will be considered in Chapter 10.

3.3. Differences due to the size of the time shift and strength of zeitgeber

The rate of adaptation to a time-zone shift is not a constant value, the initial stages generally being faster than the later ones when the amount of adaptation still required is only small. It is also found that, even though more time is required to adapt to a journey through a large number of time zones, the initial rate of adaptation is greater in the case of such a journey (Klein and Wegmann, 1979). This second result—that the size of shift of the external environment can influence the rate of adaptation of the circadian rhythm—implies some form of 'phase-response curve', which will be discussed in Chapter 12.

Of more practical importance is the observation by Klein and Wegmann (1974) that the rate of adaptation depends upon the strength of zeitgeber after the shift. They compared the rate of adaptation of a psychomotor test to both eastward and westward shifts of 6 hours. After each shift, half the subjects were allowed to take part in outdoor activities every second day while the other half were confined to their hotels, though well aware of local time. The 'outdoor' group adjusted to the time shift in either direction more rapidly than the 'hotel' group, *see Fig.* 9.5 (*bottom*). This suggests that strengthening the zeitgeber in the new time zone, such as by activity (or something associated with it) plays a considerable role in speeding up the process of adaptation. It also implies that the rate of adaptation to simulated time-zone shifts in isolation units, from which social zeitgeber from the rest of society are absent, might be abnormally slow.

3.4. Differences due to inter-individual variation

It has been observed that adaptation to a time-zone shift was accomplished less easily by older subjects (Evans et al., 1972; Klein and Wegmann, 1979); other differences that have been found are that subjects with large amplitude rhythms under control conditions adapt more slowly to simulated time-zone transitions (Aschoff, 1978) and extroverts adapt more rapidly than introverts (Colquhoun and Folkard, 1978). However, since, with all these factors, more work has been done upon shift workers, further discussion of these issues will be deferred until Chapter 10.

4. Entrainment by Partition: Masking

4.1. Entrainment by partition

As *Figs.* 9.2–9.5 have indicated, the process of adaptation can generally be described as a direct movement of the rhythm from the old to the new time zone, that is an advance of phase after an eastward shift and a delay after one in the westbound direction. This is not invariably the case and 'anomalous' directions of movement have been found.

Some examples are quoted by Aschoff (1978) and by Klein and Wegmann (1979) and come from both real and simulated time-zone shifts (*Fig.* 9.6). The anomalous shifts do not seem to occur randomly, however. Thus they have been observed in humans (a) nearly always only in association with eastward shifts, (b) especially when the number of time zones passed through is large and (c) with variables such as deep body temperature and the urinary excretion of potassium and steroid that have large endogenous components. The 'anomaly' is that adaptation takes place by the rhythm delaying by 18 hours rather than by advancing by 6 hours (in the case of a zeitgeber advance of 6 hours). The process has been called 'entrainment by partition' (Aschoff, 1978) or an 'antidromic phase response' (Klein and Wegmann, 1979).

The explanation of this phenomenon is uncertain but it has been related to the observation that the free-running rhythm of humans is greater than 24 hours. As a consequence of this there would be a tendency for phase-shifts of the endogenous component to be by delay rather than advance (Klein and Wegmann, 1979). Why in most subjects and for most variables this tendency seems not to be realized and entrainment by partition is not seen is not known. One possibility is that the process whereby the zeitgeber can shift the endogenous component of a rhythm is generally strong enough to advance it; by contrast, the process is less effective in some subjects, the result of which is that the endogenous rhythm 'free-runs' with a period of more than 24 hours. If these speculations are correct, then: (a) in those animals with a free-running period of less than 24 hours, one would expect to find that entrainment by partition occurred more frequently after westward time-zone shifts;

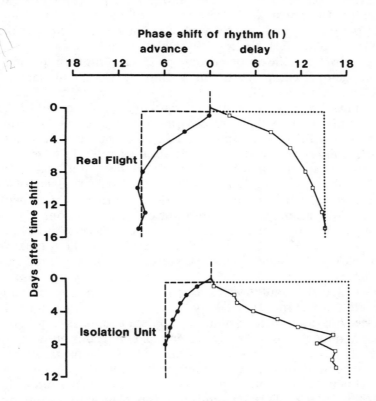

Fig. 9.6. Changes in the phases of the body temperature rhythm after transmeridian flights through 9 time zones in an eastward direction and after a phase-advance by 6 hours of the sleep times in an isolation unit. ●, Subjects in whom adaptation progressed in the 'correct' direction (phase-advance); □, subjects in whom adaptation was achieved by phase-delay of the rhythm. Dashed lines indicate course of adaptation if adaptation were immediate and by phase-advance. Dotted lines indicate course of adaptation if adaptation were immediate and by phase-delay (*From* Aschoff, 1978, Fig. 12; data of Klein et al., 1977; Wever, 1979.)

and (b) in the case of rhythms with small endogenous components, entrainment by partition would be rare because the endogenous component, even if it were adapting 'anomalously', would be more difficult to detect under nychthemeral circumstances in the presence of the large, fully adapted, exogenous component.

4.2. Masking

The fact that entrainment by partition has not been observed in those variables which have a strong exogenous component (for example, urine flow and psychometric tests) raises another difficulty in interpretation which applies to almost the whole subject of time-zone transitions. The difficulty is that the great majority of experiments has been performed in nychthemeral conditions, the result of which is that one cannot be certain of the extent to which the observed changes in rhythms measure adaptation of the endogenous component. The pattern of sleep and wakefulness is known to exert a direct effect upon many rhythms (Mills et al., 1978b) and such an effect has been termed 'masking' by Aschoff (1978). Data relevant to this problem have been collected recently in experiments in which time-zone shifts were simulated in an isolation unit (Mills et al., 1978a,c). In addition, constant routines (*see* Chapter 1) were performed on two occasions, once before the simulated shift and once a few days after it. The shifts were of 12 hours or of an advance or delay of 8 hours. Rhythms in rectal temperature and a variety of urinary constituents were sought by cosinor analysis but, in addition, the two constant routines were compared with one another by cross-correlation techniques (for details of which *see* Mills et al., 1978a). An example of the results obtained is shown in *Fig.* 9.7. This illustrates some of the main findings of this study which were:

1. The degree of adaptation measured under nychthemeral conditions overestimated the amount of adaptation to the time-zone shift that had taken place. *Fig.* 9.7 indicates almost complete adaptation by 2 days after the shift when measured under nychthemeral conditions but a comparison of the two constant routines indicates this to be far from the case.

2. The second constant routine indicated that the endogenous component had behaved 'anomalously' and become later even after an eastward shift, as in *Fig.* 9.7. This tendency for the endogenous component to become later was particularly marked after the 12-hour shifts when the great majority of rhythms were delayed. Further, comparison of the two constant routines by cross-correlation techniques confirmed that a delayed component was often present after an 8-hour phase-advance of the zeitgeber and nearly always so after an 8-hour phase-delay or a 12-hour shift (Mills et al., 1978a).

These results indicate that the endogenous component often adapts to time-zone shifts by becoming later whatever the direction of shift that is involved. If this is so, then the process of 'entrainment by partition' might be seen more often in nychthemeral conditions in those subjects in whom the endogenous component was larger or the exogenous component smaller than average. Presumably, for most subjects and variables, this component is too small to be obvious in nychthemeral conditions and requires the lack of 'masking' that a constant routine provides before it

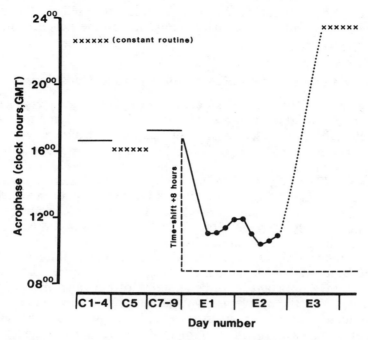

Fig. 9.7. Acrophases of the urinary urate excretory rhythm in a single subject before and after a simulated flight through 8 time zones in an eastward direction. Before time-shift, acrophases are of cosine curves fitted to days when a normal nychthemeral routine was followed (C1–4 and C7–9) and a 'constant routine' (C5) when no sleep was taken. After time-shift, acrophases were derived from cosine curves fitted to 24-hour intervals pergressively incremented by 4 hours (●, E1 and E2). On the third day after time-shift (E3), a second 'constant routine' was performed. Dashed line represents the course the acrophase would follow if adaptation were immediate and complete. On the first 2 days after time-shift, adaptation is nearly complete but the second 'constant routine' indicates that this may be due to a masking effect. (*From* Mills et al., 1978a, Fig. 6.)

can be demonstrated. The observation that even in the constant routines not all endogenous components become later suggests that in at least some cases the internal oscillator can be phase-advanced by the external environment, a point that has already been made when the results from nychthemeral studies were discussed.

Finally, it must be pointed out that the 'direction asymmetry effect', different rates of adaptation of variables and acceleration of adaptation by strengthening zeitgebers, all discussed above, have all been observed (with the exception of the study of Gerritzen and Strengers, 1974) under nychthemeral conditions with rhythmic intake of food. The extent to which the properties of the endogenous component are known to be important in these phenomena (as has been surmised in the relevant sections) is uncertain; clearly, further constant routine experiments would be valuable here.

5. Consequences of Repeated Time-zone Transitions

There remain the problems associated with those subjects, generally military or civil aircrew, who are repeatedly flying from one time zone to another. Recently, increased interest in this has been shown, especially in those aspects concerned with performance, and summaries are to be found in Klein et al. (1976), Nicholson (1978), Hartman (1979) and Klein and Wegmann (1979). As yet these studies have contributed comparatively little to our understanding of circadian rhythms and concentrated rather upon performance decrement and fatigue. In part this is because time and circumstances rarely permit detailed scientific investigations to be performed; in part this is because aircrews must deal with a complex of multiple time-zone shifts, multiple schedule changes, fatigue and stress which will render any data extremely difficult to interpret. Equally, it is not certain how important an understanding of circadian rhythms will be in solving some of the problems involved since motivation and stress seem to be so much more important in these circumstances (see below). Even so, the following account will attempt to show that some progress is being made in this field when circadian rhythms are considered.

5.1. Sleep disturbance

There is general agreement that the inability to take sufficient sleep is a major problem (Klein and Wegmann, 1979; Hartman, 1979). Preston (1970) found that all members of aircrew on long-haul flights (especially the older subjects) suffered from cumulative sleep deficit; such a deficit was found also in groups of flying personnel undergoing two simulated 8-hour westward time-zone shifts in an isolation unit (Preston, 1973; Preston et al., 1976).

Many studies, like those just referred to, rely upon a subjective assessment of sleep times as recorded in log books, but a study in which sleep time was assessed by measuring the EEG confirmed that a deficit in time spent asleep (confirmed by the log reports) had arisen (Harris et al., 1971). This sleep deficit seems to be made up during 'rest days' (Hartman, 1971).

Atkinson et al. (1970) have attributed the accumulation of a sleep deficit to the difficulty associated with attempting to sleep at 'odd' hours; presumably factors such as body temperature, the sympathetic nervous system and 'arousal' (see Chapter 6) could be important here. Ways in which the aircrews attempt to minimize the amount of cumulative sleep loss vary. In principle (Hartman, 1979), they could:

    (i) attempt to adjust to each time-zone shift;

(ii) attempt to maintain home time;

(iii) take sleep when and if possible.

In practice, many pilots seem to adopt a 'split-sleep' system with 'naps' being taken just before, and soon after, a mission (Nicholson, 1970a,b). Nicholson considers that the value of these 'naps' is to reduce fatigue acutely and to prepare for the next spell of duty.

It is noticeable, when inspecting the sleep logs of some crew members, that 'naps' are often taken either at times coincident with night time in their home time zone or at times that become progressively later, as if the desire to sleep was showing a rhythm with a period greater than 24 hours. These points are illustrated in *Fig.* 9.8A and B. Alternatively, this latter observation might reflect duty schedules and the general westward progression of the flights in accord with which view is the observation that, during an eastward circumglobal trip, sleeps were taken progressively earlier (*Fig.* 9.8C).

At this point an experiment that was performed upon a single pilot in an isolation unit is relevant (Mills et al., 1978b). The subject undertook a simulated flight in a westward direction 'stopping' at 'Los Angeles' and 'Perth' for about two days on each occasion and living on a routine in accord with local time at these places. He then returned home by the same route. Regular urine samples were taken as well as rectal temperature recordings; the rhythms were assessed by cosinor analysis. As the typical examples of urinary chloride and potassium excretion show (*Fig.* 9.9), the rhythms were not synchronized to a 24-hour day but indicated a period of more than 24 hours with times of peak becoming progressively later as the 'trip' continued. This finding suggests that irregular patterns of sleep and wakefulness are associated with 'free-running' rhythms with a period in excess of 24 hours (*see also* Minors and Waterhouse, 1980). If it can be assumed that the rhythms of fatigue and sleep—wakefulness behave similarly, then a similar explanation for the sleep—wakefulness pattern seen in *Fig.* 9.8B can be offered. Furthermore, such a suggestion might explain why circumglobal trips in an eastward direction (*Fig.* 9.8C) are unpopular with aircrews; thus, assuming that adaptation to phase-advances is more difficult than to phase-delays (*see* s. 3.1), a greater disturbance to circadian rhythms would be caused by circumglobal journeys in an eastward direction. In connection with this, the subject of *Fig.* 9.8C rated himself as very fatigued for much of the trip (Preston et al., 1970; *see also* Parrot and Petiot, 1978).

Recent isolation unit studies (Minors and Waterhouse, 1980), which will be described in more detail in Chapter 10, have shown that deep body temperature and urinary excretion rhythms can be synchronized to a period of 24 hours if some sleep is taken at the same time each day, even if another 'nap' is taken at irregular times. This result might be relevant to the finding already described (*see Fig.* 9.8A) that some pilots take 'naps' at times coincident with night time in their home time zone. A similar

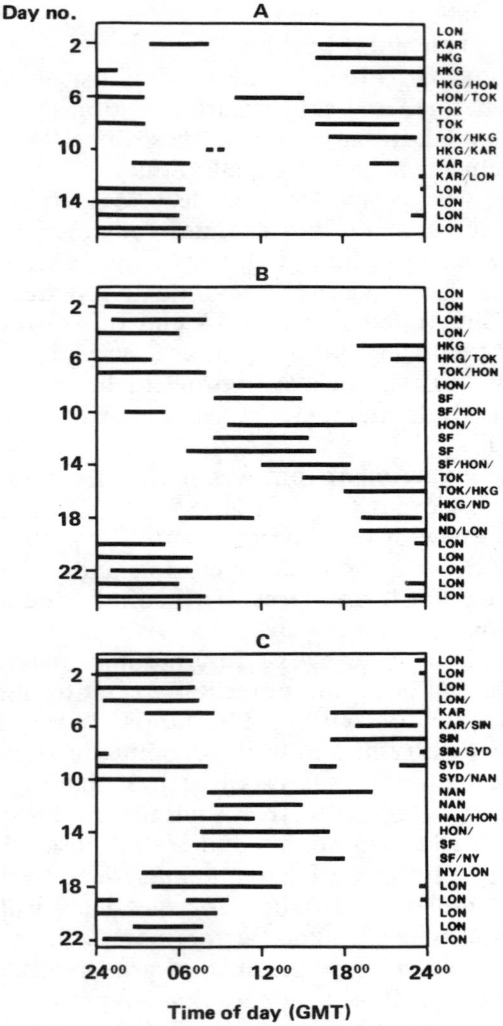

Fig. 9.8. Times of sleep (black bars) in three cockpit crews on commercial flights. The right of each plot shows the city of residence on each day. A slash indicates days on which flights were made. (Lon, London; Kar, Karachi; HKG, Hong Kong; HON, Honolulu; TOK, Tokyo; SF, San Francisco; ND, New Delhi; SIN, Singapore; SYD, Sydney; NY, New York; Nan, Nandi.) A, pilot on a 12-day tour from London/Honolulu/London. B, pilot on Far East tour, London/San Francisco/London. C, pilot on a round-the-world (eastbound) flight. Note that in B on the westbound flight (day 13 on) the times of sleep become progressively later whilst for the eastbound flight (C) times of sleep become progressively earlier. (Data of Preston, 1970; 1973.)

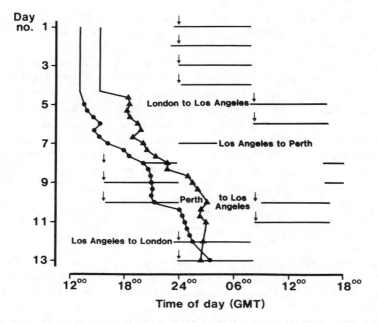

Fig. 9.9. Times of sleep (black bars) and acrophases of the urinary excretion of potassium (●) and chloride (▲) in a pilot who simulated in an
isolation unit the time-changes during a 'journey' from London to Los
Angeles to Perth and return. Arrows indicate midnight as experienced
by the pilot. Acrophases derived by fitting cosine curves to 24-hour
intervals of data pergressively incremented by 8 hours. Note the acrophases become progressively later and are not hence synchronized to
24-hour periodicity. (Data of Mills et al., 1978b.)

inference can be made in the case of data from astronauts circling the
globe or visiting the moon, whose day—as judged by sunlight—will often
be grossly abnormal. Under such conditions there is often difficulty with
sleep, but this is ameliorated when the astronauts adhere to a more
rigorous schedule of sleep and wakefulness even if this is not identical to
their earthbound home time (Nicholson, 1972).

5.2. Mental performance and other rhythms

Some tests of the speculations above could be made if data of other
rhythms from the circumglobal studies were available; unfortunately this
is not so. However, in spite of the difficulties involved (*see above*), some
data have been obtained from flights of long duration. In spite of a
cumulative sleep deficit, assessments of 'piloting skill' (Hartman and
Cantrell, 1967) and 'combat efficiency' (Caille et al., 1972) have shown
remarkably little decrement. This has been attributed to the ability of
'motivation' to overcome fatigue, at least temporarily (*see* Chapter 6),

though it has been postulated that this depletes the 'reserves' of the subjects and requires time for recovery of these reserves after the mission (Harris et al., 1971).

Measurements of oral temperature and a variety of psychometric tests in the isolation unit experiments of Preston's group (Preston, 1973; Preston et al., 1976) showed decrements in both mean temperature and performance level. A similar depression of mean oral temperature rhythm after real flights has been reported (Harris et al., 1970). This depression in mean values is generally attributed to stress and fatigue and later reports of measurements of the urinary excretion of a number of substances (adrenaline, noradrenaline, sodium, potassium, urea and 17-OHCS), in which mean values rose, have been interpreted similarly (Hale et al., 1972a, 1973). In addition, attention has been drawn to the relationship that continues to exist between fatigue and temperature in isolation unit experiments (Preston, 1973) and after real flights (Hartman et al., 1974) (c.f. Chapter 6).

In summary, crews undergoing repeated time-zone shifts show little obvious impairment in performance even though they have lost sleep. Measurements of rhythms of urinary excretion, oral temperature and psychometric performance confirm that subjects are fatigued and stressed, but what has happened to their circadian rhythmicity is not clear at the present time (see also Hale et al., 1972b).

6. Advice to Travellers

Having outlined all the problems that beset the long-distance traveller, we must now consider if any advice to him can be given! To put the matter into perspective, it is likely that the effects of 'jet-lag' are normally marginal in so far as motivation can, at least in the short term, counteract the effects of fatigue, of stress and of inappropriate phasing of circadian rhythms that follow a time-zone transition (see Alluisi and Chiles, 1967).

The advice that is given depends upon whether the traveller is staying in the new time zone for a short or a long time and whether he will then be returning home or continuing on his travels to another time zone. Thus: if a long stay in any time zone is planned, the traveller would be advised to try to adapt to it as quickly as possible; if he envisaged only a short stay and then a return home, he would be advised to remain adapted to home time; if he was undertaking a series of time zone shifts, he could either attempt to adapt to each in turn (remembering that it is easier to delay rhythms than to advance them) or to synchronize his rhythms to home time. As a result of these considerations, the following courses of action are open to him (see Conroy, 1971; Klein et al., 1976; Klein and Wegmann, 1979):

1. Accept that adaptation has not taken place and choose the occasions on both old and new times when performance would be best. Thus, after a

westward flight, an appointment in the morning of the new zone corresponds to one in late afternoon in the old zone and is acceptable; but an evening meeting, which would coincide with sleep on the old time, should be avoided. Similar considerations would suggest that merits of meetings late in the new day, and argue against morning appointments, after eastward flights.

2. Attempt to speed up adaptation as much as possible. As has been described above (s. 3.3), there is evidence that the process of adaptation is speeded up in the presence of strong zeitgeber (*see also* Chapter 12). This would argue for a whole-hearted acceptance of the new social routine and adherence to the new times of meals, zeitgeber, sleep etc. If circumstances permit, it has been suggested (Graeber et al., 1979; Cuthbert et al., 1980) that one could 'pre-adapt' to the new time zone in the day or so before the flight by altering one's regimen appropriately; clearly this is not always possible.

3. Attempt to maintain rhythms synchronized to home time. It seems that this can be achieved by taking part of one's sleep at a time coincident with night time in the home time zone (*see* s. 4.1 and *Fig.* 9.8A). Further comment on this procedure will be made in Chapter 10.

It must always be remembered that long journeys are often associated with a loss of sleep as well as time shifts and that a 'nap' might relieve some of this fatigue. However, the time of 'nap' requires some thought since it is important that it should not hinder the subject from sleeping again when it is night time in the new time zone.

Finally, attempts have been made to calculate the rest periods required for recovery after time-zone transitions (Siegel et al., 1969; Buley, 1970; Gerathewohl, 1974). These calculations have been based upon analysis of the time taken for adaptation of temperature, steroid, urinary and psychometric rhythms to time-zone shifts by healthy subjects. The information required includes the duration of the flight, the time of day when it takes place and the number of time zones crossed. In a more recent method, factors related to direction of flight and age of traveller also are required (Gerathewohl, 1974). Even though these calculations are based upon premises that can be challenged (for example, adaptation has been measured under nychthemeral rather than constant routine conditions and the assumption has been made that the rest period required equals the time taken for the rhythms to adapt to the new time zone), when tested, they have been found to be of use to both staff and administration. The value of such a pragmatic solution must not be underrated even if it lacks somewhat in theoretical backing.

References

Alluisi E. A. and Chiles W. D. (1967) Sustained performance, work-rest scheduling, and diurnal rhythms in man. *Acta Psychologica* 27, 436–42.

Aschoff J. (1978) Features of circadian rhythms relevant for the design of shift schedules. *Ergonomics* 21, 739–54.

Aschoff J., Hoffmann K., Pohl H. et al. (1975) Re-entrainment of circadian rhythms after phase-shifts of the zeitgeber. *Chronobiologia* **2**, 23—78.

Atkinson D. W., Borland R. G. and Nicholson A. N. (1970) Double crew continuous flying operations: a study of aircrew sleep patterns. *Aerospace Med.* **41**, 1121—6.

Brüner H., Jovy D. and Klein K. E. (1960) Ein objektives Messverfahren zur Feststellung der psychomotorischen Leistungsbereitschaft. *Int. Z. Angew. Physiol. Einschl. Arbeitsphysiol.* **18**, 306—18.

Buley L. E. (1970) Experience with a physiologically based formula for determining rest periods on long distance air travel. *Aerospace Med.* **41**, 680—3.

Caille E. J. P., Quideau A. M. C., Girand J. F. J. et al. (1972) Loss of sleep and combat efficiency: effects of the work/rest cycle. In: Colquhoun W. P. (ed.) *Aspects of Human Efficiency. Diurnal Rhythm and Loss of Sleep.* London, English Universities Press, pp. 177—93.

Colquhoun W. P. and Folkard S. (1978) Personality differences in body-temperature rhythm, and their relation to its adjustment to night work. *Ergonomics* **21**, 811—17.

Conroy R. T. W. L. (1971) Time zone transitions and business executives. *Trans. Soc. Occup. Med.* **21**, 69—72.

Cuthbert B. N., Graeber R. C., Sing H. C. et al. (1980) Rapid transmeridial deployment: II. Effects of age and countermeasures under field conditions. Proceedings XIV International Conference of the International Society for Chronobiology. Hanover 1979. Milan, Il Ponte (in the press).

Elliott A. L., Mills J. N., Minors D. S. et al. (1972) The effect of real and simulated time-zone shifts upon the circadian rhythms of body temperature, plasma 11-hydroxycorticosteroids and renal excretion in human subjects. *J. Physiol.* **221**, 227—57.

Endo S., Yamamoto T. and Sasaki M. (1978) Effects of time zone changes on sleep—west—east flight and east—west flight. *Jikeikai Med. J.* **25**, 249—68.

Evans J. I., Christi G. A., Lewis S. A. et al. (1972) Sleep and time zone changes. A study in acute sleep reversal. *Arch. Neurol.* **26**, 36—48.

Gerathewohl S. J. (1974) Simple calculator for determining the physiological rest period after jet flights involving time-zone shifts. *Aerospace Med.* **45**, 449—50.

Gerritzen F. and Strengers Th. (1974) Adaptation of circadian rhythms in urinary excretions to local time, after rapid air travel. In: Scheving L. E., Halberg F. and Pauly J. E. (ed.) *Chronobiology.* Tokyo, Igaku Shoin, pp. 555—9.

Graeber R. C., Cuthbert B. N., Sing H. C. et al. (1979) Rapid transmeridian deployment: I. Use of chronobiologic countermeasures to hasten time zone adjustment in soldiers. *Chronobiologia* **6**, 102.

Halberg F., Halberg E. and Montalbetti (1970) Premesse e sviluppi della chrono-farmacologia. *Quad. Med. Quantitativa* **8**, 7—54.

Halberg F. and Lee J. K. (1974) Glossary of selected chronobiologic terms. In: Scheving L. E., Halberg F. and Pauly J. E. (ed.) *Chronobiology.* Tokyo, Igaku Shoin, pp. XXXVII—L.

Halberg F. and Nelson W. (1978) Chronobiologic optimization of aging. In: Samis H. V. and Capobianco S. (ed.) *Aging and Biological Rhythms.* London, Plenum, pp. 5—56.

Halberg F., Nelson W. and Runge W. J. (1971) Plans for orbital study of rat biorhythms results of interest beyond the biosatellite program. *Space Life Sci.* **2**, 437—71.

Hale H. B., Hartman B. O., Harris D. A. et al. (1973) Physiologic cost of prolonged double-crew flights in C-5 aircraft. *Aerospace Med.* **44**, 999—1008.

Hale H. B., Hartman B. O., Harris D. A. et al. (1972a) Physiologic stress during 50-hour double crew missions in C-141 aircraft. *Aerospace Med.* **43**, 293—9.

Hale H. B., Hartman B. O., Harris D. A. et al. (1972b) Time zone entrainment and flight stressors as interactants. *Aerospace Med.* **43**, 1089—1094.

Harris D. A., Hale H. B., Hartman B. O. et al. (1970) Oral temperature in relation to inflight work/rest schedules. *Aerospace Med.* **41**, 723—7.

Harris D. A., Pegram G. V. and Hartman B. O. (1971) Performance and fatigue in experimental double-crew transport missions. *Aerospace Med.* **42**, 980–6.

Hartman B. O. (1971) Field study of transport aircrew workload and rest. *Aerospace Med.* **42**, 817–21.

Hartman B. O. (1979) Management of irregular rest and activity. In: *Sleep, Wakefulness and Circadian Rhythm.* AGARD Lecture Series No. 105, AGARD, ch. 13.

Hartman B. O. and Cantrell G. K. (1967) Sustained pilot performance requires more than skill. *Aerospace Med.* **38**, 801–3.

Hartman B. O., Hale H. B., Harris D. A. et al. (1974) Psychobiologic aspects of double-crew long-duration missions in C-5 aircraft. *Aerospace Med.* **45**, 1149–54.

Hauty G. T. and Adams T. (1966a) Phase shifts of the human circadian system and performance deficit during the periods of transition: I. East–west flight. *Aerospace Med.* **37**, 668–74.

Hauty G. T. and Adams T. (1966b) Phase shifts of the human circadian system and performance deficit during the periods of transition: II. West–east flight. *Aerospace Med.* **37**, 1027–33.

Hauty G. T. and Adams R. (1966c) Phase shifts of the human circadian system and performance deficit during the periods of transition: III. North–south flight. *Aerospace Med.* **37**, 1257–62.

Klein K. E., Brüner H., Gunther E. et al. (1972) Psychological and physiological changes caused by desynchronization following transzonal air travel. In: Colquhoun W. P. (ed.) *Aspects of Human Efficiency. Diurnal Rhythm and Loss of Sleep.* London, English Universities Press, pp. 295–305.

Klein K. E., Brüner H. and Holtmann H. (1970) Circadian rhythm of pilots' efficiency and effects of multiple time zone travel. *Aerospace Med.* **41**, 125–32.

Klein K. E., Herrmann R., Kuklinski P. et al. (1977) Circadian performance rhythms: Experimental studies in air operations. In: Mackie R. R. (ed.) *Vigilance: Theory, Operational Performance and Physiological Correlates.* New York, Plenum, pp. 111–32.

Klein K. E. and Wegmann H.-M. (1974) The resynchronization of human circadian rhythms after transmeridian flights as a result of flight direction and mode of activity. In: Scheving L. E., Halberg F. and Pauly J. E. (ed.) *Chronobiology.* Tokyo, Igaku Shoin, pp. 564–70.

Klein K. E. and Wegmann H.-M. (1979) Circadian rhythms in air operations. In: *Sleep, Wakefulness and Circadian Rhythm.* AGARD Lecture Series No. 105, AGARD, ch. 10.

Klein K. E., Wegmann H. M., Athanassenas G. et al. (1976) Air operations and circadian performance rhythms. *Aviat. Space Environ. Med.* **47**, 221–30.

Klein K. E., Wegmann H. M. and Hunt B. I. (1972) Desynchronization of body temperature and performance circadian rhythm as a result of outgoing and homegoing transmeridian flights. *Aerospace Med.* **43**, 119–32.

Lafontaine E., Lavarnhe J., Courillon J. et al. (1967) Influence of air travel east–west and *vice versa* on circadian rhythms of urinary elimination of potassium and 17-hydroxycorticosteroids. *Aerospace Med.* **38**, 944–7.

Levine H., Halberg F., Sothern R. B. et al. (1974) Circadian phase-shifting with and without geographic displacement. In: Ferin M., Halberg F., Richart R. M. et al. (ed.) *Biorhythms and Human Reproduction.* London, Wiley, pp. 557–74.

Mills J. N. (1973a) Air travel and circadian rhythm. *J. Coll. Phys. Lond.* **7**, 122–31.

Mills J. N. (1973b) Transmission processes between clock and manifestations. In: Mills J. N. (ed.) *Biological Aspects of Circadian Rhythms.* London, Plenum, pp. 27–84.

Mills J. N., Minors D. S. and Waterhouse J. M. (1978a) Adaptation to abrupt time shifts of the oscillator(s) controlling human circadian rhythms. *J. Physiol.* **285**, 455–70.

Mills J. N., Minors D. S. and Waterhouse J. M. (1978b) The effect of sleep upon human circadian rhythms. *Chronobiologia* 5, 14—27.

Mills J. N., Minors D. S. and Waterhouse J. M. (1978c) Exogenous and endogenous influences on rhythms after sudden time shift. *Ergonomics* 21, 755—61.

Minors D. S. (1975) PhD Thesis, University of Manchester.

Minors D. S. and Waterhouse J. M. (1980) Anchor sleep as a synchronizer of rhythms on abnormal schedules. *Int. J. Chronobiol.* 7, (in the press).

Moiseeva N. J., Bogolovsky M. M., Simonov M. Y. et al. (1976) Characteristics of the circadian sleep rhythm in relation to environmental factors associated with the rotation of the earth and to biological macrorhythms in man. *J. Interdiscipl. Cycle Res.* 7, 15—24.

Nicholson A. N. (1970a) Military implications of sleep patterns in transport aircrew. *Proc. R. Soc. Med.* 63, 570—2.

Nicholson A. N. (1970b) Sleep patterns of an airline pilot operating world-wide east—west routes. *Aerospace Med.* 41, 626—32.

Nicholson A. N. (1972) Sleep patterns in the aerospace environment. *Proc. R. Soc. Med.* 65, 192—4.

Nicholson A. N. (1978) Irregular work and rest. In: *Textbook of Aviation Medicine, Physiology and Human Factors.* London, Tri-Med Books, pp. 494—503.

Parrot J. and Petiot J.-C. (1978) Less than 24 hour pseudo-periodicity in work schedules of truck drivers, in relation to their sleep. *Int. Arch. Occup. Environ. Hlth* 41, 179—88.

Preston F. S. (1970) Time zone disruption and sleep patterns in pilots. *Trans. Soc. Occup. Med.* 20, 77—86.

Preston F. S. (1973) Further sleep problems in airline pilots on world-wide schedules. *Aerospace Med.* 44, 775—82.

Preston F. S., Bateman S. C., Meichen F. W. et al. (1976) Effects of time zone changes on performance and physiology of airline personnel. *Aviat. Space Environ. Med.* 47, 763—9.

Richards L. G. and Jacobson I. D. (1978) Concorde: ride quality and passenger reactions. *Aviat. Space Environ. Med.* 49, 905—13.

Sasaki T. (1964) Effect of rapid transportation around the earth on diurnal variation in body temperature. *Proc. Soc. Exp. Biol. Med.* 115, 1129—31.

Sasaki T. (1972) Circadian rhythm in body temperature In: Itoh S., Ogata K. and Yoshimura H. (ed.) *Advances in Climatic Physiology.* Tokyo, Igaku Shoin, pp. 319—33.

Sekiguchi C., Yamaguchi O., Kitajima T. et al. (1976) Effects of rapid round trips against time displacement on adrenal cortical—medullary circadian rhythms. *Aviat. Space Environ. Med.* 47, 1101—6.

Siegel P. V., Gerathewohl S. J. and Mohler S. R. (1969) Time-zone effects. Disruption of circadian rhythms poses a stress on the long-distance air traveller. *Science* 164, 1249—55.

Strughold H. (1971) *Your Body Clock.* New York, Scribner.

Wever R. A. (1979) *The Circadian System of Man. Results of Experiments under Temporal Isolation.* Berlin, Springer-Verlag.

Wever R. A. (1980) Phase shifts of human circadian rhythms due to shifts of artificial zeitgebers. *Chronobiologia* 7, 303—27.

chapter 10 *Shift Work*

1. Introduction

Whereas most of the community sleeps by night and works and takes its leisure by day, an appreciable fraction of the working population either occasionally or permanently works hours that are at variance with this. The exact proportion of the workforce involved in shift work varies between different countries, between different types of work and between workforces of different size (Sergean, 1972; Rutenfranz et al., 1976; Harrington, 1978).

The reasons that are given to justify the necessity of night work are social, technological and economic (*see* for example, Rutenfranz et al., 1977). Socially, there is a need for the continuous availability of essential services (medical, police, military, transport, electricity, etc); technologically there are some processes (in the chemical, petroleum and steel industries) that must be run continuously; economically, in a world in which profits are given high priority, expensive equipment must not be used at less than full capacity for a substantial part of the day.

There is a wide variation in the different types of shift system that are employed. (Indeed, Professor John Mills, to whose memory we have dedicated this book, once said 'There are nearly as many shift systems as shift workers'.) Some of these systems are described by Rutenfranz et al. (1976, 1977) but comparative studies upon the effects of various shifts upon the circadian rhythms are few. Shift systems can be distinguished in terms of:

(i) the number of hours that are worked. This is generally 8 hours, but sometimes 12 hours, split shifts (2 X 4 hours with a substantial break between) or workspans of more irregular duration are involved;

(ii) the time at which work is done. This can be 'morning', 'afternoon' and 'night' with 8-hour systems or 'day' and 'night' with 12-hour systems; and, most importantly as far as studies of circadian rhythms are concerned,

(iii) the frequency with which the shift changes or 'rotates'. This can range from 'rapid' (in which the shift changes every 1 or 2 days) to 'slow' (in which the change is every week or few weeks) or 'permanent' (in which the same shift is continuously worked). 'Irregular' shifts (in which no pattern is evident) are sometimes worked also.

2. Some Problems in Assessing Implications of Shift Work

Any social implications that shift work might have for shift workers and their families can be assessed only by questioning them; difficulties associated with interpreting data from such surveys have been discussed by Wedderburn (1978). To take a naive example, the answer given to the question: 'Do you like shift work?' might depend in part upon what the shift worker believes the aim of the question to be, such as 'higher wages for bad conditions' or 'remove those who complain'. Some of these problems can be overcome if the anonymity of the worker is respected (a single-blind test), but then the researchers are at a disadvantage if, later on, they wish to pursue some of the answers or choose some of the workforce for further studies. Since the answer to many general questions will probably involve balancing the advantages (financial, promotional) with the disadvantages (*see below*), it is preferable to ask more specific questions (*see Table* 10.1). Unfortunately, such a course of action can run the risk of making the questionnaire too complex or time-consuming. As Wedderburn (1978) has pointed out, when a number of questions are involved it is desirable to have a certain amount of overlap between them so that the degree of 'internal consistency' between answers can be checked. This procedure has certain similarities with the way in which Thayer's Activation—Deactivation checklist is compiled and scored (*see* Chapter 6).

When attempts are made to assess any medical problems that might arise in shift workers, the use of a questionnaire will give rise to difficulties similar to those already discussed in the case of social problems. If factory medical records are studied, the problem arises that, if workers choose to visit their own general practitioner instead, these factory records might underestimate any health problems that might exist. Further, if it is commonly held by the shift worker (whether correctly or not) that shift work is associated with a particular ailment, will this increase, decrease or have no effect upon his tendency to consult medical staff about such an ailment?

In previous chapters, the arguments in favour of assessing circadian rhythms in the laboratory rather than in the field have been put forward. On the other hand, as Chapter 9 has already indicated, valuable contributions to our understanding of these rhythms in humans have been made after real time-zone transitions. Similarly, as Reinberg and his colleagues have argued (Reinberg, Vieux et al., 1979c) there is a need for studies on shift workers whilst they are performing shift work; in brief, there is no substitute for the 'real thing'. As this chapter will show, a number of valuable studies have been performed and, as Reinberg's group points out, the immense amount of cooperation and planning that is required on the parts of experimenters, workers and management in order that field studies should be successful must not be underestimated.

3. Complaints Associated with Shift Work

In spite of all these difficulties, there is a large measure of agreement between different groups when their general findings are considered (*see* Rutenfranz et al., 1976; Harrington, 1978; Wedderburn, 1978). *Table* 10.1, taken from a study by Wedderburn, shows the results from a questionnaire involving a group of workers on rotating shifts, so enabling the same subjects to compare their attitudes to 'morning', 'afternoon' and 'night' shifts. *Table* 10.1 bears out the generally held view that the night

Table 10.1. Description of each shift type
'Think of work on—shift. How well does each of the following words describe what it is like for you on—shift?'*

| | Percentage 'yes' response about | | |
Descriptive item	Morning shift	Afternoon shift	Night shift
Quickly over	84	38	29
Seems a longer shift	12	40	64
Gives me indigestion	7	4	21
Tiring	58	15	80
Disturbs my sleep	36	2	52
More friendly atmosphere	48	51	21
Makes me irritable	25	18	48
Peaceful	49	62	58
More responsibility at work	28	16	24
More independent	35	26	42
Good for family life	73	19	17
Gives me more spare time	88	15	49
Wastes the day	10	68	55
Starts too early	53	7	20
Restricts my social life	17	79	77
Sexless	13	11	44

From Wedderburn (1978).

shift, and to a lesser extent the morning shift, is unpopular among the workforce. This unpopularity seems to derive from three main causes: it reduces the worker's social life, it fatigues him and it renders him prone to disorders of the gastrointestinal tract—the last two conditions, it will be recalled, being very similar to those that constitute the 'jet-lag' syndrome.

Each of these complaints will be considered briefly.

3.1. Unsociability

It is obvious that to work on the night or afternoon shifts will restrict the person's ability to play a full role in social events; equally, the need to sleep after night work or to be absent in the early evening when on the afternoon shift will restrict family life. The result is that shift work affects all members of the household. Yet the shift worker often has spare time in the daytime and blocks of days off. For some, especially those with solitary spare-time pursuits, these advantages outweigh the disadvantages. Indeed, Harrington (1978) concludes that, whereas some workers dislike shift work, most put up with it and about 10 per cent actually like it; all of which demonstrates the interpretative problems associated with the simple answer 'Yes' or 'No' to the broad question: 'Do you like shift work?'

3.2. Health deterioration

This is an aspect of shift work which has provoked considerable controversy and about which there is still not agreement. Recent reviews include those by Rutenfranz et al. (1976, 1977), by Harrington (1978) and by Winget et al. (1978). As Harrington points out, it is possible to make a case for or against an association between a particular disorder and shift work according to the references chosen! As the cited reviews indicate, there are not strong cases in favour of the hypotheses that shift work is associated with increased mortality rates, absenteeism or frequency of cardiovascular and neurological disorders. Indeed, absenteeism has often been found to be less in shift workers, attributed by some (for example, Aanonsen, 1964) to an increased job-motivation in this group.

However, an increased incidence of a number of ill-defined symptoms, grouped together as 'gastrointestinal disorders' has often been claimed and further comment upon this is necessary. Rutenfranz et al. (1976) have noted that the increased incidence of gastrointestinal disorders was less marked or absent in studies performed after 1945, by which time compulsory shift work associated with wartime had been discontinued. The inference that can be made is that those who continue shift work are relatively symptom-free and continue voluntarily, whereas those who develop symptoms of gastrointestinal disorder leave shift work. There are two important implications of this hypothesis. The first is that some workers are better able than others to cope with shift work and possible means by which such an advantage could be predicted will be discussed later. The second implication is that shift workers of some experience might represent a self-selected group that is more healthy than the workforce as a whole. If this were correct, then decreased absenteeism (*see above*) and the lower incidence of disorders found in some studies (for example, Taylor, 1976) might be a reflection on the workforce rather than shift work itself.

There is considerable evidence (reviewed by Harrington, 1978; Reinberg, Vieux et al., 1979b) for the view that some workers are less able than others to adapt to shift work; in accord with this, the health of subjects who have been forced to give up shift work is worse than that of workers continuing with shift work (Aanonsen, 1964). A recent study by Koller et al. (1978) divided subjects into day workers, night workers and 'drop-outs' (that is workers who have decided against further night work). When questioned, this last group indicated most difficulties with social and health matters and the day workers least; when assessed by doctors, the health of day workers ranked higher than that of drop-outs and night workers when gastrointestinal disorders were considered. Similarly, Angersbach et al. (1980) have found not only a higher incidence of gastrointestinal (but not cardiovascular) disorders in shift workers when compared with day workers, but also a general increase in ill-health in 'drop-outs' who have transferred from shift work to day work.

The possible process of 'self-selection' that might be manifest by shift workers introduces a difficulty of interpretation into transverse studies that can be ameliorated by longitudinal studies. Recently, two such studies have been reported. In the first (Meers et al., 1978), 104 subjects were studied before and after starting a slow rotating shift system (1 week working 08^{00} to 16^{00}; 1 week 16^{00} to 24^{00}; 1 week 24^{00} to 08^{00}; 1 week off; always Sundays off). The subjects were asked to assess a number of aspects of their health. The authors noted that after 6 months of shift work, there was an increased number of complaints about ill-health when comparisons were made with control conditions in the same workers before the shift work began. These trends continued in the 95 subjects who were questioned again after 4 years 4 months. Of those who had stopped shift work in the meanwhile, there was evidence that their general health had stabilized rather than deteriorated like that of their shift-working colleagues. However, the authors also point out that no serious illness attributable to shift work was found (*see above*).

The second study (Akerstedt and Torsvall, 1978) was performed upon 400 subjects before and after some of them were affected by changed shift schedules that had resulted from an overall decrease in shift working. The shifts being worked before the changes were: shift 1, day work only; shift 2, morning and evening shifts, alternating each week; shift 3, morning, evening and night shifts alternating each week; shift 4, morning, evening and night shifts, alternating each few days. After the changes the workforce could be divided into those who worked the same shifts as before (and so acted as controls) and those who worked different shifts. This latter group was subdivided into: those changing from shift 4 to shift 3; those changing from shifts 3 or 4 to shift 2; those changing from shifts 3 or 4 to shift 1. Subjects were questioned on a number of matters including sleep, mood complaints, gastrointestinal and social complaints before, and 1 year after, the changes had been instituted. Some of the results are shown in *Fig.* 10.1. This indicates very few differences for those workers whose shifts were unchanged, but improvements (often statistically significant) in workers no longer doing night shifts (changing from shifts 3 or 4 to shifts 1 or 2). There was also some evidence that

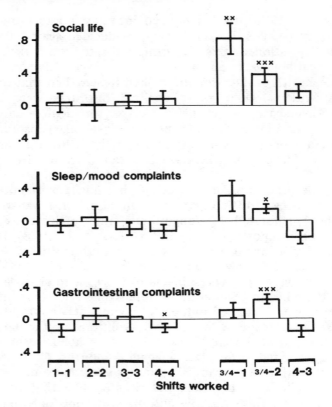

Fig. 10.1. Mean changes in ratings of 'social complaints', 'sleep/mood complaints' and 'gastrointestinal complaints' (± 1 s.e.) for different shifts worked. Figures above the zero lines indicate an improvement and figures below a deterioration. The abscissa indicates the shift previously worked (figure before dash) and the shift worked after shift systems were changed (figure after dash). On the left, workers for whom there was no change in shift worked and, on the right, for workers for whom there was a change of shift. Shift 1 = daywork; shift 2 = morning and evening shifts on alternating weeks; shift 3 = morning, evening and night shifts with 1 week on each shift; shift 4, as shift 3 except the shift rotation was faster (2 or 3 days on each shift). Significance levels of changes: x, $P<0.05$; xx, $P<0.01$; xxx, $P<0.001$. For further details *see* text. (*From* Akerstedt and Torsvall, 1978; Fig. 1.)

control workers staying on shifts involving night work (4–4) showed increases in gastrointestinal complaints at the end of the year.

Both these longitudinal studies provide evidence in favour of the hypothesis that shift work is associated with some medical disorders, especially gastrointestinal disturbances.

As has already been described, the questionnaire results might be influenced by the general beliefs of the worker and this might even modify the stress that the worker will place upon any health problems they might have. Presumably, medical examinations need not be influenced as much by such factors (see the slight differences between subjective and objective assessments in the study of Koller et al., 1978) but single-blind studies seem rare (Harrington, 1978) and the resultant preponderance in the literature of subjective assessments of health is undesirable. A criticism of longitudinal studies is that the sample ages, and, as will be discussed later, ageing might be associated with changes independent of shift working. This does not seem always to be important as, in the study of Akerstedt and Torsvall (1978), the control subjects showed few changes in incidence of health problems even though, of course, they aged as much as the workers whose shifts changed.

The cause of the increase in gastrointestinal complaints cannot be attributed to a different daily calorie intake since this has been shown not to alter (Rutenfranz et al., 1977; Reinberg, Migraine et al., 1979). The distribution of meals is changed, with breakfast often being missed by night shift workers and being replaced by 'nibbling'—generally of carbohydrate—throughout the night (Reinberg, Migraine et al., 1979). In addition, there are reports of increased smoking and caffeine or alcohol consumption (Rutenfranz et al., 1977), any or all of which might be implicated in the aetiology of gut disorders.

3.3. Fatigue

Loss of sleep and fatigue during the waking period has been a regular complaint from shift workers; both in studies in which the time spent asleep was logged and when electroencephalographic (EEG) monitoring of sleep was performed, data have been obtained that confirm this belief (see reviews by Mills, 1967; Rutenfranz et al., 1976, 1977; Harrington, 1978; Winget et al., 1978; Knauth et al., 1980). Thus Foret and Lantin (1972) found that 10 train drivers achieved about 8 hours sleep on rest days but just under 6½ hours when on night shift and sleeping during the daytime. In other studies, different values for the absolute amount of sleep were found but the essential result—that on shift work the worker becomes sleep-deprived—is clear enough.

The reasons for this deprivation are both environmental and physiological. After a night shift, workers need to sleep during the daytime and are far more likely to be woken then by external disturbances. Not surprisingly, many of their complaints refer to 'noise', whether this be due to children, to traffic or to telephones, etc. Should they wake temporarily (normal enough even during nocturnal sleep) these conditions render it more difficult for them to get back to sleep.

There are physiological reasons too, based upon the slow adaptation of the endogenous components of rhythms to a changed schedule, as a result of which the worker will be trying to sleep while his body is preparing for a new day. (Some of these rhythms will be described in more detail in the

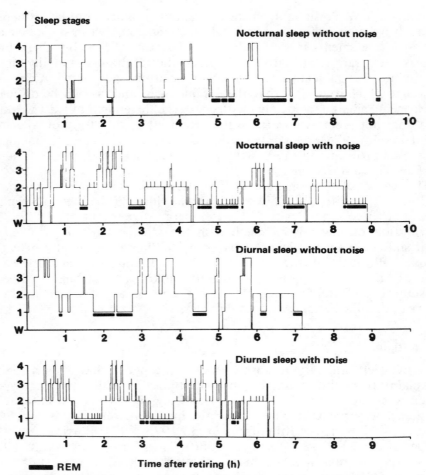

Fig. 10.2. EEG sleep changes in a single subject sleeping either in the day-time or at night and with or without noise in the sleeping room. Brief awakenings indicated by upward spikes. (*From* Rutenfranz et al., 1976, Fig. 1.)

next section.) Thus the worker is liable to be woken by the need to micturate and there will be a general increase in 'arousal'—as assessed by catecholamine and temperature measurements—all of which will militate against his obtaining adequate sleep. The result is that workers often take more naps than controls in addition to their main sleep; however, this seems not to be completely successful since subjects sleep longer during their days of rest, presumably since they are still sleep-deprived (Tune, 1969; Foret and Lantin, 1972; Smith, 1979). If the inverse relationship between temperature and fatigue holds during shift work (*see also* s. 4.2) not only sleep-deprivation but also inappropriately phased temperature rhythms will contribute to the sensation of fatigue. There is evidence that subjects differ in their ability to take sleep at unusual hours; this aspect also will be discussed later.

Not only the quantity of sleep but its quality too is different. EEG studies are rare but those that exist indicate changes in both the total amount of the different stages and in sleep 'infrastructure' in shift workers undergoing rapid rotation (Foret and Benoit, 1978, 1979; Webb and Agnew, 1978). *Fig.* 10.2 (from Rutenfranz et al., 1976) illustrates this changed profile together with the effects of noise during daytime sleep. The effect of noise is to reduce the amount of stage 4 sleep (one component of slow-wave sleep) and increase the number of awakenings (compare the top two records). Sleep during the daytime results in a shorter sleep (compare the first and third record) and REM frequency declines rather than increases as the sleep progresses (*see also* Knauth and Rutenfranz 1975). A possible explanation of this, based upon a circadian rhythm of REM frequency, has already been given in Chapter 5.

If shift workers react to changed routines in the same way as do laboratory subjects, then the release of hormones during their diurnal sleep will show a changed pattern (*see* Chapter 7). The implications that all these changes in sleep and endocrine profiles have for the general well-being of the shift worker are completely unknown, but, for some speculation on this, *see* Oswald (1978).

4. Performance during Shift Work

Although the views and health of the workforce are important factors, it is necessary also to consider whether or not any differences in safety and production efficiency exist between various shifts.

4.1. Field studies of complex tasks

One way in which to assess the efficiency of different shifts is to consider those studies in which performance in factory tasks has been measured directly in the place of work. These have been reviewed recently by Hildebrandt et al. (1975) and Harrington (1978).

There is not agreement as to whether there is a variation in the number of accidents with time of day (Rutenfranz et al., 1977; Harrington, 1978; Winget et al., 1978) but the difficulty in deciding what constitutes an 'accident' might well be a deciding factor. One method which provides evidence that can be used for assessments of both safety and performance is the measurement of the number of errors that are made at different times of the day. *Fig.* 10.3 summarizes some of the results that have been obtained in this way. This figure shows that the frequency of errors tends to be high during the night, lower during the daytime and shows a post-lunch dip in performance (*see* Chapter 6) of varying prominence from

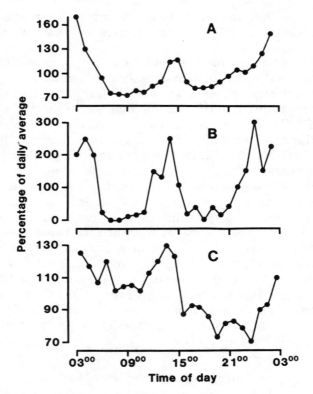

Fig. 10.3. Mean daily course of errors in performance expressed as per-
centages of the daily average. A, frequency of errors in reading gas meters
(data of Bjerner and Swensson, 1953), means of 62 000 values; B, frequency
of falling asleep while driving a car (data of Prokop and Prokop, 1955),
means of 500 values; C, frequency of automatic compulsive brakings due
to driver omission in train drivers (data of Hildebrandt et al., 1974), means
of 2238 values.

about 13⁰⁰ until 15⁰⁰. Similarly, the delay by switchboard operators
to respond to calls (Browne, 1949) suggests a poorer nocturnal performance. In all cases, there is a striking parallelism between these results and
those from simple psychometric tests and temperature (*see* Chapter 6).
The inference often made is that the temperature rhythm underlies these
performance rhythms, though neither simultaneous temperature recordings
nor other measurements of 'arousal' were made in these studies.

However, Meers (1975) describes an investigation in which a comparison
between body temperature and performance was made. The investigation
involved a group of shift workers in a sugar refinery working a three-shift
system that rotated each week. The task investigated—crystallizing
sugar—could be assessed by the quality of the end-product and entailed a
number of complex decisions requiring considerable skill and experience.

Further, there was no feedback, that is, it was not until the very end of the operation that its success could be judged. This study indicated that: (a) the quality of sugar produced on the night shift was up to 20 per cent lower than that produced by the same workers on the day shift; (b) as the week progressed, there was some improvement in the quality of product; (c) oral temperature was lower during work in night workers but tended to rise as the week progressed; (d) the temperature rhythm reverted to its normal phasing after a diurnal routine had been adopted for only one day. Meers concluded from these results that:

(a) night workers were initially unadapted to their hours of work and performed with less efficiency;

(b) continuing on the same shift led to some adaptation of physiological rhythms and improvement in performance;

(c) adjustment from a nocturnal to a diurnal routine was very rapid. As will be seen, other studies with simpler measures of performance lead to similar conclusions where a fairly slow (weekly) rotation of shifts is concerned.

However, it is important to realize some of the difficulties of interpretation that attend field work of this complexity; some of these have been discussed by Meers (1975) in relation to his study just described. In brief, the difficulties arise because there are dissimilarities between nocturnal and diurnal working conditions (for example, in supervision, motivation, noise, lighting and fatigue) and any or all of these might complicate the interpretation of the results obtained (*see also Table* 10.1).

A clear illustration of some of these problems comes from the other workers studied by Meers (1975). These were wire-drawers and their task differed from that of the sugar-crystallizers inasmuch as it was machine-paced and the operator had feedback, that is a knowledge of how the process was progressing. Compared with the day shift, there was a 5 per cent decrease in the quantity of wire produced by the night shift. However, this could not unambiguously be directly attributed to a nocturnal decrease in the operators' efficiency since there were less staff to deal with machine defects at night and so breakdowns lasted longer. When correction for this difference was made, the two shifts were the same. The fact that the task was machine-paced would seem in this case to have rendered the overall efficiency of the task less dependent upon any circadian rhythms of the operators.

Further difficulties in interpretation have been discussed by Hildebrandt et al. (1975) in connection with their own study (*see Fig.* 10.3). In this, they measured the frequency with which a warning light in the cabin of a locomotive was not extinguished by the driver, the result of which was that automatic braking ensued. The observed data were considered by Hildebrandt and his colleagues to result from three factors:

1. A circadian rhythm (which they related to temperature).
2. Visibility of the warning light (this was less during daylight hours).

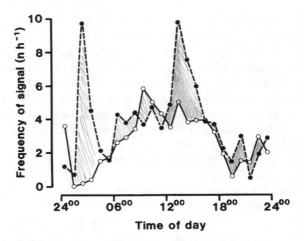

Fig. 10.4. Daily course of the mean hourly frequency of the sounding of a warning signal due to driver omission in train drivers. ○—○ readings from the first to third hour of work only; ●－－● readings from the fourth to sixth hours of work only. (Data of Hildebrandt et al., 1974.)

3. Fatigue of the locomotive driver.

The first factor has been commented upon already. The second factor results in the daylight frequencies being raised and those during darkness being decreased, an influence, it will be noted, that will tend to oppose the effects of a normally-phased temperature rhythm. The third factor was fatigue, which was investigated in two ways. First, Hildebrandt et al. (1975) established that there was a direct relationship between the number of brakings between the times 13^{00} and 15^{00} and the number of hours that the driver had been on duty. Second, when the data were replotted (*Fig.* 10.4) after dividing them into those from the first or second half of the driver's shift, the post-lunch and nocturnal decrements in performance were far greater in the presence of fatigue than in its absence. The effects of fatigue and motivation upon rhythms in mental performance in laboratory-based tests have been discussed in Chapter 6; the results from this and other field studies support the general conclusions reached there, namely that rhythms are accentuated by fatigue, by overloading the subject or by boring tasks and that they can be reduced if the tasks are interesting and the subject's motivation is high (*see also* Harrington, 1978).

In summary, the interpretation of performance rhythms in complex tasks is beset by problems and Meers (1975) concluded 'at the end of this field study it became clear that the difficulties in collecting, analysing and interpreting data are almost insurmountable'. Nevertheless, the conclusions from these studies making use of complex tasks are essentially similar to those derived from field studies in which simpler psychometric tests and other physiological rhythms have been assessed, and it is to those that we now turn.

4.2. Field studies with simpler markers

Advantages of simple markers

As was discussed in Chapter 6, a parallelism, if not a causal link, exists between the rhythms of three variables, namely simple psychometric tests, body temperature and urinary catecholamines. These three variables can be studied under field conditions rather more easily than the complex tasks just described. This gain must be offset against the artificial nature of the tests employed, of course, but the recent observation that similar results are achieved with simple tests in the laboratory and in the field and with actual 'on-job' performance suggests that simple tasks can be a useful, as well as convenient, marker (Monk and Embrey, 1981). As will be seen, the further assumption that temperature and catecholamine rhythms are markers for performance in complex tasks is substantially true, but it must be remembered that the parallelism need not hold invariably. Therefore, the recent development of a portable device for measuring simple and choice reaction times and storing the results for later analysis in the laboratory is a promising one (Glenville and Wilkinson, 1979) since it should enable the researcher to collect more data from subjects at their workbench (and hence under more natural conditions) than has previously been possible.

Temperature as a marker

Knauth and his colleagues (Knauth and Ilmarinen, 1975; Knauth et al., 1978) measured rectal temperature in a group of student volunteers before, during and after a 3-week span of continuous night work. Throughout the study, subjects were in contact with normal social influences but they slept in the laboratory and had to spend their leisure time in the vicinity. The average temperatures of all subjects on different days of the study are shown in *Fig.* 10.5. The control day (*Fig.* 10.5 *top left*) showed a steady climb through the work period to a peak just before retiring; after retiring, temperature continued to drop steeply to a minimum value at about mid-sleep (*see also* Chapter 2). During the first night shift (*Fig.* 10.5 *middle left*) the temperature rhythm was unchanged, as a result of which the subjects worked while temperature was falling and slept while it was rising. Later days on the night shift (*Fig.* 10.5 *bottom left* and *top right*) resulted in the (work-time) night temperature rising and the (sleep) day temperature falling by comparison with controls. Whilst this happened, the rhythm was 'flattened', that is, its amplitude was decreased. Subsequently (*Fig.* 10.5, *middle* and *bottom right*), the amplitude recovered, diurnal sleep became associated with ('appropriate') low

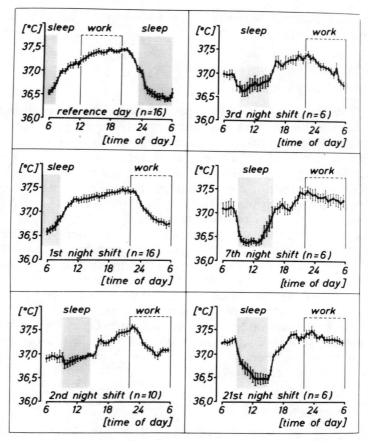

Circadian rhythm of rectal temperature $[\bar{x} \pm SE]$

Fig. 10.5. Circadian variations in rectal temperature in experimental subjects working diurnally (reference day) and nocturnally. (*From* Knauth et al., 1978, Fig. 1.)

temperatures and the waking period with high temperatures. These changes were almost complete by day 7 and by day 21 (*Fig.* 10.5 *bottom right*) stability seemed to have been achieved.

Was adaptation to night work complete? The answer depends upon the criterion used. Thus, adaptation was complete inasmuch as the temperature changes during sleep were normal but it was incomplete inasmuch as the maximum temperature did not occur at the end of work as on the control days. If cosine curves were fitted to the data (Knauth et al., 1978), when stability has been achieved, the time of minimum was related to mid-sleep in the same way as during the control days, but the acrophase had not changed by the same number of hours as mid-work. This last point will be discussed further below.

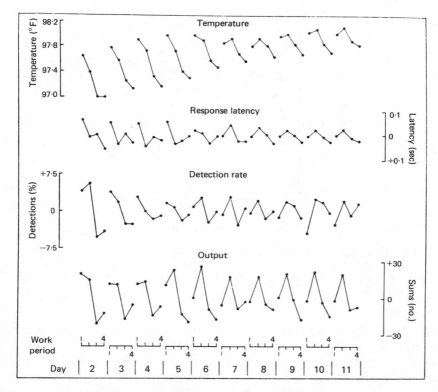

Fig. 10.6. Three-day moving averages of mean body temperatures and of mean performance scores from 10 subjects working on an 8-hour night shift. Measurements were made during each of four work periods during a single night's work. The work periods were separated by two 10-minute 'tea breaks' and one 1-hour meal break. Data from 12 consecutive sessions of night work are shown. (With permission *from* Colquhoun, 1971. Copyright by Academic Press Inc. (London) Ltd.)

The relationship between temperature and a variety of psychometric tests during shift work performed by naval personnel was investigated by Colquhoun and his colleagues (Colquhoun et al., 1968a,b; 1969; Colquhoun, 1971). Their subjects worked a number of shift systems throughout which regular measurements of oral temperature and of performance in auditory discrimination vigilance and calculation tests were made. With all shift systems (whether rotating or stable) a strong similarity (but not identity) between temperature and psychometric performance was found. This is illustrated by *Fig.* 10.6 in which the process of adaptation to an 8-hour night shift is shown. Note that, with successive night shifts, temperature adjusts and that the changing shape of the temperature rhythm was matched by the changing rhythms in the psychometric measurements. However, this matching was not exact

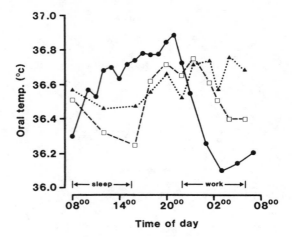

Fig. 10.7. Mean circadian variations in body temperature of 31 subjects before and during night work. ●—● before; □––□ on the sixth day of night work; ▲...▲ on the twelfth day of night work. (*From* Colquhoun et al., 1968b, Fig. 5.)

(indeed the figure suggests that the rhythm in ability to perform calculations changed slightly ahead of that of oral temperature). In addition, in the 12-hour shifts, in which performance fell off towards the end, an additional effect of fatigue seemed to be present (Colquhoun et al., 1969). When, on an 8-hour night shift, the temperature rhythms for the whole nychthemeron were considered (*Fig.* 10.7) the slow and partial adaptation of the rhythm and its decrease in amplitude—'flattening'—are clearly seen as in the study of Knauth and his colleagues (*above*).

Catecholamines as a marker

The relationship between self-rated alertness and catecholamines has been investigated by Akerstedt and his group, and reference to this has already been made (*see* Chapter 6). The use of catecholamines as a marker for the processes of adaptation to shift work also has been made by this group (Akerstedt and Fröberg, 1975; Akerstedt, 1977, 1979; Akerstedt et al., 1977). In a study upon railway workers (Akerstedt and Fröberg, 1975; Akerstedt et al., 1977) temperature, alertness and urinary catecholamines were measured during the week before the 3 weeks of nightwork, the first and third week of nightwork itself and the recovery week afterwards. The results for urinary adrenaline are shown in *Fig.* 10.8. These results indicate: (*a*) a normal rhythm during the control week (low nocturnal sleep values and high diurnal work values, *Fig.* 10.8A); (*b*) poor adaptation during the first week of night work (high diurnal sleep values and low nocturnal work values, *Fig.* 10.8B); (*c*) partial adaptation during the third week (sleep

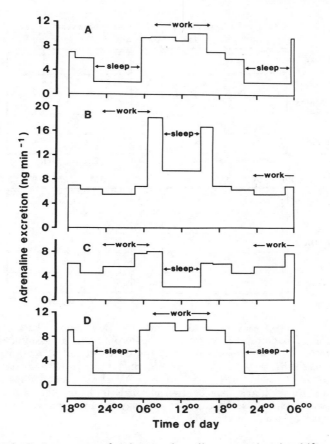

Fig. 10.8. Daily course of urinary adrenaline excretion in shift workers during: A, the last week of day work; B, the first week of night work; C, the third week of night work; D, the first week of return to day work. Data means from 15 workers. For clarity data have been repeated beyond 24 hours. (*From* Akerstedt and Froberg, 1975, Fig. 3.)

values are now low but work values are still not as high as control values, *Fig.* 10.8C). Note also that the average level was raised in the first week (*Fig.* 10.8B), a result indicating that some form of 'stress' was associated with the change in routine, and that the fall during sleep was immediate which strongly suggests that it was caused by a direct masking effect of sleep rather than adaptation of an internal clock. By the third week, the 'stressful' rise was over, but the amplitude of the rhythm was less, that is, 'flattening' had taken place. Temperature and alertness rhythms showed similar time courses; with noradrenaline, the changes were far more rapid and this would accord with the view that this rhythm has a more marked exogenous component.

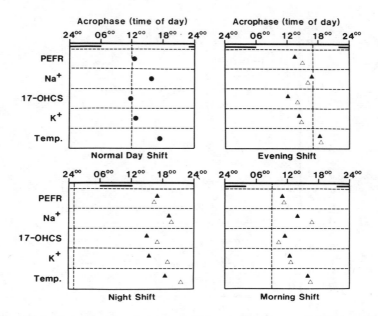

Fig. 10.9. Acrophases of peak expiratory flow rate (PEFR), urinary excretion of sodium, 17-hydroxycorticosteroids and potassium, and oral temperature in shift workers working the different shifts. Working hours: normal day shift, 07^{45} to 16^{30}; morning shift, 06^{00} to 13^{00}; evening shift, 13^{00} to 21^{00}; night shift, 21^{00} to 06^{00}. Vertical dashed lines indicate mid-work times. Black bars below time scale indicate time in bed. Acrophases shown are those on the first shift day (▲) and the seventh shift day (△). (Data of Chaumont et al., 1979.)

Changes in other variables

Even though this has not been investigated at all fully, it would seem likely that the process of adaptation to shift work takes place at different rates when different variables are considered. Claims have been made that this is the case (*see*, for example, Monk et al., 1978), but by far the most extensive comparison of the rates of adaptation of different variables to shift work has been made by Reinberg and his colleagues (Reinberg et al., 1975; Chaumont et al., 1979; Vieux et al., 1979). They studied workers on a weekly rotating shift system and collected data on the 1st and 7th day of each shift. They found that the acrophases of some variables (for example, PEFR and urinary sodium excretion, rapidly adapted to night shift whereas the acrophases of others (for example, temperature, 17-hydroxycorticosteroids and urinary potassium excretion) adjusted more slowly (*see Fig.* 10.9). These results all support the view that rhythms with large exogenous components adjust more rapidly than those with large endogenous components, a conclusion essentially the same as that reached in Chapter 9 when the response to time-zone transitions was considered.

Adjustment back to day work

A further problem associated with shift work is illustrated by *Fig.* 10.8D. It shows that the rhythms adjusted back to a diurnal existence far more rapidly than they adjusted to night work in the first place, an observation made by Meers (1975) also (*see* s. 4.1). Undoubtedly most shift workers will adopt a diurnal existence during their days off (but *see also* s. 6.2) but the *rapid* change of rhythms requires further comment. One implication of this finding is that after a few days of recreation any adaptation to night work will probably have been lost and the process of adaptation must begin again.

Shift work and zeitgeber

In summary, therefore, the data indicate that, when shifts are rotated slowly (every 7 or 21 days in the studies cited), adaptation of rhythms is slow and (in some respects) incomplete and that, at least transiently, rhythms tend to 'flatten'; further, during rest days, the rhythms rapidly adjust to become appropriate for diurnal living. Why should this be so?

Two factors might be important. First, unlike time-zone transitions, shift work is not associated with a change in all zeitgeber. In certain respects, therefore, the circumstance is similar to that obtained when subjects try to reverse their habits in the laboratory or hospital but are still aware of the normal nychthemeron. In these experiments, as in shift work, adjustment of physiological and psychological variables is slow and incomplete (*see* Aschoff et al., 1975). Presumably it is the non-coincidence, and even antagonism, between the sleep–wakefulness cycle and other external rhythmicities that act as zeitgeber in man that accounts for the 'flattening' of rhythms and slows their adjustment. By contrast, during days off, zeitgeber would act in concert to adjust rhythms to a phasing appropriate for a diurnal existence. Further comment on this asynchrony between zeitgeber will be made in Chapter 12.

Secondly, there is a change in the distribution of work and leisure in a night worker as compared to a day worker (Mills, 1967). Thus, the conventional sequence is sleep–work–leisure, whereas the night worker will have the sequence sleep–leisure–work. Accepting that many physiological rhythms rise during the first half and fall during the second half of activity, then the day worker will be working before his rhythms have peaked whereas the night worker will be working after his peak has been passed. This effect is clearly shown in the large study by Reinberg and his colleagues upon oil refinery workers (Reinberg et al., 1975; Chaumont et al., 1979). A large number of physiological and psychometric variables was assessed on the 1st and 7th day of each of four 7-day shifts ('day', 'morning', 'evening' and 'night'). Some of the results are shown in *Fig.* 10.9; not only did some variables adapt more rapidly than others (as has already been described) but the acrophases (shown in *Fig.* 10.9) bore different relationships to mid-work on the different shifts. Thus, the acrophases were after mid-work on the day and morning shifts, slightly before on the evening shift and considerably before on the night shift. On the other hand, the phase of the adapted rhythms with respect to mid-sleep (and hence mid-activity) was similar on all shifts (compare with the temperature data of Knauth and his colleagues, s. 4.2).

In summary, it would seem that the slow rate of adaptation to shift work is a reflection of the competing zeitgeber influences whereas the different leisure—work and work—leisure sequences in the waking period account for the abnormal relationship between mid-work and acrophase that is found on some shifts when stability has been attained. It would be informative to know what would happen to the rhythms of a workforce on day shift that took its leisure before it worked or of a workforce on night shift that took its leisure, rather than its sleep, after work. Intuitively it is likely that these arrangements would be socially unacceptable, but they would enable the importance of the abnormal sequence of sleep—leisure—work in determining the phase of rhythms during shift work to be assessed.

How successful has the use of simple markers been?

This is a very difficult problem to assess. On the one hand, through the use of simple markers a considerable amount of data has been accumulated; this has enabled some idea of the changes in circadian rhythms to be described and has even enabled certain models of what is believed to be happening in the body during shift work to be proposed. To this extent, the use of simple markers has been successful. On the other hand, the contributions that those markers have made to our understanding of the wider problems of performance and efficiency in the worker are less clear. Thus, even though a general parallel between psychometric performance and temperature has been found in shift work (*see* the results of Colquhoun and his colleagues, s. 4.2), as already mentioned, the parallel was not exact. In addition, as the studies of Mann et al. (1972) have shown (these were described in Chapter 6, s. 4.1), temperature can be a poor predictor of performance in certain psychometric tests in shift workers. When performance in complex ('real') tasks in the field is considered, the data of Meers (1975) (s. 4.1) are some of the very few that can be cited; it cannot be decided from these how closely rhythms of temperature and the performance of complex tasks are related. In all of these respects, the usefulness of simple markers is subject to a certain amount of reservation.

However, these simple markers have been used also to assess changes on other shift schedules as well as to give some indication of the relative merits of these different schedules together with possible means of distinguishing between workers who are tolerant or intolerant of shift work. It is to these studies that we now turn.

5. Results and Problems with Other Shift Systems

Even though the data so far described have come from investigations of slowly rotating schedules, some studies of the circadian rhythms of subjects on rapidly rotating and irregular systems have been performed and will now be outlined.

5.1. Rapidly rotating versus slowly rotating shift systems

Rapidly rotating shifts

Fig. 10.10 (from Knauth and Rutenfranz, 1976) shows the temperature records of 2 subjects who underwent a '2–2–2' shift system (i.e. a rapidly rotating system of 2 mornings, 2 afternoons and 2 nights followed by some days off). The temperature rhythm is consistent from day to day in accord with the view already discussed that adaptation to a changing schedule is slow. Quantitatively, the acrophase of the best-fitting cosine curve was stable from day to day. Similar results were found with a '1–1–1' system.

Rapid or slow rotation of shifts?

The essential difference between rapidly and slowly rotating shift systems, therefore, is that in the latter, adaptation is at least partial and so is associated with some shifting of the circadian rhythms whereas, in the former, adaptation is much less, with the result that, except for rhythms with large exogenous components, very little change in timing of the rhythms is observed.

Is one shift system better than the other? This is a question that has caused considerable debate (*see*, for example, Folkard and Monk, 1981). The argument that shift work has a deleterious effect upon health has already been advanced; a distinction between the effects produced by slowly rotating and rapidly rotating shifts is difficult to make since rarely has only one type of shift system been practised by any group of workers for a period long enough to allow differential effects upon health to become manifest. The views of shift workers themselves and their families (Harrington, 1978; Reinberg, Vieux et al., 1979c) suggest that different types of shift system have different advantages. Thus, the advantage of the slow rotation of shifts is that social life is easier to organize in advance; the advantage of rapidly rotating systems is that large numbers of consecutive night shifts are not worked. As far as researchers are concerned, those

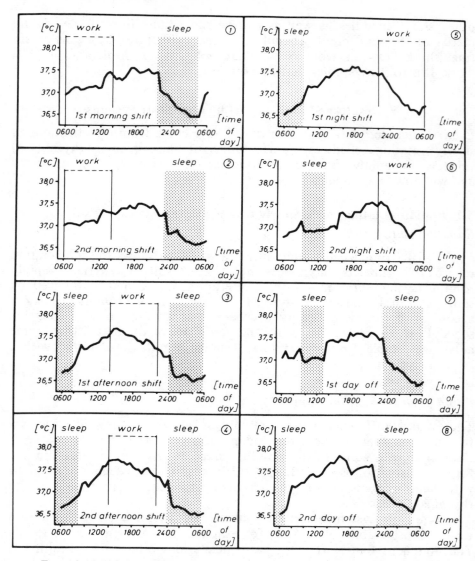

Fig. 10.10. Mean courses of the daily temperature variations measured half-hourly in 4 experimental subjects during a 2—2—2 rotating shift system. (*From* Knauth and Rutenfranz, 1976, Fig. 8.)

in favour of slow rotation of shifts (for example, Winget et al., 1978) believe that an 'appropriate' phasing of circadian rhythms to the work schedule is important and that this requires time (*see Fig.* 10.5). Those in favour of rapid rotation of shifts (for example, Rutenfranz et al., 1977, Reinberg, Vieux et al., 1979b) stress the incomplete adaptation of rhythms to changed work schedules and argue instead for a (rapidly rotating) system which does not allow substantial adjustment to take place (*see Fig.* 10.10).

Indirect evidence in favour of this latter view comes from recent studies (Andlauer and Reinberg, 1979; Andlauer et al., 1979; Reinberg, Vieux et al., 1979a; Reinberg, Andlauer et al., 1980) in which workers on both slowly and rapidly rotating shifts were described as 'tolerant' or 'intolerant' of shift work on the basis of whether or not they suffered from digestive troubles, persistent fatigue and sleep disturbance. The amplitude of the oral temperature rhythm (assessed both by cosine curve fitting and by the difference between minimum and maximum) was higher in the 'tolerant' group. It is not known whether the small amplitudes in the intolerant group existed prior to their starting night work or whether they were, in some way, a consequence of working at night (but *see* s. 6.2). However, as has been commented upon in Chapter 9 and will be elaborated upon later, the rate of adaptation of circadian rhythms to shifts in zeitgeber is inversely proportional to their amplitude (Aschoff, 1978).

Andlauer and Reinberg (1979) conclude from their studies that subjects with large amplitude circadian rhythms will adjust more slowly to shift work and manifest more stable rhythms; it is, they assert, this stability that is associated with tolerance to shift work. In another study with workers on a rapidly rotating shift system, those with 'major' difficulties relating to tolerance of shift work manifested oral temperature rhythms of lower amplitude than subjects matched for age and shift work experience who had 'minor' or 'no' difficulties (Reinberg, Vieux et al., 1979a; Reinberg, Andlauer et al., 1980).

Two points seem worthy of comment here. The first is that it might be that the differences in amplitude of rhythms as between 'tolerant' and 'intolerant' workers are the result of some other factor (*see* s. 6.2). Secondly, it must be remembered that one disadvantage with rapidly rotating shift systems, if circadian rhythms are stable, is that, assuming that temperature is an indicator of psychometric performance, then the workers' rhythms will be inappropriately phased for night work on every occasion that this takes place.

5.2. Irregular and non-24-hour shift systems and the need for stability

Results

Recently, reports have appeared of studies on personnel working irregular shift schedules (Colquhoun et al., 1978, 1979). Oral temperature readings were taken hourly during a shift system that consisted of 4 hours duty followed by 4, 8 or 12 hours off duty; the complete cycle lasted 3 days and was repeated for the 48 days of the experiment. As the experiment progressed, the amplitude of the temperature rhythms decreased and there was a general loss of 24-hour rhythmicity as assessed by cosine curve fitting. In some cases the data became irregular, in others there were also 8- and 4-hour components. Such a result can be explained if it is accepted, first, that in the absence of regular zeitgeber the endogenous rhythm will 'free-run' (see Chapter 9, s. 5) and, secondly, that sleep has a direct effect upon temperature (see Chapter 2, s. 4).

Recently (Schaefer et al., 1979), a report has appeared on submariners on a regular work–rest schedule of 18 hours (6 hours work, 12 hours rest) while social and feeding influences continued on a 24-hour pattern. Rhythms of temperature, pulse, respiration and blood pressure measured under nychthemeral conditions before the shift system showed marked 24-hour components. At the beginning and end of the 18-hour cycles, rhythmic components with periods of 12, 36 and 48 hours were found in all variables; again this probably indicates the exogenous effect of the schedule. In addition, a 24-hour component was still present (though its size was diminished) but, as the mission proceeded, an increasing number of rhythms with a period between 25 and 28 hours was found. These results agree with those of Colquhoun et al. (1978, 1979) and suggest also that a residual 24-hour periodicity of meals and social influences is not sufficient to entrain rhythms to a 24-hour period in the face of non-24 hour sleep–wakefulness schedules. Experiments carried out in an isolation unit (Minors and Waterhouse, 1980) have confirmed that irregularly taken sleeps are associated with rhythms of rectal temperature and urinary excretion that 'free-run' with a period of about 25 hours even though meal times were taken in accord with a 24-hour clock and the subjects were aware of 'real' time (see also Chapter 12).

Conferring stability on irregular schedules

If stability of rhythms is an important factor when acceptability of shift systems is concerned, then a series of experiments (Minors and Waterhouse, 1980) that has already been outlined in Chapters 2 and 9 is again relevant. These experiments showed that if 4 hours of sleep were taken regularly— 'anchor sleep'—even though another 4 hours were still taken randomly, then the rhythms showed a stable period of 24 hours. These results might have application to shift workers. Thus, people on irregular schedules who wished to stabilize their rhythms, or workers on slowly rotating schedules who did not wish their circadian rhythms to lose adaptation might benefit by taking part of their sleep at the same time each day (both on a leisure and a work day). In Chapter 9 it was mentioned that some pilots on circumglobal navigation appeared to try to 'nap' at home night time during the course of their journey and, as will be mentioned later, there is some evidence that permanent shift workers adjust their days off to suit the requirements of their shift schedules. In all cases it seems possible that 'anchor sleeps' are being incorporated into the daily routines (see also Wedderburn, 1972). Presumably rhythms of shift workers on rapidly rotating schedules will be stabilized if a 'normal' sleep time appears often enough to act as 'anchor sleep' (*see Fig.* 10.10).

5.3. Two further problems

Two further problems exist when different shift systems are compared.

1. Time-zone transitions and shift work both produce a dissociation between the individual and his environment, a shifting of circadian rhythms and an internal dissociation between the different rhythms. Even if the proposed importance to shift workers of stability of their circadian rhythms is confirmed in other studies, the possible effects of internal dissociation and 'inappropriate' phasing of rhythms (both of which would seem to be marked on night shifts during rapidly rotating systems, *see* Vieux et al. (1979) and *Figs.* 10.9, 10.10) will still require elucidation.

2. A problem that has not been investigated at all is one that has been discussed with respect to time-zone transitions, namely the extent to which the endogenous rhythms have been masked by exogenous influences (*see Fig.* 9.7). The use of constant routines in shift work studies has not been made though its potential importance has been described (Akerstedt, 1979). There is some evidence that adaptation of the endogenous component of a rhythm to a slowly rotating but not a rapidly rotating schedule does take place. Thus, after 3 weeks on a night shift, the temperature acrophase (which was shifted compared with control values) had not fully adjusted back to control values after 2 normal days whereas, with rapidly rotating shifts, the acrophase was hardly affected by the 1−1−1 shift system and so was immediately phased appropriately for days off (Knauth and Rutenfranz, 1976). Such a result is very similar to that obtained by Lafontaine and his colleagues (1967) (Chapter 9, s. 2.3) and enables the same inferences to be drawn.

6. Permanent Shift Workers and Adaptability to Shift Work

Earlier, reference has been made to the fact that some shift workers 'drop out' because of social and medical difficulties. Accordingly, those that remain—permanent shift workers and, especially, night workers—might be better able to cope with the problems of shift work and a study of them might indicate why this is so. Such an approach naturally leads to a more general discussion of why shift work is more acceptable to some workers than others.

6.1. Permanent night workers

It is generally found that permanent night workers have rhythms better adapted to their work than do rotating shift workers. In the newspaper industry, where night work is normal and permanent, a study found that

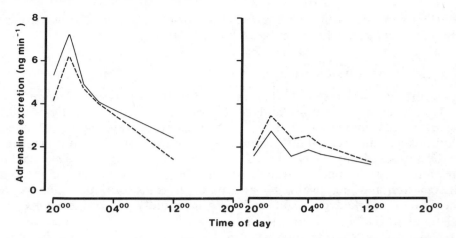

Fig. 10.11. Mean levels of adrenaline excretion in: *left,* permanent night workers; *right,* occasional night workers. Data plotted are the mean rates of excretion on the first night of night work (solid lines) and the fifth night of permanent night workers (seventh night for occasional night workers) (dashed lines). (*From* Pátkai et al., 1975, Fig. 1.)

plasma 11-hydroxycorticosteroid levels were high on waking at 14^{00} and low at the end of work at 06^{00} (Conroy et al., 1970). Clearly such a result must be interpreted with caution since the sampling frequency was so low; nevertheless, it suggests that the steroid rhythm, one with a high endogenous component, had adapted to a nocturnal routine.

Folkard et al. (1978) compared nocturnal temperatures and subjective estimates of well-being from 22^{00} until 06^{00} in two groups of nurses who had been on night work for at least 30 months. One group (full time) had worked 4 nights per week and the other (part time) 2 nights per week. Rhythms unadapted to night work would show a fall of temperature and sense of well-being during much of the shift and adaptation to night work over the course of the week was assessed by decreases in the falls of temperature and well-being as the shift progressed. When the first two nights were compared, adaptation of the full-time group was better than that of the part-time group; by contrast, adaptation of alertness was not different, emphasizing yet again that motivation can overcome the effects of circadian rhythms.

Akerstedt and his group (Pátkai et al., 1975, 1977; Akerstedt, 1979) have compared urinary catecholamine rhythms in two groups of workers in the newspaper industry. One group alternated between night and day work, the other worked nights permanently. *Fig.* 10.11 shows some of the results for adrenaline excretion. Again better adaptation is found in permanent night workers in so far as the rhythms show a higher amplitude, that is the process of 'flattening' (*see above*) does not seem to be so marked. Similar results for self-rated activation and for deep body temperature were obtained.

6.2. Adaptability to shift work

Why should permanent night workers adapt to night work better and why are some workers intolerant of shift work?

Clearly, as Harrington (1978) and Rutenfranz et al. (1977) have argued, there are strong social factors that influence the workforce. Nevertheless, as this brief account has indicated, there is evidence to indicate that some workers 'drop out' often as a result of declining health. Recently, much interest has been shown in the possibilities that workers who will be intolerant of shift work and that circumstances that predispose to difficulties with shift work can be predicted. Some of these possibilities will now be discussed.

Do experience and age speed up the process of adaptation?

To test this hypothesis, a longitudinal study is required but such data are not yet available. As Reinberg et al. (1975) and Klein and Wegmann (1979) point out, there is some evidence in man and mice that ageing *decreases* the ability to adjust to time-zone shifts. A transverse study by Adum (1975) investigated the percentage of professional truck drivers in different age groups who had difficulty in night driving. The results (*Table* 10.2) can be interpreted to indicate an increasing difficulty with age (after 31 years) and a decreasing difficulty with experience (compare the two youngest groups). However such an interpretation is very speculative, ignoring as it does many social factors that are likely to have been different between the age groups.

Table 10.2. Relationship between age and difficulty experienced in night driving*

Age group	Difficulty in night driving
26–30 yr	20%
31–35 yr	5%
36–40 yr	12%
41–45 yr	12%
46–50 yr	20%
over 50 yr	50%

*From Adum, 1975.

Are the amplitudes of circadian rhythms relevant?

Wever (1979) has shown that the rate of entrainment of subjects to a time shift in an isolation unit was inversely proportional to the amplitude of the temperature rhythm measured during control conditions. In field studies upon oil refinery workers, Reinberg's group (Reinberg, Vieux et al., 1978a,b, 1979d) have shown that the same relationship holds when adjustment to the first night of shift work is considered for temperature, peak expiratory flow rate and urinary 17-hydroxycorticosteroid rhythms. The significance of this relationship is not certain but clearly it is closely related to another controversy that has already been commented upon, namely is it better for a worker's rhythm to remain in phase with a diurnal life-style or to adjust to the new shift whenever night work is done (*see* s. 5.3).

Again, at a speculative level, since ageing is believed to be associated with a decreased amplitude of circadian rhythms (*see* Chapter 8), older people would be disadvantaged for shift work if stability of circadian rhythms were important; further, aged subjects should adjust more easily to time-zone transitions, a prediction that does not accord with the data (*see* Klein and Wegmann, 1979).

Is the phase of the temperature rhythm important?

The distinction between 'larks' and 'owls' and their temperature rhythms has already been made (*see* Chapter 6, s. 4). It is generally accepted that 'owls' prefer the night shift and are better able to adapt to it and that 'larks' are better able to cope with the morning shift (*see*, for example, Ostberg, 1973; Colquhoun and Folkard, 1978; Horne et al., 1980). In part this probably relates to the earlier rise of temperature in 'larks' and the later fall in 'owls' and to the parallelism that exists between core temperature, alertness and performance. In addition, some relation between temperature and sleep is indicated by the work of Breithaupt et al. (1978). Thus they measured the latency to sleep and duration of sleep in 3 'morning' and 3 'evening' females ('larks' and 'owls' respectively) after the subjects had been sent to bed at 21^{00}, 23^{00}, 01^{00} or 03^{00}. Their results showed that the 'morning' types were better able to get to sleep (even at the earlier time) had lower temperatures on retiring and woke earlier than 'evening' types; as a result, the 'morning' types incurred sleep deficits if their bedtime was delayed. In contrast to this, the 'evening' types were able to 'lie in' in the mornings and thereby did not incur as much sleep deficit; however, they had greater difficulty in falling asleep, especially with an early bedtime. The implications of these findings for shift workers —that 'morning' types would suffer from sleep deficits after delayed retiring times, that they would be more able to gain sleep before 'morning' shifts and that 'evening' types would have difficulty in sleeping before the morning shift but not in sleeping after a night shift—all concur with the conclusions of Ostberg (1973) and Colquhoun and Folkard (1978).

Further evidence has been obtained recently in studies by Hildebrandt and Stratmann (1979) upon nurses. A number of circadian rhythms (temperature, heart-rate and subjective assessment of vigilance) were measured after between 7 and 18 days night shift and after about 10 days of nychthemeral recovery. Adaptation to the night shift was assessed from a number of 'indices' that monitored changes in shape, phase and ultradian components in the circadian rhythms. Extroverts complained less about night work and adjusted to it by a flattening of their circadian rhythms; by contrast, introverts complained more and manifested greater disturbances in their circadian rhythms including an increased amplitude of circadian rhythms. These results have a number of implications. First they support the view that night shifts are more easily performed by extroverts (evening types, 'owls'). Secondly, they suggest that the amplitude of the circadian rhythms determines the type of shift system for which the worker will be better suited; with high-amplitude rhythms, the associated stable phasing renders the worker more suited to rapidly rotating shifts but with low-amplitude rhythms the associated lability and propensity for rapid adaptation render him more suitable for slowly rotating shifts (see above). Thirdly, they imply that the attitude of the shift worker is an important consideration.

Does the attitude of the shift worker contribute?

A study dealing with the attitude of workers rather than their circadian rhythms has been made by Folkard et al. (1979). They have assessed the 'flexibility' of groups of nurses on different types of shift work and have concluded that 'inflexible' subjects, that is, those who were unable to take daytime 'naps', unable to sleep through noise and unable to overcome feelings of drowsiness, were less suited to night work; the importance of daytime sleep and 'naps', of a suitably peaceful neighbourhood and of an accommodating family have all been stressed by Lortie et al. (1979) also. The *motivation* or *commitment* of permanent night workers has been emphasized by others repeatedly (Pátkai et al., 1977; Folkard et al., 1978). In other words, a permanent night worker must be able and willing to adjust his social as well as his working life to meet the demands of his work schedule, and he must be prepared to tolerate the consequences of the diurnal habits of the rest of society. There is even some evidence that during his days off, the permanent night worker will adopt schedules that are influenced by his night work experience (Folkard et al., 1978) and this attempt to 'standardize' (at least partially) his life style might contribute to the observation (s. 5.1) that the 'tolerant' shift worker shows rhythms of higher amplitude. Such an adjustment is easier where shift work is 'universal, accepted and lifelong' (Conroy et al., 1970) or in isolated communities in which the shift worker is not the odd one out (Lobban, 1965); it is particularly difficult in the case of night workers who are expected to run a diurnal household as well (for example, part-time nurses who are also mothers).

7. Conclusions

In conclusion, there is considerable evidence that shift work has certain disadvantages, the main ones being social isolation and the sensation of fatigue, this latter probably attributable to a deficit in sleep. Claims that health deteriorates have been substantiated at least partly only in the case of gastrointestinal disorders. In some tasks, performance is poorer, especially during night work and when the work is of a boring or repetitive nature.

When the circadian rhythms of shift workers are investigated, there is clear evidence that they can be disorganized in a manner similar to that seen after a time-zone transition. As yet, the aspect of disorganization that is most responsible for the sense of general malaise is unknown (or indeed if there is any connection). Accordingly, it is difficult to be certain whether rapidly or slowly rotating shift systems are to be preferred; however, an intermediate rate of rotation of shifts would seem to be an undesirable compromise since there is not enough time for adaptation and yet too much for the rhythm to be unaffected. Evidence is growing that permanent shift workers are 'better' adapted to shift work than shift workers in general and 'drop-outs' in particular, but, although various suggestions have been made as to why these differences between individuals exist, again much further investigation is required.

References

Aanonsen A. (1964) *Shift Work and Health.* Oslo, Universitetsforlaget.

Adum O. (1975) Shiftwork in professional drivers. In: Colquhoun P., Folkard S., Knauth P. et al. (ed.) *Experimental Studies of Shiftwork.* Opladen, Westdeutscher Verlag, pp. 273–276.

Akerstedt T. (1977) Inversion of the sleep wakefulness pattern: effects on circadian variations in psychophysiological activation. *Ergonomics* **20**, 459–74.

Akerstedt T. (1979) Altered sleep/wake patterns and circadian rhythms. Laboratory and field studies of sympathoadrenomedullary and related variables. *Acta Physiol. Scand.* Suppl. 469, 1–48.

Akerstedt T. and Fröberg J. E. (1975) Work hours and 24h temporal patterns in sympathetic-adrenal medullary activity and self-rated activation. In: Colquhoun P., Folkard S., Knauth P. et al. (ed) *Experimental Studies in Shiftwork.* Opladen, Westdeutscher Verlag, pp. 78–93.

Akerstedt T. and Torsvall L. (1978) Experimental changes in shift schedules—their effects on well-being. *Ergonomics* **21**, 849–56.

Akerstedt T., Pátkai P. and Dahlgren K. (1977) Field studies of shift work: II. Temporal patterns in psychophysiological activation in workers alternating between night and day work. *Ergonomics* **20**, 621–31.

Andlauer P. and Reinberg A. (1979) Amplitude of the oral temperature circadian rhythm and tolerance to shift work. *Chronobiologia* Suppl. 1, 67–73.

Andlauer P., Reinberg A., Fourré L. et al. (1979) Amplitude of the oral temperature circadian rhythm and the tolerance to shift-work. *J. Physiol. (Paris)* **75**, 507–12.

Angersbach D., Knauth P., Loskant H. et al. (1980) A retrospective cohort study comparing complaints and diseases in day and shift workers. *Int. Arch. Occup. Environ. Hlth* **45**, 127–40.

Aschoff J. (1978) Features of circadian rhythms relevant for the design of shift schedules. *Ergonomics* 21, 739–54.

Aschoff J., Hoffmann K., Pohl H. et al. (1975) Re-entrainment of circadian rhythms after phase-shifts of the zeitgeber. *Chronobiologia* 2, 23–78.

Bjerner B. and Swensson A. (1953) Shiftwork and rhythm. *Acta Med. Scand.* Suppl. 278, 102–107.

Breithaupt H., Hildebrandt G., Döhre D. et al. (1978) Tolerance to shift of sleep, as related to the individual's circadian phase position. *Ergonomics* 21, 767–74.

Browne R. C. (1949) The day and night performance of teleprinter switchboard operators. *Occup. Psychol.* 23, 1–6.

Chaumont A.-J., Laporte A., Nicolai A. et al. (1979) Adjustment of shift workers to a weekly rotation (study 1). *Chronobiologia* Suppl. 1, 27–34.

Colquhoun W. P. (1971) Circadian variations in mental efficiency. In: Colquhoun W. P. (ed.) *Biological Rhythms and Human Performance.* London, Academic Press, pp. 39–107.

Colquhoun W. P., Blake M. J. F. and Edwards R. S. (1968a) Experimental studies of shift work I: A comparison of rotating and stabilised 4 hour shift systems. *Ergonomics* 11, 437–53.

Colquhoun W. P., Blake M. J. F. and Edwards R. S. (1968b) Experimental studies of shift work II: Stabilised 8 hour shift systems. *Ergonomics* 11, 527–46.

Colquhoun W. P., Blake M. J. F. and Edwards R. S. (1969) Experimental studies of shift work III: Stabilised 12 hour shift systems. *Ergonomics* 12, 865–82.

Colquhoun W. P. and Folkard S. (1978) Personality differences in body-temperature rhythm, and their relation to its adjustment to night work. *Ergonomics* 21, 811–17.

Colquhoun W. P., Paine M. W. P. H. and Fort A. (1978) Circadian rhythm of body temperature during prolonged undersea voyages. *Aviat. Space Environ. Med.* 49, 671–8.

Colquhoun W. P., Paine M. W. P. H. and Fort A. (1979) Changes in the temperature rhythms of submariners following a rapidly rotating watch-keeping system for a prolonged period. *Int. Arch. Occup. Environ. Hlth* 42, 185–90.

Conroy R. T. W. L., Elliott A. L. and Mills J. N. (1970) Circadian excretory rhythms in night workers. *Br. J. Indust. Med.* 27, 356–63.

Folkard S. and Monk T. H. (1981) Individual differences in the circadian response to a weekly rotating shift system. In: Reinberg A., Vieux N. and Andlauer P. (ed.) *Night and Shift Work, Biological and Social Aspects.* Oxford, Pergamon Press, pp. 367–74.

Folkard S., Monk T. H. and Lobban M. C. (1978) Short and long-term adjustment of circadian rhythms in 'permanent' night nurses. *Ergonomics* 21, 785–99.

Folkard S., Monk T. H. and Lobban M. C. (1979) Towards a predictive test of adjustment to shift work. *Ergonomics* 22, 79–91.

Foret J. and Benoit O. (1978) Étude du sommeil de travaileurs à horaires alternants: adaptation et récupération dans le cas de rotation rapide de poste (3–4 jours). *Eur. J. Appl. Physiol.* 38, 71–82.

Foret J. and Benoit O. (1979) Sleep recordings of shift workers adhering to a three-to-four-day rotation (Study 2). *Chronobiologia* Suppl. 1, 45–53.

Foret J. and Lantin G. (1972) The sleep of train drivers: an example of the effects of irregular work schedules on sleep. In: Colquhoun W. P. (ed.) *Aspects of Human Efficiency.* London, English Universities Press, pp. 273–82.

Glenville M. and Wilkinson R. T. (1979) Portable devices for measuring performance in the field: the effects of sleep deprivation and night shift on the performance of computer operators. *Ergonomics* 22, 927–33.

Harrington J. M. (1978) *Shift Work and Health. A Critical Review of the Literature.* London, HMSO.

Hildebrandt G., Rohmert W. and Rutenfranz J. (1974) 12 & 24 h rhythms in error frequency of locomotive drivers. *Int. J. Chronobiol.* 2, 175–80.

Hildebrandt G., Rohmert W. and Rutenfranz J. (1975) The influence of fatigue and rest period on the circadian variation of error frequency in shift workers (engine drivers) In: Colquhoun P., Folkard S., Knauth P. et al. (ed.) *Experimental Studies on Shiftwork.* Opladen, Westdeutscher Verlag, pp. 174–87.

Hildebrandt G. and Stratmann I. (1979) Circadian system response to night work in relation to the individual circadian phase position. *Int. Arch. Occup. Environ. Hlth* **43**, 73–83.

Horne J. A., Brass C. G. and Pettitt A. N. (1980) Circadian performance differences between morning and evening types. *Ergonomics* **23**, 29–36.

Klein K. E. and Wegmann H.-M. (1979) Circadian rhythms in air operators. In: *Sleep, Wakefulness and Circadian Rhythm.* AGARD Lecture Series No. 105, AGARD, ch. 10.

Knauth P. and Ilmarinen J. (1975) Continuous measurement of body temperature during a three-week experiment with inverted working and sleeping hours. In: Colquhoun P., Folkard S., Knauth P. et al. (ed.) *Experimental Studies of Shiftwork.* Opladen, Westdeutscher Verlag, pp. 66–73.

Knauth P. and Rutenfranz J. (1975) The effects of noise on the sleep of nightworkers. In: Colquhoun W. P., Folkard S., Knauth P. et al. (ed.) *Experimental Studies of Shiftwork.* Opladen, Westdeutscher Verlag, pp. 57–65.

Knauth P. and Rutenfranz J. (1976) Experimental shift work studies of permanent night, and rapidly rotating, shift systems. I. Circadian rhythm of body temperature and re-entrainment at shift change. *Int. Arch. Occup. Environ. Hlth* **37**, 125–37.

Knauth P., Rutenfranz J., Herrmann G. et al. (1978) Re-entrainment of body temperature in experimental shift-work studies. *Ergonomics* **21**, 775–83.

Knauth P., Rutenfranz J., Schulz H. et al. (1980) Experimental shift work studies of permanent night, and rapidly rotating, shift systems. *Int. Arch. Occup. Environ. Hlth.* **46**, 111–25.

Koller M., Kundi M. and Cervinka R. (1978) Field studies of shift work at an Austrian oil refinery. I: Health and psychosocial wellbeing of workers who drop out of shiftwork. *Ergonomics* **21**, 835–47.

Lafontaine E., Lavarnhe J., Courillon J. et al. (1967) Influence of air travel east-west and *vice versa* on circadian rhythms of urinary elimination of potassium and 17-hydroxycorticosteroids. *Aerospace Med.* **38**, 944–7.

Lobban M. C. (1965) Dissociation in human rhythmic functions. In: Aschoff J. (ed.) *Circadian Clocks.* Amsterdam, North-Holland, pp. 219–27.

Lortie M., Foret J., Teiger C. et al. (1979) Circadian rhythms and behaviour of permanent nightworkers. *Int. Arch. Occup. Environ. Hlth* **44**, 1–11.

Mann H., Pöppel E. and Rutenfranz J. (1972) Untersuchungen zur Tagesperiodik der Reaktionszeit bei Nachtarbeit. III. Wechselbeziehungen zwischen Körpertemperatur und Reaktionszeit. *Int. Arch. Arbeitsmed.* **29**, 269–84.

Meers A. (1975) Performance on different turns of duty within a three-shift system and its relation to body temperature—two field studies. In: Colquhoun P., Folkard S., Knauth P. et al. (ed.) *Experimental Studies of Shiftwork.* Opladen, Westdeutscher Verlag, pp. 188–205.

Meers A., Maasen A. and Verhaegen P. (1978) Subjective health after six months and after four years of shift work. *Ergonomics* **21**, 857–9.

Mills J. N. (1967) Circadian rhythms and shift workers. *Trans. Soc. Occup. Med.* **17**, 5–7.

Minors D. S. and Waterhouse J. M. (1980) Anchor sleep as a synchronizer of rhythms on abnormal schedules. *Int. J. Chronobiol.* **7**, (in the press).

Monk T. H. and Embrey D. E. (1981) A field study of circadian rhythms in actual and interpolated task performance. In: Reinberg A., Vieux N. and Andlauer P. (ed.) *Night and Shift Work, Biological and Social Aspects.* Oxford, Pergamon Press, pp. 473–80.

Monk T. H., Knauth P., Folkard S. et al. (1978) Memory based performance measures in studies of shiftwork. *Ergonomics* 21, 819—26.

Ostberg O. (1973) Interindividual differences in circadian fatigue patterns of shift workers. *Br. J. Indust. Med.* 30, 341—51.

Oswald I. (1978) Editorial. Sleep and hormones. *Eur. J. Clin. Invest.* 8, 55—6.

Pátkai P., Akerstedt T. and Pettersson K. (1977) Field studies of shift work: I. Temporal patterns in psychophysiological activation in permanent night workers. *Ergonomics* 20, 611—19.

Pátkai P., Pettersson K. and Akerstedt T. (1975) The diurnal pattern of some physiological and psychological functions in permanent night workers and in men working on a two-shift (day and night) system. In: Colquhoun P., Folkard S., Knauth P. et al. (ed.) *Experimental Studies of Shiftwork.* Opladen, Westdeutscher Verlag, pp. 131—41.

Prokop O. and Prokop L. (1955) Ermüdung und Einschlafen am Steuer. *Dtsch. Z. Gerichtl. Med.* 44, 343—55.

Reinberg A., Andlauer P., Guillet P. et al. (1980) Oral temperature, circadian rhythm amplitude, ageing and tolerance to shift work. *Ergonomics* 23, 55—64.

Reinberg A., Chaumont A.-J. and Laporte A. (1975) Circadian temporal structure of 20 shift workers (8-hour shift-weekly rotation): an autometric field study. In: Colquhoun P., Folkard S., Knauth P. (ed.) *Experimental Studies of Shiftwork.* Opladen, Westdeutscher Verlag, pp. 142—65.

Reinberg A., Migraine C., Apfelbaum M. et al. (1979) Circadian and ultradian rhythms in the eating behavior and nutrient intake of oil refinery operators (Study 2). *Chronobiologia* Suppl. 1, 89—102.

Reinberg A., Vieux N., Andlauer P. et al. (1979a) Oral temperature, circadian rhythm amplitude, ageing and tolerance to shift-work (study 3). *Chronobiologia* Suppl. 1, 77—85.

Reinberg A., Vieux N., Andlauer P. et al. (1979b) Concluding remarks. Shift work tolerance: perspectives based upon findings derived from chronobiologic field studies on oil refinery workers. *Chronobiologia* Suppl. 1, 105—110.

Reinberg A., Vieux N., Chaumont A.-J. et al. (1979c) Aims and conditions of shift work studies. *Chronobiologia* Suppl. 1, 7—23.

Reinberg A., Vieux N., Ghata J. et al. (1978a) Circadian rhythm amplitude and individual ability to adjust to shift work. *Ergonomics* 21, 763—6.

Reinberg A., Vieux N., Ghata J. et al. (1978b) Is the rhythm amplitude related to the ability to phase-shift circadian rhythms of shift-workers? *J. Physiol. (Paris)* 74, 405—9.

Reinberg A., Vieux N., Ghata J. et al. (1979d) Consideration of the circadian amplitude in relation to the ability to phase shift circadian rhythms of shift workers. *Chronobiologia* Suppl. 1, 57—63.

Rutenfranz J., Colquhoun W. P., Knauth P. et al. (1977) Biomedical and psychosocial aspects of shift work. A review. *Scand. J. Work Environ. Hlth* 3, 165—82.

Rutenfranz J., Knauth P. and Colquhoun W. P. (1976) Hours of work and shiftwork. *Ergonomics* 19, 331—40.

Schaefer K. E., Kerr C. M., Buss D. et al. (1979) Effect of 18-h watch schedules on circadian cycles of physiological functions during submarine patrols. *Undersea Biomed. Res.* Submarine Suppl. S81—S90.

Sergean R. (1972) A note on current trends in the arrangement of working hours in the UK. In: Colquhoun W. P. (ed.) *Aspects of Human Efficiency.* London, English Universities Press, pp. 283—8.

Smith P. (1979) A study of weekly and rapidly rotating shiftworkers. *Int. Arch Occup. Environ. Hlth* 43, 211—20.

Taylor P. J. (1976) Occupational and regional association of death, disablement and sickness absence among Post Office staff 1972—1975. *Br. J. Indust. Med.* 33, 230—5.

Tune G. S. (1969) Sleep and wakefulness in a group of shift workers. *Br. J. Indust. Med.* **26**, 54–8.

Vieux N., Ghata J., Laporte A., Migraine C. et al. (1979) Adjustment of shift workers adhering to a three- to four-day rotation. (Study 2.) *Chronobiologia* Suppl. 1, 37–42.

Webb W. B. and Agnew H. R. (1978) Effects of rapidly rotating shifts on sleep patterns and sleep structure. *Aviat. Space Environ. Med.* **49**, 384–9.

Wedderburn A. A. I. (1972) Sleep patterns on the 25 hour day in a group of tidal shiftworkers. Studia Laboris et Salutis, Report no. 11, Stockholm, *Nat. Inst. Occup. Hlth* pp. 101–106.

Wedderburn A. A. I. (1978) Some suggestions for increasing the usefulness of psychological and sociological studies of shiftwork. *Ergonomics* **21**, 827–33.

Wever R. A. (1979) *The Circadian System of Man. Results of Experiments under Temporal Isolation.* Berlin, Springer-Verlag.

Winget C. M., Hughes L. and LaDou J. (1978) Physiological effects of rotational work shifting: a review. *J. Occup. Med.* **20**, 204–10.

chapter 11 *Clinical Implications*

It must be admitted at the outset that the usefulness of circadian rhythms to the practice of human medicine is disappointingly limited at the present time. In part the reason is that most of the drug studies have necessarily been performed upon animals and the application to human patients of information gained by these means must proceed with caution. However, another reason is that in many cases the advantages of a 'chronobiological' approach are small or even marginal; it is to be hoped that this position will change as a result of all the work in this field that is being carried out.

1. Normal Values and Circadian Rhythms

Diagnosis of an illness often requires an estimation to see whether some value is outside a range associated with health and the success of a treatment can be assessed by determining whether extreme values move to within the normal range. The use of the range rather than a single value acknowledges the differences that exist for any variable within a healthy population and such a normal range often has a statistical validation, for example, it can be the range from two standard deviations below the population mean up to two standard deviations above it. The variation which is found in a population derives from many sources, such as sex, age, height, weight etc. and any scheme whereby these factors can be taken into account—as with the use of nomograms—will decrease the range of any variable and thereby improve the chance of diagnosing an 'abnormal' value.

However, even after all these corrections have been taken into account, we are left with a 'homeostatic' rather than a 'circadian' model of what is normal and healthy. As, it is hoped, this book has made clear, few, if any, variables are constant throughout the nychthemeron and, since such rhythmicity generally has an endogenous component, measuring it under 'constant' or 'standard' conditions will not remove that rhythmicity. Accordingly, with reference to the cardiovascular system, what is an acceptable value at one time of the day might be unacceptably high or low at another (for example, Delea, 1979; Pickup et al., 1979). This point has been eloquently argued by Halberg and his colleagues on a number of

245

occasions (Halberg, Halberg et al., 1973; Halberg, Haus et al., 1973; Halberg and Ahlgren, 1979).

Yet this is not always the case. Sometimes the timing of samples for diagnostic purposes is not as critical as this because the amplitude of the rhythm is so small (for example, those of plasma sodium concentration and urinary creatinine excretion, *see* Chapter 4). It may be argued also that conditions are 'standardized' in those cases where samples are taken on waking and when the patient is under 'basal conditions'. Provided that the normal range also was derived from samples taken from healthy subjects at this time, such an argument has some strength, but, even then, waking might not be the most appropriate time for sampling. Thus the rhythm of plasma phosphate concentration changes most rapidly at about the time of waking (*see Fig.* 4.8), so comparatively small differences between times of sampling might produce misleading information.

The existence of these circadian changes has implications in diagnosis, the assessment of treatment and the performance of medical personnel.

1.1. The time of diagnosis

To a certain extent there has been some recognition of the implications of circadian rhythms in diagnosis for some time. Thus, it is widely accepted (Conroy, 1969; Mills, 1974) that the end of a febrile phase of some disease is more readily assessed in the morning (when cutaneous heat loss is minimal) than in the evening (when the circadian rhythm of heat loss is naturally greater, *see* Chapter 2) and that rhythms of renal excretion are inverted in some forms of renal disease. Some other examples where the time of diagnosis is important are:

1. Two disorders of adrenocortical function are Cushing's disease, when there is hypersecretion, and Addison's disease, when secretion is low. The nychthemeral rhythm of cortisol has already been described (Chapter 7); it is one of considerable amplitude and comparatively uninfluenced by exogenous factors. Ideally, changes in plasma concentration should be assessed by sampling frequently throughout the 24 hours; however, a sample taken just before retiring would be most appropriate for the diagnosis of Cushing's disease and just after waking in the case of Addison's disease (Ernest, 1966; Conroy, 1969; Mills, 1974).

2. Many species of parasite must be in the bloodstream of their victim when their vector is flying so that they can be transmitted to another host. In some cases, part of their life-span is spent in other regions of the body; with certain microfilarial parasites, when they accumulate in the lungs their concentration in the peripheral blood decreases almost to zero. The intriguing story of how it is believed the rhythms of vector and parasite are matched in different microfilariae is told elsewhere (Hawkins, 1973) but the important point here is that appropriate timing of blood samples would be required to assess correctly the extent of infection.

3. The evidence that there is a diurnal rhythm of fetal movement with peak values in the evening was given in Chapter 8 (s. I.2.2.). Since, in addition, abnormally low rates of fetal movement are associated with poor fetal outcome (Sadovsky and Yaffe, 1973; Pearson and Weaver, 1976; Spellacy et al., 1977), assessment in the evening (when movement should be high) would be of more use than early in the day.

1.2. The time of assessment of treatment

When the efficacy of treatment is being assessed, two of the many points that must be considered are:

1. The return to normal of an abnormal value. The comments already made relating to diagnosis will apply equally here of course.

2. Adverse side-effects should not become too great. Thus, one form of cancer therapy involves the use of immunosuppressive drugs, an adverse side-effect of which is to decrease the number of white blood cells (WBC). This form of treatment is often arranged so that the WBC count does not fall below a value of $2-4 \times 10^9$/l. Such a lower limit does not take into account the nychthemeral variation in leucocytes that is believed to be produced by the cortisol rhythm (Conroy and Mills, 1970). That is, the normal rhythm of leucocytes shows a maximum just before midnight and a minimum just before midday (the inverse of the cortisol rhythm); as a result, a 'low' value of leucocytes at midnight would seem to be a far more potent warning sign than one at midday (Halberg and Nelson, 1978).

1.3. The performance of medical personnel

As Chapter 6 has indicated, rhythms of human performance exist, values during the night generally being lowest. The implications of such findings for shift workers in general were considered in Chapter 10. As that account indicated, nurses have provided some data though results from other members of the medical profession are far fewer. Nevertheless, in a study upon junior doctors (Friedman et al., 1971), errors were determined in a vigilance task in which the subjects were asked to identify cardiac arrhythmias in an electrocardiogram recording. Errors of identification were twice as frequent when the doctors had previously taken less than 2 hours sleep than when they had slept for 7 hours. The poorer performance at night, even in 'permanent' night workers, is presumably a reflection of fact that man is a 'diurnal' animal. However, a distinction might exist between performance in repetitive and routine tasks and that in stimulating or emergency situations. As has been stressed in Chapters 6 and 10, motivation can temporarily overcome the poor performance due to fatigue or the effects of circadian rhythms in some situations, but whether or not this is the case in clinical circumstances is not known. In one study on an intensive care unit for the newborn, the number of deaths was higher at night (Tyson et al., 1979). The authors attributed this result to a number of factors (*see* s. 5.2), one of which was a decreased nocturnal performance of the staff involved.

2. Autorhythmometry

2.1. What is autorhythmometry?

Having outlined the possible application of circadian rhythms to clinical practice, the topic of autorhythmometry, keenly proposed by Halberg (Halberg et al., 1972; Halberg, 1973; Halberg, Halberg et al., 1973; Halberg et al., 1979), can be considered. The usefulness of autorhythmometry is based upon the observation that individuals differ not only when mean values taken at any single time of the day are considered but also when their nychthemeral rhythms are considered in detail (Halberg and Ahlgren, 1979). A group of American high school children performed a series of self-measurements, including peak expiratory flow rate, blood pressure, oral temperature and mood (Halberg et al., 1972). Measurements were taken between six and nine times per day for as many days as possible. The results were then assessed by fitting cosine curves, and an example—systolic blood pressure—is shown in *Fig.* 11.1. These results indicate not only that statistically significant cosine fits can be made to data obtained in this way but also that each individual has a personal record of his (or her) mesor plus its 95 per cent confidence interval under conditions when he (or she) is presumed to be healthy. Note that if the mean value for systolic blood pressure for the group as a whole were calculated from these data, the standard error of the mean would be considerably increased. Of particular interest is the result that the mesors for the male subjects 14 and 15 differed significantly from each other: a value of 115 mmHg would be abnormally high for one and abnormally low for the other; yet neither subject would fall outside the 95 per cent confidence limits of the mean calculated for the group as a whole. A similar battery of tests was performed for a larger number of days by a single subject, in addition to which frequent sampling of blood pressure was done by an automatic non-invasive technique. For all variables, it was possible to describe the data in terms of the mesor, amplitude, acrophase and period of the best-fitting cosine curve (Scheving et al., 1974).

2.2. The potential usefulness of autorhythmometry

The usefulness of such an approach is that the individual would have a record—a 'chronobiological profile'—of his own rhythms during conditions of presumed health. It is then argued (Halberg, Halberg et al., 1973): (*a*) that changes in this record might be signs of a developing illness or malady and enable preventative measures to be taken before serious consequences arose; (*b*) that autorhythmometry could be used to monitor

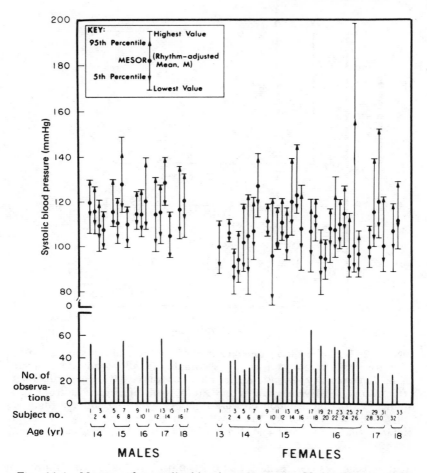

Fig. 11.1. Mesors of systolic blood pressures in 50 healthy students, illustrating that no mean blood pressure can be usefully allocated to a population and that individualized ranges must be obtained. Note in particular that highest values obtained in male students 15 and 4 were lower than the lowest value for male student 14. Similarly the highest values obtained in female students 1, 2, 3, 19, 27 and 28 were lower than the lowest value obtained by female students 8 and 15. (*From* Halberg, Halberg et al., 1973, Fig. 2.)

the effectiveness of treatment; and (*c*) that continuing recording could be used to check that a disorder remained under control.

An example which illustrates the usefulness of continuous monitoring is illustrated in *Fig.* 11.2. This figure shows the diastolic blood pressure of a 4-year-old girl during recovery from renal nephritis. A significant fall can be seen in the mesor of the rhythm of diastolic blood pressure as recovery (assessed independently) proceeded; more importantly, as far as the

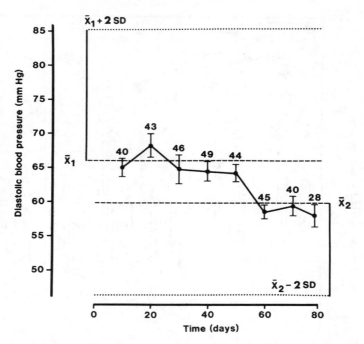

Fig. 11.2. Diastolic blood pressure in a 4-year-old child during and after nephritis. Daily mesors ± 1 s.e. are shown; figures above indicate numbers of readings per day. Despite higher blood pressures during nephritis than after recovery all values are within the normal range derived from two studies: $\bar{x}_1$ the mean blood pressure in 50 4–5-year-old children (data of Allen-Williams, 1945); $\bar{x}_2$ the mean blood pressure in 93 3–5-year-old children (data of Faber and James, 1921). (*After* Halberg and Nelson, 1978.)

present discussion is concerned, at no time did any value for the mesor fall outside the 'normal' range of the population as a whole. This indicates the advantage of considering values for each patient individually rather than as a member of a 'normal population', even when the values for the normal population have been obtained under 'standard' conditions. This point is very similar to that made in considering *Fig.* 11.1. In the same way, the potential value of home monitoring of blood pressure in patients being treated for hypertension has been stressed by Raftery (1975). However, as Raftery points out, a problem exists when the tediousness of the repetitive measurements required is considered. It remains to be seen whether such problems will limit the use of autorhythmometric techniques by the population in general.

The potential of autorhythmometry in the prediction and diagnosis of illness (*see above*) is one of considerable promise. Such an approach assumes that circadian rhythms are changed in illness and it is this proposition that must now be considered in some detail.

3. Changed Circadian Rhythms in Illness

To assess accurately changes in circadian rhythmicity in illness, large amounts of data are required but this is often ethically unacceptable in patients who are unwell. Additional problems are encountered when psychiatric or mentally retarded patients are considered due to their lack of understanding or motivation (Sollberger, 1974). As a result the changes in rhythmicity reported in the literature often have to be described as 'irregular' or 'abnormal' rather than in terms of changed values of cosine curve parameters (period, mesor, amplitude and acrophase).

Noteworthy exceptions are the data that have come from studies upon temperature measurements of the female breast under control conditions and with a variety of tumours and conditions necessitating surgery (Gautherie and Gros, 1977; Simpson, 1978; Haus et al., 1979; Halberg et al., 1979). In one case (Halberg et al., 1979) the temperature data were obtained autorhythmometrically and indicated a rise in mesor and fall in amplitude of the circadian rhythm of the fibrocystic breast when compared with the contralateral breast which was assumed to be normal. More data can conveniently be obtained automatically from a 'chronobra' that has recently been developed (Simpson, 1978). Only the present means of data storage seems to limit the frequency and span of measurement that is possible; preliminary results from these studies indicate the appearance in cancerous breasts of rhythms of abnormal timing, frequency, amplitude and mesor (Gautherie and Gros, 1977; Simpson, 1978; Haus et al., 1979).

Inspection of the literature relating to circadian rhythms in illness indicates that (when cosinor analysis has been performed) changes in any rhythm parameter can take place. Reviews that cover these changes in rhythms according to the type of illness can be found as follows: for metabolic disorder in diabetes, Alberti et al. (1975), Asplin et al. (1979); for sleep—wakefulness disorder, Weitzman and Pollack (1979); for respiratory disorder, Connolly (1979); for renal disorder, Aslanian et al. (1978), Wesson (1979), Wesson and Simenhoff (1979); for mental retardation, Quay and Guth (1975); for psychiatric disorders, Atkinson et al. (1974), Sollberger (1974), Kripke et al. (1978; 1979); for oncology, Simpson (1978), Haus et al. (1979); and for endocrines, the reviews cited for each hormone in Chapter 7 and Weitzman (1976, 1979).

3.1. Do changed circadian rhythms act as 'markers' for illness?

A central issue when changed circadian rhythms are considered is to what extent these changes are a reliable indicator of the illness (and therefore act as an aid to diagnosis) rather than a secondary symptom (which need not always be manifest). Data enabling such a distinction to be made are not plentiful, but there are two approaches that can be used to infer that a direct link between an illness and a changed circadian rhythm exists. These two approaches are as follows.

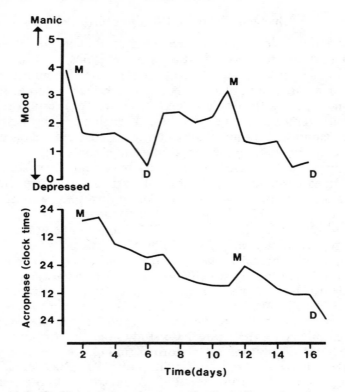

Fig. 11.3. Cyclic variations in the self-rating of mood and temperature acrophases in a female manic depressive. M and D indicate occurrences of mania and depression respectively. Note that the temperature rhythm has a period of about 21 hours and that occurrences of mania and depression occur when the phase of the temperature rhythm is most abnormal with respect to clock time. (Data of Kripke et al., 1978.)

The severity of the disorder is matched by the severity of the rhythm abnormality

Examples where this relationship is believed to hold are:

1. Luteinizing hormone release in anorexia nervosa (Pirke et al., 1979).

2. Thyrotrophin release in endogenous depression (Weeke and Weeke, 1978).

3. Cases are known in which the mood of the patient changes cyclically, sometimes with a regular period of two or three days, in other cases less regularly (Atkinson et al., 1974; Nikitopoulou and Crammer, 1976; Kripke et al., 1978). Nikitopoulou and Crammer (1976) observed that the days on which their patients were most depressed coincided with those days on which the circadian rhythm of temperature was most abnormal. Another example is shown in *Fig.* 11.3 taken from a female patient who showed a regular, 10-day cycle in mood. It has been proposed that cyclic

CLINICAL IMPLICATIONS 253

illness results from the interaction between two oscillators of different period (*see* Atkinson et al., 1974). If this is so then a 10-day cycle would result from the interaction of oscillators of 24- and 21·8-hour periods; as *Fig.* 11.3 indicates, the temperature rhythm advanced by an amount consistent with a 21·8-hour period. The important finding for present purposes is that the patient's mood was most extreme on those days when the acrophase of her temperature rhythm, like that for the rhythms of activity and urinary excretion, was placed most inappropriately for a diurnal existence (between midnight and 06⁰⁰ when she was depressed and between 18⁰⁰ and midnight when manic). However, a more recent study of a larger group of manic depressives (Kripke et al., 1979) did not reveal abnormal phase relationships of their circadian rhythms at a time when they were hypomanic.

Do changes in circadian rhythms approach normal as the disorder is treated?

Some data relating to this question have been obtained from the group of manic depressives already described in connection with *Fig.* 11.3 (Kripke et al., 1978). Lithium ameliorates the symptoms of depression in some but not all patients. There is some evidence that those who respond to lithium possess a circadian rhythm that runs fast (*see Fig.* 11.3) but is slowed by the lithium treatment, so that a normal phase relationship between the environment and the circadian rhythms becomes possible (*see also* Atkinson et al., 1974). Interestingly enough, 2 patients who did not respond to lithium showed rhythms with a period greater than 24 hours when tested under nychthemeral conditions before treatment (Kripke et al., 1978).

Some further examples in which recovery of circadian rhythms and symptoms are believed to be coincident are plasma tryptophan and tyrosine levels in depression (Niskanen et al., 1976); plasma adreno-corticotrophic hormone and cortisol levels in Cushing's disease (Schteingart and McKenzie, 1980); blood pressure and hypertension (Engel et al., 1979); mitotic rhythms of cancerous breast tissue (Garcia-Sainz and Halberg, 1966); and luteinizing hormone release in anorexia nervosa (Pirke et al., 1979). However, Jaquet et al. (1980) have recently studied a group of patients who had been treated for acromegaly by removal of the adenohypophysis. Three of their 8 subjects had abnormal growth hormone rhythms (*see* Chapter 7) even though their treatment (assessed independently) was presumed to be successful.

In summary, there is some evidence that, in some illnesses, changed circadian rhythms accompany the illness often to a sufficient extent for them to be of diagnostic worth. However, such a link cannot be used to deduce either the site of the internal clock or origin of the illness; once again, one must remember that correlation does not prove causality.

3.2. Can adjustment of circadian rhythms ameliorate symptoms of illness?

That the symptoms of some manic depressive patients are most marked when the phase relationship between the environment and their circadian rhythms is most abnormal, and that their internal clock runs fast (*see Fig.* 11.3) might explain the observation (Atkinson et al., 1974) that manic depressives prefer jet travel in an eastward (requires an advance of internal clock) rather than a westward (requires delay) direction. Related to this are the results that have been obtained upon a single individual who, when he was coming out of a depressed phase, underwent a simulated time-zone transition of 6 hours advance (eastward direction). The patient showed remission of symptoms for about 2 weeks; there was limited evidence that some of his circadian rhythms were temporarily phased more appropriately after this advance of his environment. However, even though such a procedure was effective after a second phase-advance, it was unsuccessful on a third or fourth occasion (Wehr et al., 1979). These findings, that the link between an abnormal phase relationship and the symptoms of depression is not immutable, argue that the link is not causal or that, at the least, other factors can intrude.

3.3. Do changes in circadian rhythms precede the illness?

Evidence that circadian rhythms become abnormal *before* the illness develops more conventional symptoms has not yet been found in humans. Such a concept has aroused particular interest in the field of oncology, the hope being that rhythmic changes would accompany the 'pre-cancerous' stage and thereby raise the possibility that effective preventive measures could be taken (Simpson, 1978; Halberg et al., 1979). Clearly, to obtain evidence to test such a hypothesis, a combination of autorhythmometry and a longitudinal study is required. So far, the only case that has been reported is that of a female who has had a small calcification of the left breast removed prophylactically (Halberg et al., 1979). Prior to partial mastectomy, temperature data from both breasts were obtained, those from the affected breast having a raised mesor and a decreased amplitude when compared with the control breast, a result found also in cancerous breasts (*see* Simpson, 1978; Haus et al., 1979). Whether or not ultradian components were present in the rhythms is not stated; nor is it known whether the changed mesor reflects a specific change or rather a non-specific response of the breast to 'trauma'. Finally, the question whether the remaining tissue will turn malignant cannot be answered and so the considerable potential of this technique is not yet certain. It is clear that, even if it turns out that changes in circadian rhythms precede the cancerous state, such an approach would require long and tedious measurement by the subject; furthermore, such an approach is not obviously suited to many other forms of cancer unless it turns out that a changed circadian rhythm in the pre-cancerous condition, specific to a particular organ or tissue, can be identified and easily monitored autorhythmometrically.

3.4. Causes of the changed rhythms in illness

The explanation of the changed circadian rhythms in different illnesses is generally quite unknown. An exception to this statement is that the inverted rhythms of water and electrolyte loss in congestive heart failure are generally attributed to an exaggerated response to changes in posture because of increased extracellular fluid volume (Conroy and Mills, 1970); thus the rhythm does not show the marked nocturnal peak if the patient maintains the same posture throughout the nychthemeron (Borst and De Vries, 1950). Related to this is the observation that, in healthy human subjects and primates, the renal responses to changes in venous return (as might be produced by changes in posture) are less marked during the night (*see* Kass et al., 1980). Consideration of the changed rhythms in illness indicates that many of the abnormal rhythms can be associated with physiological processes that are mediated by the hypothalamus. Further, the observation that the circadian rhythm of skin temperature shows ultradian components with breast cancer has been speculated to be evidence that the normal control mechanisms have broken down (Simpson, 1978). These speculations will be elaborated in Chapter 12 when the possible site of the 'master internal clock', together with the concepts of a causal nexus and a hierarchy of internal oscillators will be considered.

4. The Effect of Time of Drug Administration—Chronotherapy

So far we have considered the implications that circadian rhythm studies can have when attempts are made to diagnose or to assess treatment of illness. Now we shall discuss a whole new area in which the time of day might exert some effect, namely the efficacy of drug treatment. Reinberg et al. (1974) compared a number of rhythms in healthy controls with a group of patients with adrenal insufficiency. The patients were treated in two ways: first, the 24-hour dose of steroid was divided into three equal parts and given at meal times; secondly, between two-thirds and three-quarters of the dose were given on rising and the rest on retiring. The results from this experiment are shown in *Fig.* 11.4 in which the 'homeostatic' schedule (A) is seen to change the acrophase of certain rhythms whereas the 'chronotherapeutic' schedule (B) is associated with rhythms indistinguishable from healthy controls.

Such a result is perhaps not that surprising, but it does indicate that the effect of a drug or treatment depends in part upon its dosage schedule. This field is called chronopharmacology or chronotherapeutics; predictably, it has been studied far more in animals than in man (*see*, for example, Reinberg and Halberg, 1971; Halberg, Haus et al., 1973) but reviews of

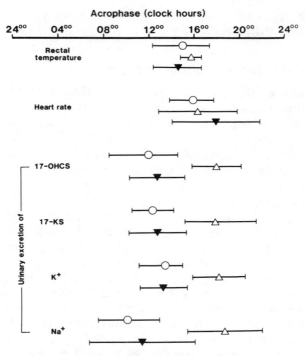

Fig. 11.4. Acrophases of several physiological variables in healthy subjects (○, *n* = 7) and patients with adrenal insufficiency receiving cortisol either at 08⁰⁰, 13⁰⁰ and 20⁰⁰ (△, *n* = 5, cortisol given in three equal doses) or at 07⁰⁰ and 23⁰⁰ (▼, *n* = 5, cortisol dose at 07⁰⁰ two or three times that at 23⁰⁰). Note the urinary excretory rhythms of patients receiving three doses of cortisol per day are phase-delayed by about 6 hours from controls. By contrast the phase of the other patients is unshifted from controls. (Data of Reinberg et al., 1974.)

the field as applied to humans have appeared (Reinberg, 1973, 1974, 1976).

The work owes much to the studies of Reinberg and he has outlined four main ways in which the time-dependence of a drug can be considered (Reinberg, 1976, 1978). These are:

1. Chronopharmacokinetics. This is a measure of the time course of the concentration of the drug in the system after administration at different times of the nychthemeron.

2. Chronesthesy. This is a measure of the susceptibility of a system to a given dose of drug at different times of the nychthemeron.

3. Chronergy. This is the effectiveness of the drug upon the system as a whole at different times of the nychthemeron; it can be considered as depending in turn upon chronesthesy and chronopharmacokinetics.

4. Side effects of drug administration.

Each of these concepts will be illustrated by examples taken mainly from the reviews just cited.

4.1. Chronopharmacokinetics

Much of the work in this field has been concerned with the fate of ingested doses of ethanol (Reinberg, 1976; Sturtevant, 1976; Minors and Waterhouse, 1980). After ingestion, ethanol levels in the plasma or urine rise and then

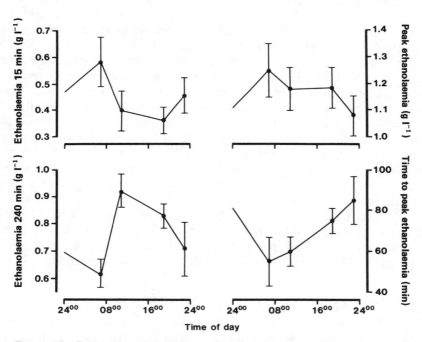

Fig. 11.5. Some chronopharmacokinetic parameters of ethanol in man. Means from 6 males ± 1 s.e. Subjects ingested ethanol 0·67 grammes per kilogramme weight orally at 07⁰⁰, 11⁰⁰, 19⁰⁰ or 23⁰⁰. Venous blood was sampled immediately before and 15, 30, 60, 90, 120, 240 and 480 minutes after ingestion. *Upper left,* blood ethanol concentration 15 minutes after ingestion; *lower left,* blood ethanol concentration 240 minutes after ingestion; *upper right,* peak blood ethanol concentration attained; *lower right,* time to reach peak blood ethanol concentration. (Data of Reinberg, 1976.)

fall, the fall being linear until low levels are reached (Sturtevant, 1976). However, as *Fig.* 11.5 indicates (from Reinberg, 1976), the rate of uptake into the blood was most rapid at about the time of rising from sleep, with peak plasma values being higher, and the 'time-to-peak' being least, at this time. The rate of removal of ethanol has been studied in more detail upon a group of 8 subjects (Minors and Waterhouse, 1980). Again maximal rates of ethanol removal, assessed by the gradient of the linear portion of the decline in urinary ethanol concentration (known to be closely related to that in plasma), were found at about the time of rising; when a cosine curve was fitted to the data, the acrophase was at about 06^{00}.

The explanation for these effects of time of ingestion is not certain, but it has been speculated that the rate of rise is some function of uptake from the gut and that the rate of elimination is related to both metabolic removal and loss in the urine.

Other explanations have been put forward to account for the chrono-pharmacokinetics of other drugs (Moore-Ede, 1973). For example, it has been shown (Reinberg et al., 1967) that the elimination of salicylate in the urine takes longer after a dose given at 07^{00} than one at 19^{00}, and Beckett and Rowland (1964) found that the elimination of amphetamines was highest in the early morning. These results can be explained if one accepts that nocturnal urine is more acid than diurnal urine (*see* Chapter 4) and that the drugs can be reabsorbed from the renal tubules only when in a non-ionized form. This hypothesis has been tested in the case of amphetamine elimination by maintaining an acid urine (when about 60 per cent of methylamphetamine, pK = 9·93, was excreted in 16 hours) or an alkaline urine (when only 2 per cent was excreted in the same period). The pK of salicylic acid is 2·97 and so decreased excretion and increased reabsorption would take place in an acid urine.

Metabolism of the drug is another means of removal and, at least in rodents, it has been shown that there are circadian variations in a number of liver enzymes involved in metabolic breakdown, these rhythms possibly being mediated by adrenal steroids (Radzialowski and Bousquet, 1968).

4.2. Chronesthesy

Intra-amniotic injections of prostaglandin $F_{2\alpha}$ have been found to induce abortion more rapidly when injected in the late afternoon (Smith et al., 1973) and the amount of halothane–nitrous oxide–oxygen mixture used to anaesthetize patients was significantly lower between midnight and 06^{00} than in other 6-hour periods (Fukami et al., 1970). In both cases, this is believed to indicate circadian variations in susceptibility of the humans to the substance.

The most elaborate series of experiments on susceptibility so far performed in humans has been that by Reinberg and his colleagues (Reinberg et al., 1965; Reinberg et al., 1969). They have measured the size of erythema and wheal in healthy, non-allergic humans in response to intradermal injections of histamine and of the histamine liberator 48/80 at different times of the day. They made similar measurements in allergic subjects in response to house dust extract and penicillin and, in asthmatic subjects, in response to histamine and house dust extract. Simultaneously, samples of urine were obtained and analysed for a number of constituents including 17-hydroxycorticosteroids (17-OHCS). In all studies a significant circadian rhythm was obtained, the acrophases for the wheal and erythematous response occurring just before midnight, a time corresponding to the minimum rate of excretion of 17-OHCS. In other experiments, the time of maximum sensitivity of the respiratory tract to histamine has been found to be between midnight and 04^{00} (De Vries et al., 1962).

It is tempting to infer from these results that there is a causal link between the steroid circadian rhythm and rhythms in sensitivity, bearing in mind the anti-inflammatory action of corticosteroids. Other evidence in favour of the importance of such a link has already been discussed when the circadian rhythm of asthma attacks was considered (Chapter 3) and further data will be considered below (s. 4.4).

4.3. Chronergy

Further investigation of the reaction of the skin to histamine has been made after ingestion of different antihistamine agents (Reinberg and Sidi, 1966; Reinberg, 1978). These experiments have shown that the decrease in wheal and erythematous response produced by the antihistamine agents was less marked, but persisted longer, when they were taken at 07^{00} rather than at 19^{00}. Thus the effect of antihistamine drugs upon the body is time-dependent; but it is not possible at the moment to determine whether the effect is due to the kinetics of uptake and loss (chronopharmacokinetics) or the susceptibility of the histamine-releasing system to the drugs (chronesthesy).

The work upon ethanol pharmacokinetics has already been described; in Reinberg's study (1976) the subjects also rated their feeling of 'drunkenness'. This too showed a circadian rhythm, peak values now occurring at about midnight, a time close the the *minimum* value for peak ethanol concentration in blood (*see Fig.* 11.5). Thus the chronergy and chronopharmacokinetics appear oppositely phased for this drug with the implication that there is a rhythm in susceptibility of the brain to ethanol (chronesthesy) of considerable amplitude, peak values occurring at about midnight.

4.4. Side effects of drug administration

Just as the therapeutic effect of a drug is influenced by its time of administration, so too can undesirable side-effects be time-dependent. The ratio of these two effects is called the 'therapeutic index' or 'therapeutic/toxic ratio'. In animals this can be defined as the ratio of the dose that kills half the animals (LD_{50}) to that which exerts the desired effect upon half of them (ED_{50}), that is the LD_{50}/ED_{50} ratio (Moore-Ede, 1973). In humans, such a concept involves a comparison of the desirable with the undesirable (non-lethal) effects.

There is some evidence from studies upon humans that the therapeutic index shows circadian rhythmicity and examples to illustrate this will now be given.

Corticosteroid therapy and the immune response

When synthetic corticosteroids are administered, one side effect is the suppression of endogenous steroid production by the pituitary—adrenal axis, a suppression that can take some time to recover. Moore-Ede (1973), Reinberg (1974) and Smolensky (1974) summarize the evidence that indicates that suppression of endogenous corticosteroid secretion is least when the exogenous steroid is given at a time corresponding to the acrophase of the circadian rhythm, more when administration is as equal doses distributed throughout the nychthemeron and greatest when administration is as a single dose and takes place in the evening. As has been described in Chapter 7, this is believed to reflect circadian changes in the sensitivity of the pituitary gland to circulating corticosteroids (*see also* Nichols et al., 1965).

As a result of these findings the time of administration of corticosteroids can be adjusted in accord with the effect required. If it is required not to suppress endogenous production (when the anti-inflammatory or immunosuppressive actions are required, for instance) then administration in the morning is most appropriate: if suppression is required (as in treatment of adrenal hyperplasia) then evening medication is preferable (Moore-Ede, 1973).

A similar point has been made by Knapp and Pownall (1980) in their review of circadian rhythms in the immune system. Thus they describe circadian rhythms in the numbers and activity of different circulating blood cells associated with the immune response in lymphocyte activity *in vitro* (Tavadia et al., 1975) and in the immune response *in vivo* (Cove-Smith et al., 1978). This last group administered tuberculin intradermally into healthy subjects at different times of the nychthemeron and measured the immune response (the area of induration) 48 hours after the injection. A circadian rhythm with an acrophase at 07[00] was found—a time, the authors noted, when conventional immunosuppressive therapy would result in lowest drug levels. The same group have speculated that such factors would account for their claim that the time of day when renal allograft rejection was most likely to take place showed an acrophase at 06[00] (Knapp et al., 1979), although a recent survey by this group suggests that the position is more complex (Knapp et al., 1980). In all cases, a relationship is believed to exist between these different aspects of the immune response and levels of endogenous corticosteroid, even though such a relationship does not seem always to have been tested.

In some cases of corticosteroid therapy the best time of administration is not simply decided. Thus the peak expiratory flow rate (PEFR) of asthmatics is raised by the administration of corticosteroid, but, when the steroid is administered at different times, the effects are not identical. This was investigated by comparing the mesor, acrophase and amplitude of the circadian changes in PEFR on a control day and on four experimental days when the steroid prednisone was administered at 01^{00}, 07^{00}, 13^{00} or 19^{00} (Reindl et al., 1969). Results indicate that, compared with control days, the greatest increase in the mesor of PEFR (desirable) as well as the greatest change in acrophase (undesirable) occurred with the 15^{00} dose. Improvements in the therapeutic index of corticosteroids can be achieved by using a combination of corticosteroids with different half-lives and potencies (Moore-Ede, 1973; Reinberg, 1974).

Hydrochlorthiazide therapy

Two large studies have been performed upon healthy human subjects (Mills et al., 1977a, b; Simpson, 1979). Both have compared the effect of time of day upon the urinary excretion of sodium, chloride and potassium. The desirable (therapeutic) effect is saliuresis but undesirable ('toxic') effects are dizziness and kaliuresis.

In the study of Mills and his colleagues, differences in urine volume and electrolyte loss due to time of administration of the drug were small and inconclusive. However, there were indications that plasma sodium and potassium were lower after 16^{00} and 20^{00} drug administration, since nearly all incidents of cramp and dizziness were associated with the 16^{00} administration and the highest values for the therapeutic/toxic ratio tended to be after morning rather than evening administration. This group concluded that it was marginally preferable to administer hydrochlorthiazide in the morning.

On the other hand, Simpson (1979) found a significantly longer saliuresis and higher therapeutic/toxic ratios after drug administration later in the day. The explanation of these differences is not known, but the two studies did differ in some respects:

1. Simpson's study considered six rather than four times of drug administration and the subjects' diet was controlled substantially more than in the study by Mills et al.

2. The saliuresis was assessed over consecutive 4-hour periods after drug administration (Simpson) or over the whole of the succeeding 24 hours (Mills et al.). The former study indicated that the effect was sometimes transient, that is, a decreased rate of excretion developed subsequent to the initial increase. Since hydrochlorthiazide therapy is generally chronic, assessment of its efficacy over as long a period as possible seems preferable.

3. The therapeutic/toxic ratio was calculated for the 72 hours after drug administration in the study of Mills et al., for 24 hours afterwards in that of Simpson.

These comments highlight the problems encountered in comparing such studies; in addition, in neither study were the data obtained from hypertensive subjects or after prolonged administration of the drug. Clearly, considerably more work is needed with this type of drug before the best time of administration can be confidently recommended.

Cancer therapy

Interest in this field is strong (*see*, for examples, Halberg, Haus et al., 1973; Reinberg, 1974; Wilson, 1974; Haus et al., 1979) but, not surprisingly, most work has again been performed upon animals. The concept of a therapeutic/toxic ratio in cancer therapy is particularly important where the drugs and other treatments used are toxic to both the host and the malignant cells. As a result, slight differences in susceptibility of either at a particular time of the nychthemeron might prove of critical worth to the efficacy of treatment (Simpson, 1978; Halberg et al., 1979). In mice, improved prognosis has been achieved in cases of leukaemia when the timing of administration of the anti-metabolite arabinose-C has been in accord with 'chronotherapeutic principles' (Halberg, Haus et al., 1973). In humans, very few data have been reported so far. Gupta and Deka (1975) and Halberg et al. (1977) have described a slight increase in the regression of tumours of the head and neck when they were treated by irradiation at the time of peak oral temperature rather than at other times, and Focan (1976, 1979) has compared the response of tumours to the administration of anti-neoplastic drugs either nocturnally or diurnally; a difference between the two times was found.

5. Implications for the Availability of Medical Services

Many hospital procedures, whether of diagnosis or treatment, are understandably geared to a routine that suits the personnel involved and therefore is diurnal. As earlier sections of this chapter have indicated, such a routine is not always ideal as far as chronological considerations are concerned, but the disadvantages are comparatively minor. However, in two important components of hospital work—births and deaths—circadian rhythms that peak during the night have been described, the implication being that the night staff (with all the disadvantages attendant upon night work, *see* Chapter 10) have a greater load to bear. In contrast to many other studies of circadian rhythms, one factor that soon emerges is the very large number of subjects that is involved. Thus, in one study by Kaiser and Halberg (1962), over 600 000 births were considered, and in a review by Smolensky et al. (1972) over 400 000 deaths and 2 000 000 births!

5.1. Onset of labour and births

In *Fig.* 11.6 are shown the circadian rhythms in onset of spontaneous labour and in natural (that is, non-induced) birth. For onset of spontaneous

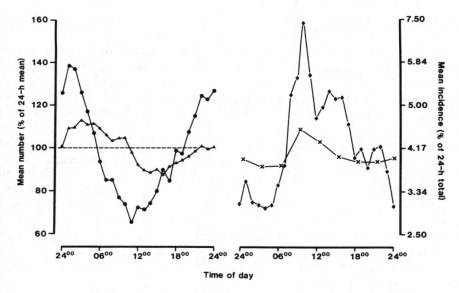

Fig. 11.6. Circadian changes in human labour and birth. Mean rates in all cases are shown. (●) hourly frequency of initiation of spontaneous labour based on 207 918 labours; (▲) hourly frequency of parturition based on 2 082 453 natural births; (◆) hourly frequency of parturition based on 30 493 induced births; (x) average incidence of 12 081 stillbirths summarized for 3-hour spans. (*From* Smolensky et al., 1972, Fig. 134.)

labour (spontaneous initiation of painful contractions and/or rupture of fetal membranes) there is a marked circadian variation. The acrophase is at about 01⁰⁰ and the frequency at the peak is about twice that at the trough, a finding in accord with the results of others (for example, Kaiser and Halberg, 1962; Breart and Rumeau-Rouquette, 1979). The circadian rhythm in births has a smaller amplitude and an acrophase at around 04⁰⁰, about 3 hours later than that for the onset of labour; the acrophases are very similar whether the first or later births are considered.

These results raise two questions. First, why should the acrophase of the onset of labour precede that of birth by about 3 hours when the average duration of labour is longer? A number of contributory factors have been advanced (Kaiser and Halberg, 1962; Smolensky et al., 1972): for example, that there is a wide variation in the duration of labour and that the duration varies according to its time of onset, labour starting during the daytime lasting longer than that starting during the night; in addition, the moment of onset of labour is not always clearly defined. As a result of these factors, the relationship between the time of onset of labour and that of birth is not clear-cut. Furthermore, the variability associated with the duration of labour will result in the amplitude of the circadian rhythm for births being less than that for labour onset (*see Fig.* 11.6). This last point has been discussed previously when the decreased amplitude of circadian rhythms in the aged was considered (*see* Chapter 8).

The second question raised is whether the observed rhythms are endogenous or exogenous. The influence of maternal posture upon fetal movement has already been mentioned (*see* Minors and Waterhouse, 1979, and Chapter 8); this could be an important factor, though data on time of onset of labour in mothers who must remain recumbent during the later stages of pregnancy or whose pattern of activity is abnormal do not seem to be available. Other influences might be any of the hormones involved in maintaining the fetus and initiating parturition; the possibility of an influence of darkness too has been raised (*see* discussion in Kaiser and Halberg, 1962). Data are not available that would enable the relative importance of these factors to be assessed and, as with considerations of zeitgeber or the causal nexus, it seems unlikely that only a single factor would be involved. Even though experiments to investigate these factors could be devised (making use of approaches described in Chapters 1 and 12), ethical considerations would preclude rapid and complete acquisition of data in this field.

Fig. 11.6 also shows the circadian rhythms of still-births and induced births (Kaiser and Halberg, 1962; Smolensky et al., 1972). The acrophases are similar to each other but almost in antiphase to that for natural birth (compare with *Fig.* 11.6, *left*). A likely explanation of the diurnal peak for induced births is that it reflects the obstetric policy of inducing labour during the daytime (Breart and Rumeau-Roquette, 1979), one advantage of which would be that medical care would be more readily available during daytime hours. However, the greater number of still-births that is observed diurnally (*Fig.* 11.6, *right*) might indicate that there are problems associated with delivery at 'unusual' times. Some evidence for this is considered by Smolensky et al. (1972) and Breart and Rumeau-Rouquette (1979), who found decreased Apgar scores and an increased frequency of neonatal distress in babies born between 14^{00} and 16^{00}. The daytime peak might result in part also from deaths following long and difficult labour (*see above*). Even if a full explanation of the diurnal peak in still-births is still awaited, one possible implication of this finding is that the obstetrician contemplating induction of birth might have to offset the increased thoroughness and efficiency of his team looking after the mother and neonate against the disadvantage that seems to exist when birth does not take place at the 'usual' time.

5.2. Mortality

In their extensive review of the field, Smolensky et al. (1972) provide a summary of results from 49 studies involving over 400 000 deaths. When the frequency of time of death throughout the nychthemeron is considered, a prominent peak at about 06^{00} with a general decrease to a minimum at about midnight is found; a lesser peak at about 16^{00} is also present. Even though such a distribution is not an exact cosine curve (the maximum and minimum differ by only 6 hours) a cosine curve fits the data significantly better than a straight line, the acrophase falling at about 07^{00}. Further analysis by these authors showed that for both adults and infants there were no significant differences in acrophase when the data were sub-divided according to sex or to cause of death. Acrophases for all the individual studies are shown in the polar plot (*Fig.* 11.7). When represented in this way, 78 per cent of the acrophases fall between midnight and midday and two-thirds between 04^{00} and 11^{00}. The cause of the secondary peak at 16^{00} is not known; it does not seem to arise from a marked incidence of mortality from any particular cause at this time even though some studies do show peak times of death during the daytime, a recent study being that of Eltringham and Dobson (1979).

The cause of the increased nocturnal mortality is generally not certain, though in certain cases, for example respiratory arrests in asthmatics, there might be a relationship between the most likely time of death (night time) and the circadian rhythm of airway resistance (Chapter 2 and Hetzel et al., 1977). However, between the hours of 04^{00} and 11^{00} (where the figure indicates that two-thirds of acrophases fall) there are marked increases of blood pressure and autonomic nervous system activity (Chapters 3 and 6) as well as rapid changes in plasma cortisol concentration (Chapter 7) and in the neural activity associated with the change from sleep to wakefulness (Chapter 5). Furthermore, there are alterations in the external environment (light and noise) together with, in a hospital environ-ment, the complications that the night staff are either less readily available or tired at the end of their work shift and that the day staff are starting the new daily routine; in such cases, it is possible that staff attention to patients will be lessened (*see* Tyson et al., 1979). All or some of these exogenous and endogenous factors might contribute to the observed rhythmicity.

6. Conclusions

As this chapter has indicated, there is a limited application of knowledge of human circadian rhythms to clinical practice. This application is most obvious at the moment when the availability of certain medical services (for example, maternity staff) is considered. In diagnosis also, both when deviations from the circadian rhythms of healthy controls are considered and when altered circadian rhythms are involved (especially in cardio-vascular, renal and endocrine disorders), a knowledge of circadian rhythms can play a role. As yet, the possibility that abnormalities of circadian rhythms will act as predictors of illness has not been realized in the human. Similarly, the important concept that different forms of treatment might be of most benefit and show least adverse side-effects at certain

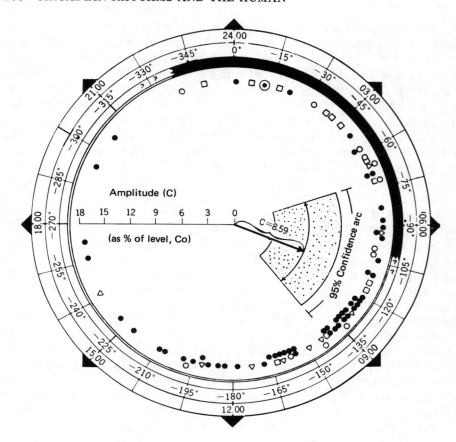

Mortality category	No. of cases	No. of (C, φ)	Symbol for acrophase
Pediatric	21,673	12	○
Postsurgical	500	1	⊙
Cardiovascular	7,644	9	△
Pulmonary	9,357	12	□
Cause unspecified	393,718	54	•
Cosinor of above	432,892	88	

Fig. 11.7. Cosinor summary of the susceptibility to 'death' in humans. Different symbols represent deaths from different causes as shown in the key. Each symbol on the polar plot represents the vector (acrophase and amplitude) for one time series. Note that the majority of the acrophases occur between 04⁰⁰ and 11⁰⁰. Inner shaded area shows 95% confidence arc derived by group cosinor analysis of all time series. (*From* Smolensky et al., 1972; Fig. 138.)

times of the nychthemeron is still at a rudimentary stage. No doubt much work will be performed in the future in many of these fields, particularly those of chronotherapeutics and oncology; some speculations concerning possible applications of circadian rhythms to hospital practice in the future have been made by Mills (1974).

References

Alberti K. G. M. M., Dornhorst A. and Rowe A. S. (1975) Metabolic rhythms in normal and diabetic man. *Israel J. Med. Sci.* **11**, 571–80.

Allen-Williams G. M. (1945) Pulse-rate and blood pressure in infancy and early childhood. *Arch. Dis. Child.* **20**, 125–8.

Aslanian N. L., Assatrian D. G., Bagdassarian R. A. et al. (1978) Circadian rhythms of electrolyte excretion in hypertensive patients and healthy subjects. *Chronobiologia* **5**, 251–62.

Asplin C. M., Hartog M., Goldie D. J. et al. (1979) Diurnal profiles of serum insulin, C-peptide and blood intermediary metabolites in insulin treated diabetics, their relationship to the control of diabetes and the role of endogenous insulin secretion. *Q. J. Med.* **190**, 343–60.

Atkinson M., Kripke D. F. and Wolf S. R. (1974) Autorhythmometry in manic-depressives. *Chronobiologia* **2**, 323–35.

Beckett A. H. and Rowland M. (1964) Rhythmic urinary excretion of amphetamine in man. *Nature* **204**, 1203–4.

Borst J. G. G. and De Vries L. A. (1950) Three types of a 'natural' diuresis. *Lancet* **2**, 1–6.

Breart G. and Rumeau-Roquette C. (1979) Spontaneous and induced rhythms in the onset and course of labour and of birth. Consequences for the artificial onset of labour. *J. Interdiscipl. Cycle Res.* **10**, 195–205.

Connolly C. K. (1979) Diurnal rhythms in airway obstruction. *Br. J. Dis. Chest* **73**, 357–66.

Conroy R. (1969) Circadian rhythms. *J. R. Coll. Surg. Irl.* **5**, 43–54.

Conroy R. W. T. L. and Mills J. N. (1970) *Human Circadian Rhythms*. London, Churchill.

Cove-Smith J. R., Kabler R., Pownall R. et al. (1978) Circadian variation in an immune response in man. *Br. Med. J.* **2**, 253–4.

Delea C. S. (1979) Chronobiology of blood pressure. *Nephron* **23**, 91–7.

DeVries G., Goei J. T., Booy-Noord H. et al. (1962) Changes during 24 hours in the lung function and histamine hyper-reactivity of the bronchial tree in asthmatic and bronchitic patients. *Int. Arch. Allergy* **20**, 93–101.

Eltringham R. J. and Dobson M. G. (1979) Cardiorespiratory arrests—a diurnal variation? *Br. J. Anaesth.* **51**, 72.

Engel R., Halbert F., Nelson W. et al. (1979) Rhythmometry gauges treatment of mesor-hypertension in the seventh and eighth decades of life. *Int. J. Chronobiol.* **6**, 163–78.

Ernest I. (1966) Steroid excretion and plasma cortisol in 41 cases of Cushing's syndrome. *Acta Endocrinol.* **51**, 511–25.

Faber H. K. and James C. A. (1921) The range and distribution of blood pressures in normal children. *Am. J. Dis. Child.* **22**, 7–28.

Focan C. (1976) Circadian rhythm and chemotherapy for cancer. *Lancet* **2**, 638–9.

Focan C. (1979) Sequential chemotherapy and circadian rhythm in human solid tumours. A randomised trial. *Cancer Chemother. Pharmacol.* **3**, 197–202.

Friedman R. C., Bigger J. T. and Kornfield D. S. (1971) The intern and sleep loss. *N. Engl. J. Med.* **285**, 201–3.

Fukami N., Kotani T., Shimoji K. et al. (1970) Circadian rhythm and anesthesia. *Jap. J. Anesthesiol.* **19**, 1235–8.

Garcia-Sainz M. and Halberg F. (1966) Mitotic rhythms in human cancer, re-evaluated by electronic computer programs—evidence for chronopathology. *J. Natl Cancer Inst.* **37**, 279–92.

Gautherie M. and Gros C. (1977) Circadian rhythm alteration of skin temperature in breast cancer. *Chronobiologia* **4**, 1–17.

Gupta B. D. and Deka A. C. (1975) Application of chronobiology to radiotherapy of tumour of oral cavity. *Chronobiologia* Suppl. 1, 25.

Halberg F. (1973) Laboratory techniques and rhythmometry. In: Mills J. N. (ed.) *Biological Aspects of Circadian Rhythms.* London, Plenum, pp. 1–26.

Halberg F. and Ahlgren A. (1979) Chronobiology—1979. *Int. J. Chronobiol.* **6**, 145–62.

Halberg F., Gupta B. D., Haus E. et al. (1977) Steps towards a cancer chronopoly-therapy. In: *Proc. XIVth Int. Congr. Ther. France.* L'Expansion Scientifique Francaise, pp. 151–96.

Halberg E., Halberg F., Cornelissen G. et al. (1979) Towards a chronopsy: part II. A thermopsy revealing a symmetrical circadian variation in surface temperature of human female breasts and related studies. *Chronobiologia* **6**, 231–57.

Halberg F., Halberg J., Halberg F. et al. (1973) Reading, 'riting, 'rithmetic—and rhythms: a new 'relevant' 'R' in the educative process. *Perspectives Biol. Med.* **17**, 128–41.

Halberg F., Haus E., Cardoso S. S. et al. (1973) Towards a chronotherapy of neoplasia: tolerance of treatment depends upon host rhythms. *Experientia* **29**, 909–34.

Halberg F., Johnson E. A., Nelson W. et al. (1972) Autorhythmometry-procedures for physiologic self-measurements and their analysis. *The Physiology Teacher* **1**, 1–11.

Halberg F. and Nelson W. (1978) Chronobiologic optimization of ageing. In: Samis H. V. and Capobianco S. (ed.) *Aging and Biological Rhythms.* London, Plenum, pp. 5–56.

Haus E., Halberg F., Scheving L. E. et al. (1979) International cancer research workshop on 'Chronotherapy of Cancer—A Critical Evaluation'. *Int. J. Chronobiol.* **6**, 67–107.

Hawking F. (1973) Circadian rhythms of parasites. In: Mills J. N. (ed.) *Biological Aspects of Human Circadian Rhythms.* London, Plenum, pp. 153–88.

Hetzel M. R., Clark T. J. H. and Branthwaite M. A. (1977) Asthma: analysis of sudden deaths and ventilatory arrests in hospital. *Br. Med. J.* **1**, 808–11.

Jaquet P., Guibout M., Jaquet C. et al. (1980) Circadian regulation of growth hormone secretion after treatment in acromegaly. *J. Clin. Endocrinol. Metab.* **50**, 322–8.

Kaiser I. H. and Halberg F. (1962) Circadian periodic aspects of birth. *Ann. NY Acad. Sci.* **98**, 1056–68.

Kass D. A., Sulzman F. M., Fuller C. A. et al. (1980) Renal responses to central vascular expansion are suppressed at night in conscious primates. *Am. J. Physiol.* **239**, F343–51.

Knapp M. S., Byrom N. P., Pownall R. et al. (1980) Time of day of taking immuno-suppressive agents after renal transplantation: a possible influence on graft survival. *Br. Med. J.* **281**, 1382–5.

Knapp M. S., Cove-Smith J. R., Dugdale R. et al. (1979) Possible effect of time on renal allograft rejection. *Br. Med. J.* **1**, 75–7.

Knapp M. S. and Pownall R. (1980) Chronobiology, pharmacology and the immune system. *Int. J. Immunopharmacol.* **2**, 91–3.

Kripke D. F., Mullaney D. J., Atkinson M. et al. (1978) Circadian rhythm disorders in manic-depressives. *Biol. Psychiat.* **13**, 335–51.

Kripke D. F., Mullaney D. J., Atkinson M. L. et al. (1979) Circadian rhythm phases in affective illnesses. *Chronobiologia* **6**, 365–75.

Mills J. N. (1974) The usefulness of chronobiological studies. *Chronobiologia* 1, 145—50.

Mills J. N., Waterhouse J. M., Minors D. S. et al. (1977a) A chronopharmacological study on hydrochlorthiazide. *Int. J. Chronobiol.* 4, 267—94.

Mills J. N., Waterhouse J. M., Minors D. S. et al. (1977b) A chronotherapeutic trial of hydrochlorthiazide. *Proceedings of XII International Conference of the International Society for Chronobiology*, Washington, 1975. Milan, Il Ponte, pp. 369—74.

Minors D. S. and Waterhouse J. M. (1979) The effect of maternal posture, meals and time of day on fetal movements. *Br. J. Obstet. Gynaecol.* 86, 717—23.

Minors D. S. and Waterhouse J. M. (1980) Aspects of Chronopharmacokinetics and chronergy of ethanol in healthy man. *Chronobiologia* 7, 465—80.

Moore-Ede M. C. (1973) Circadian rhythms of drug effectiveness and toxicity. *Clin. Pharmacol. Ther.* 14, 925—35.

Nichols T., Nugent C. A. and Tyler F. G. (1965) Diurnal variations in suppression of adrenal function by glucocorticoids. *J. Clin. Endocrinol.* 25, 343—49.

Nikitopoulou G. and Crammer J. L. (1976) Change in diurnal temperature rhythm in manic-depressive illness. *Br. Med. J.* 1, 1311—14.

Niskanen P., Huttunen M., Tamminen T. et al. (1976) The daily rhythm of plasma tryptophan and tyrosine in depression. *Br. J. Psychiat.* 128, 67—73.

Pearson J. F. and Weaver J. B. (1976) Fetal activity and fetal well being: an evaluation. *Br. Med. J.* 1, 1305—7.

Pickup A. J., Braithwaite R., Dinsdale J. et al. (1979) Diurnal variation in systolic time interval. *Lancet* 1, 616—17.

Pirke K. M., Fichter M. M., Lund R. et al. (1979) Twenty-four hour sleep—wake pattern of plasma LH in patients with anorexia nervosa. *Acta Endocrinol.* 92, 193—204.

Quay W. B. and Guth S. (1975) Chronobiology in mental retardation research: progress and prospects. *Chronobiologia* 2, 243—64.

Radzialowski F. M. and Bousquet W. F. (1968) Daily rhythmic variation in hepatic drug metabolism in the rat and mouse. *J. Pharmacol. Exp. Ther.* 163, 229—38.

Raftery E. B. (1975) Home blood pressure recording. *Lancet* 1, 259—60.

Reinberg A. (1973) Chronopharmacology. In: Mills J. N. (ed.) *Biological Aspects of Circadian Rhythms*. London, Plenum Press, pp. 121—52.

Reinberg A. (1974) Chronopharmacology in man. In: Aschoff J., Ceresa F. and Halberg F. (ed.) *Chronobiological Aspects of Endocrinology*. Schattauer Verlag, pp. 305—37.

Reinberg A. (1976) Advances in human chronopharmacology. *Chronobiologia* 3, 151—66.

Reinberg A. (1978) Clinical chronopharmacology, an experimental basis for chronotherapy. *Drug. Res.* 28, 1861—7.

Reinberg A., Ghata J., Halberg F. et al. (1974) Treatment schedules modify circadian timing in human adrenocortical insufficiency. In: Scheving L. E., Halberg F. and Pauly J. E. (ed.) *Chronobiology*. Tokyo, Igaku Shoin, pp. 168—73.

Reinberg A. and Halberg F. (1971) Circadian chronopharmacology. *Ann. Rev. Pharmacol.* 11, 455—92.

Reinberg A. and Sidi E. (1966) Circadian changes in the inhibitory effects of an antihistaminic drug in man. *J. Invest. Derm.* 46, 415—19.

Reinberg A., Sidi E. and Ghata J. (1965) Circadian reactivity rhythms of human skin to histamine or allergen and the adrenal cycle. *J. Allergy* 36, 273—83.

Reinberg A., Zagula-Mally Z. W., Ghata J. et al. (1967) Circadian rhythm in duration of salicylate excretion referred to phase of excretory rhythms and routine. *Proc. Soc. Exp. Biol. Med.* 124, 826—32.

Reinberg A., Zagula-Mally Z., Ghata J. et al. (1969) Circardian reactivity rhythm of human skin to house dust, penicillin, and histamine. *J. Allergy* 44, 292—306.

Reindl K., Falliers C., Halberg F. et al. (1969) Circadian acrophase in peak expiratory flow rate and urinary electrolyte excretion of asthmatic children; phase shifting of rhythms by prednisone given in different circadian system phases. *Rass. Neur. Veg.* **23**, 5–26.

Sadovsky E. and Yaffe Y. (1973) Daily fetal movement recording and fetal prognosis. *Obstet. Gynecol.* **41**, 845–50.

Scheving L. A., Sheving L. E. and Halberg F. (1974) Establishing reference standards by autorhythmometry in high school for subsequent evaluation of health status In: Scheving L. E., Halberg F. and Pauly J. E. (ed.) *Chronobiology.* Tokyo, Igaku Shoin, pp. 386–93.

Schteingart D. E. and McKenzie A. K. (1980) Twelve hour cycles of adrenocorticotropin and cortisol secretion in Cushing's disease. *J. Clin. Endocrinol. Metab.* **51**, 1195–8.

Simpson H. W. (1978) An outline of mammary chronophysiology and pathology. A new tumour 'marker'. In: Griffiths K., Neville A. M. and Pierrepoint C. G. (ed.) *Tumour Markers.* Sixth Tenovus Workshop. Cardiff, Alpha Omega Publishing, pp. 317–339.

Simpson H. W. (1979) Hydrochlorthiazide diuresis in healthy man: review of the circadian mediation. *Nephron* **23**, 98–105.

Smith I. D., Shearman R. P. and Korda A. R. (1973) Chronoperiodicity in the response to intra-amniotic injection of prostaglandin $F_{2\alpha}$ in the human. *Nature* **241**, 279–80.

Smolensky M. S. (1974) Rationale for circadian-system phased glucocorticoid management. In: Scheving L. E., Halberg F. and Pauly J. E. (ed.) *Chronobiology.* Tokyo, Igaku Shoin, pp. 197–201.

Smolensky M., Halberg F. and Sargent F. (1972) Chronobiology of the life sequence. In: Ito S., Ogata K. and Yoshimura H. (ed.) *Advances in Climatic Physiology.* Tokyo, Igaku Shoin, pp. 281–318.

Sollberger A. (1974) Chronobiology and rhythms in psychiatry and psychology. In: Scheving L. E., Halberg F. and Pauly J. E. (ed.) *Chronobiology.* Tokyo, Igaku Shoin, pp. 515–16.

Spellacy W. N., Cruz A. C., Gelman S. R. et al. (1977) Fetal movements and placental lactogen levels for fetal-placental evaluation. A preliminary report. *Obstet. Gynecol.* **49**, 113–15.

Sturtevant F. M. (1976) Chronopharmacokinetics of ethanol. I. Review of the literature and theoretical considerations. *Chronobiologia* **3**, 237–62.

Tavadia H. B., Fleming K. A., Hume P. D. et al. (1975) Circadian rhythmicity of human plasma cortisol and PHA-induced lymphocyte transformation. *Clin. Exp. Immunol.* **22**, 190–3.

Tyson J., Schultz K., Sinclair J. C. et al. (1979) Diurnal variation in the quality and outcome of newborn intensive care. *J. Pediatr.* **95**, 277–80.

Weeke A. and Weeke J. (1978) Disturbed circadian variation of serum thyrotropin in patients with endogenous depression. *Acta Psychiat. Scand.* **57**, 281–9.

Wehr T. A., Wirz-Justice A., Goodwin F. K. et al. (1979) Phase advance of the circardian sleep–wake cycle as an antidepressant. *Science* **206**, 710–13.

Weitzman E. D. (1976) Biologic rhythms and hormone secretion patterns. *Hospital Practice* **11**, 79–86.

Weitzman E. D. (1979) Sleep stage organization: neuroendocrine relations. In: *Sleep, Wakefulness and Circadian Rhythm.* AGARD Lecture Series No. 105, AGARD, ch. 3.

Weitzman E. D. and Pollack C. P. (1979) Disorders of the circadian sleep–wake cycle. *Med. Times.* **107**, 83–94.

Wesson L. G. (1979) Diurnal circadian rhythms of renal function and electrolyte excretion in heart failure. *Int. J. Chronobiol.* **6**, 109–17.

Wesson L. G. and Simenhoff M. L. (1979) Circadian rhythms of renal function and electrolyte excretion in surgical hypopituitarism with observation of postural influence. *Int. J. Chronobiol.* **6**, 31–7.

Wilson W. L. (1974) Iatrogenic toxicity from cancer chemotherapy—can it be reduced by timing treatment according to rhythms? *Int. J. Chronobiol.* **2**, 171–3.

chapter 12 *Circadian Rhythm Mechanisms*

1. A Hierarchy of Rhythms and the Causal Nexus between Them

As Chapters 2–8 of this book have illustrated, circadian rhythmicity in man seems ubiquitous; indeed a variable that did not show circadian rhythmicity under nychthemeral circumstances would be rare! In many cases the rhythmicity can be attributed mainly to exogenous causes (*see*, for example, Chapter 3) but often at least some endogenous influence is present. Accepting this, the problem is raised of whether the different endogenous components are connected in some way or rather they are independent of each other. These concepts of a hierarchy of clocks and the existence between them of a causal nexus have been mentioned in Chapter 1 and considered in detail in humans elsewhere (Mills, 1966; 1973).

Fig. 12.1 outlines some of the main links that have been surmised at different parts of the text. These links have been inferred generally from the knowledge that one variable affects another (for example, the autonomic nervous system influences the heart rate) together with a consideration of the appropriateness of the timing of the two rhythms. However, as has often been stated, correlation does not prove causality and more elaborate criteria have been laid down (Mills, 1966; 1973). Thus, a variable, A, is the sole and sufficient cause of another, B, if:

1. A can induce B at any time of the nychthemeron.
2. B is no longer rhythmic if A is held constant.
3. Changes in phase or period of A must immediately be followed by equal changes in B.
4. The physiological circadian changes in A must be sufficient to induce the circadian changes observed in B.

In practice not enough data are available to apply all of these criteria to any suspected link. One case where the evidence is rather more complete is that of the relationship between plasma cortisol and numbers of circulating blood eosinophils. Briefly, the cell number is depressed by cortisol and the main pieces of evidence supporting the link as far as circadian rhythms are concerned are (*see* reviews by Mills, 1966; Conroy and Mills, 1970):

   i. eosinophil numbers decrease from a maximum at about 04^{00} to a minimum at about 15^{00}, an observation which indicates that the cortisol rhythm is 'appropriately' phased;

ii. the rhythm is absent after adrenalectomy and in Addison's disease (*see* criterion 2); and

iii. the rhythm can be reproduced by injection of a physiological dose of cortisone at 06⁰⁰ but not by injection of divided doses spread equally throughout the nychthemeron (*see* criterion 4).

However, the effect of steroid injections at physiologically 'unusual' times—*see* criterion 1—together with the effect of time-zone transitions, shift work and non-24-hour schedules upon both rhythms—*see* criterion 3—do not seem to have been investigated at all fully (but *see* Sharp, 1960a).

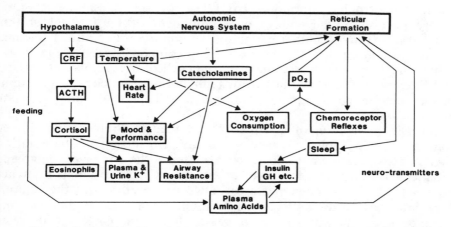

Fig. 12.1. Possible hierachial nexus between different circadian rhythms. (*See* text for further details.)

Of course, even though it is time-consuming to prove that all the criteria hold for a particular link, it is much easier to disprove a 'sole and sufficient' connection by a single set of data. When the causal nexus possibly responsible for renal excretory rhythms was considered (Chapter 4), this procedure was used on a number of occasions. By way of an example, there is considerable evidence that plasma cortisol is responsible for the rhythm of potassium excretion; however, the observation that cortisol and urinary potassium rhythms could be dissociated in some night workers indicated that other factors can be involved, at least in these circumstances.

The lack of success in establishing a 'sole and sufficient' link between circadian rhythms no doubt results from the naive assumption that the link between any two rhythms should be single, uni-directional and immutable; consideration of other physiological systems indicates that such a position is most unlikely. Similarly, when factors influencing performance in psychometric tests were considered (Chapter 6), the rhythms of temperature and catecholamine levels at least seemed associated in some way; and when factors influencing airway resistance were considered (Chapter 3), both the autonomic nervous system and plasma corticosteriods were suggested to play some role. In both cases, this extra complexity has been indicated in *Fig.* 12.1.

In addition, the search for a 'sole and sufficient cause' does not easily accommodate the role played by the rhythmic influences of the external world—the exogenous component—which has been the subject of much of the book. In other words, it would seem more useful to consider that the exogenous and endogenous components of a circadian rhythm are both compound. Thus, in the sequence, corticotrophin-releasing factor, adrenocorticotrophic hormone (ACTH), cortisol, plasma potassium, urinary potassium, the endogenous component of the cortisol rhythm would result from a number of factors, an important one of which would be the rhythm of ACTH; the exogenous component also could derive from more than one source. In turn, the cortisol rhythm would act as a contributory (endogenous) factor for a number of other circadian rhythms such as plasma potasssium and phosphate (Chapter 4) and eosinophil numbers (*above*).

Another implication of a sequence of linked rhythms was considered from a different point of view when the 'circadian amplitude ratios' (CAR) of temperature and substances excreted in the urine were compared (Chapter 4). Since, at each stage of a sequence, an exogenous influence can exist, as the number of stages between a rhythm and the internal clock increases, so the endogenous component will tend to diminish and the exogenous component will increase, that is the CAR will decrease in value. For example, the CAR would be greater for cortisol than for plasma potassium in the sequence already described and, by reference to *Fig.* 12.1, the CARs for temperature and ACTH would be greater than those for oxygen consumption and cortisol respectively.

One inference from such an approach is that the rhythms of temperature and urinary 17-hydroxycorticosteroids are 'closer' to the internal clock. Such an inference implies that the hypothalamus would be a promising area of the brain to consider when the site of the internal clock is being sought (*see* s. 3.2). A similar conclusion relating to the importance of the hypothalamus was drawn in Chapters 3 and 6, but it must be made clear that such an approach does not indicate whether the hypothalamus is in turn driven by another oscillator (reticular formation?, *see* Chapters 3 and 6), or whether different parts of the hypothalamus can have different clock-like functions.

Futhermore, some idea of the potential complexity of the system can be appreciated when it is realized that the 'control' areas possibly involved—the hypothalamus, autonomic nervous system and reticular formation—need not be independent of each other and that there is every reason to believe that other circadian rhythms (for examples, those of temperature, PO_2, plasma amino acid concentrations, *see below*) can also exert an effect upon these areas (*see Fig.* 12.1). In summary, it seems that the concept of a causal sequence or chain is better replaced by one of a causal network or nexus. Also, not all components of this nexus possess endogenous components of equal size (those at the 'top' of the hierarchy possess the largest endogenous components) and interactions can take place between components at all levels. It is this balance between a hierarchy and an interacting nexus that is intended by the use of the term 'hierarchical nexus', one that has already been used in Chapter 4. (*See also* Moore-Ede et al., 1976; Moore-Ede and Sulzman, 1977.)

2. How Many Clocks?

When compared with the nychthemeral state, there is a change in the relative phasing of different circadian rhythms in free-running conditions (for examples, *see* Chapters 2, 5 and 7) and transiently after time-zone transitions (Chapter 9) or shift work (Chapter 10). This process of change in the relative phases might indicate the presence of a number of internal oscillators each of which was responding to the changed conditions independently. However, the preferred explanation is that such a phase change results from the process of entrainment of a rhythm by a zeitgeber and depends upon the relative strengths of the oscillator and zeitgeber (*see below*), further, the different rates of adjustment to a time-zone shift or shift work are generally taken to reflect for any rhythm the different balance between exogenous and endogenous components that is believed to exist, rather than separate adjustment of different oscillators. (This is the view that has been adopted by the present authors.) In other words, the phenomenon of internal dissociation, whilst a predictable consequence of having more than one oscillator, is not proof of the same.

2.1. Can the rhythm of sleep—wakefulness be independent of body temperature?

In humans the case for considering there to be more than one internal oscillator has often rested on whether the sleep—wakefulness rhythm can be satisfactorily described as an oscillatory process and, if so, whether this oscillator can act independently of the oscillator that dominates the rhythm of deep body temperature.

Some of the evidence that the sleep—wakefulness rhythm behaves like an oscillator has already been described. Thus, it has been claimed that (*a*) there is an inverse relationship between length of sleep and length of prior wakefulness (this has been commented upon in Chapter 5) and that (*b*) as the zeitgeber period changes, the phase of the sleep—wakefulness cycle alters (this is illustrated in *Fig.* 4.6 and has been discussed further in Chapter 5; it will be elaborated upon when the process of entrainment is considered).

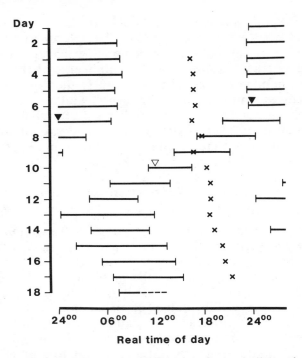

Fig. 12.2. The sleep–wakefulness rhythm (black bars indicate time in bed) and rhythm of rectal temperature (x indicates daily acrophase) in a single female subject studied in an isolation unit. Initially the subject lived on 24-hour time but at the point marked ▼ the clock in the unit was adjusted to run fast so that when it indicated the passage of 24 hours, 21 hours had actually elapsed. At the point marked ▽ the clock was stopped and the subject allowed to free-run. For clarity, the time scale represents over 24 hours. (Data of Minors and Waterhouse, 1981.)

Further evidence is given in *Fig.* 12.2 which shows the sleep–wakefulness and temperature rhythms of a single subject in an isolation unit who was instructed to live a regular regimen of sleep, waking and meal times in accord with a clock (Minors and Waterhouse, 1981). This clock originally ran at the correct rate (control period) but later—unknown to the subject—was altered to a 21-hour day for 4 apparent days. At the end of this time, when real and apparent midnight were 12 hours apart, the clock was stopped and the subject instructed to 'continue estimating time as well as you can'. As the *Fig.* 12.2 shows, the sleep–wakefulness rhythm continued to advance with a period of about 21 hours for some days after the clock was stopped, whereas the temperature oscillator continued quite independently with a period of 25 hours; this result is easier to explain if the hypothesis that the sleep–wakefulness rhythm was determined by another oscillator is accepted.

The major evidence that the sleep—wakefulness oscillator can be independent of that controlling the temperature rhythm comes from the process of spontaneous internal desynchronization (*see Fig.* 2.3), but other data that can be interpreted similarly are:

1. As *Fig.* 12.2 illustrates, the sleep—wakefulness cycle and the temperature rhythm must be independent, at least during days 7 to 12.

2. As *Fig.* 4.6 indicates, even though the rhythms of sleep—wakefulness and temperature both show phase changes indicative of the presence of an endogenous component, that for temperature is far greater. One possible explanation is that different oscillators control the two variables.

3. The phenomenon of 'partial entrainment' has been described in which some, but not all, rhythms are entrained by a particular zeitgeber. Thus Aschoff (1978) and Wever (1979a) describe subjects whose temperature rhythms were entrained to a light—dark zeitgeber but whose sleep—wakefulness rhythm free-ran independently. Note that the opposite result—an entrained sleep—wakefulness and a free-running temperature rhythm—would have been far less powerful evidence, since the sleep—wakefulness pattern might have been forced upon the subjects by the alternation of light and dark in the external environment (*see* Chapter 5).

In summary, therefore, there is evidence to indicate that there is more than one oscillator in humans, a view that has been enthusiastically proposed by Aschoff and Wever (Aschoff, 1969, 1976, 1978, 1979; Wever, 1972, 1975a; Aschoff and Wever, 1976; Wever, 1979a), and more than one oscillator in squirrel monkeys, as has been suggested by Moore-Ede and his colleagues (Moore-Ede et al., 1976; Moore-Ede and Sulzman, 1977).

The search for more than one oscillator can be broadened in two different ways, by considering sub-primate species and by considering individual tissues. Thus, although it is not intended to survey the large amount of work performed in this field, recent reviews (*see,* for examples, Block and Page, 1978; Menaker et al., 1978) have indicated that many species show more than one oscillator. Often these are paired (as in the optic lobes of the cockroach or the eyes of *Aplysia*) and it has been suggested that a number of functions are controlled by different oscillators that are normally coupled together. Therefore an evolutionary precedent for the presence of more than one clock exists, but we are quite ignorant of whether or not the oscillators in humans (wherever they might be) have homologues in other parts of the animal kingdom. (*But see* Lydic et al., 1980.)

2.2. Cellular rhythmicity and its control

Isolated organs and tissues, and even unicellular animals, all exhibit circadian rhythmicity (*see* accounts by Moore-Ede et al., 1976; Palmer, 1976; Moore-Ede and Sulzman, 1977; Saunders, 1977; Block and Page, 1978). Such an observation not only enables the search for the circadian clock to be directed to a cellular level (see below) but raises the problem of how, under normal circumstances, the cellular clocks are rendered subservient to some 'master oscillator'. (Such a problem is related to one already raised, namely how the balance betweeen a causal nexus and a hierarchy can be achieved.) With respect to the control of cellular clocks, the answer to this dilemma is not known, but since some circadian rhythms with considerable endogenous components (those of temperature, plasma adrenaline and cortisol, for instance) exert widespread effects upon many rhythms within the body, it seems quite possible that it is through them that the 'lesser' oscillators are controlled. It is conceivable that a breakdown of the efficacy of such a mechanism would go some way to explaining the presence of ultradian rhythms in some forms of cancer (see Chapter 11).

2.3. Interactions between the two master oscillators

The circadian rhythm system of the human acts as though the two master oscillators are normally coupled to each other (when the subject is 'internally synchronized') with a free-running period that averages 25·0 hours

Table 12.1. Periods of the temperature rhythm (hours) before and after spontaneous internal desynchronization in subjects studied in isolation

	Group A	Group B
Internally synchronized	25·55 ± 0·46 (n = 15)	24·47 ± 0·15 (n = 11)*
After desynchronization — Temperature	24·85 ± 0·30 (n = 15)	24·88 ± 0·13 (n = 11)
After desynchronization — Activity	Lengthened	Shortened

Group A, subjects in whom activity period lengthened (see for example, Fig. 2.3); Group B, subjects in whom activity period shortened. (Data from Aschoff and Wever, 1976.)
Results as mean ± s.e.
*Significantly different from Group A.

(Wever, 1979a). When the two oscillators become uncoupled, generally the sleep—wakefulness oscillator manifests a longer period, and the 'temperature oscillator' (that is, the oscillator which has a dominant effect upon the circadian rhythm of deep body temperature) shortens its period slightly (see Fig. 2.3); when the sleep—wakefulness oscillator shortens its period, then the 'temperature oscillator' lengthens its period slightly. Remarkably enough (Table 12.1), during internal desynchronization, the temperature rhythms have indistinguishable periods even though the sleep—wakefulness periods differ greatly. Aschoff and Wever (1976) have suggested that these results indicate that the free-running period during the internally synchronized state is a compromise due to the coupling of two oscillators of different periods. Support for this view

comes from the observation (*Table* 12.1) that, before internal desynchronization, the coupled period of those subjects in whom the
sleep—wakefulness cycle lengthened with internal desynchronization was
slightly longer than the period of the uncoupled 'temperature oscillator'
and vice versa for those subjects in whom the sleep—wakefulness period
shortened. Why some subjects have sleep—wakefulness rhythms with
periods that are much longer rather than shorter than those of their
'temperature oscillators' is not known.

Furthermore, since the coupled period of the mutually entrained
oscilllators is far closer to that of the temperature rhythm than that of
the sleep—wakefulness rhythm during the desynchronized state (*Table*
12.1), it is inferred that the oscillator dominating the temperature rhythm
is stronger (by about 15 times) than that controlling the sleep—wakefulness
rhythm (Wever, 1979a). Further evidence in favour of the two oscillators
having different strengths will be presented when zeitgeber are considered.

Even when the two oscillators are desynchronized, they continue to
exert effects upon each other, as is indicated by the following observations:

1. The phenomenon of 'relative co-ordination' is shown (*see Fig.* 2.3
and Aschoff, 1969). That is, when the two rhythms are almost in phase,
their periods temporarily become similar and when they are out of phase
the periods have more marked differences; as a result, the two variables do
not free-run smoothly but in an irregular manner. The effect is more
marked for the sleep—wakefulness oscillator than for that controlling
temperature, since the latter is much stronger than the former and so less
influenced by it.

2. 'Naps' are taken during the subjective day at a time corresponding to
minimum body temperature and the objective assessment of activity
shows a 25-hour component that is equivalent to the period of the
temperature rhythm (*see* Chapter 5).

3. There is a small component in the temperature rhythm with a period
equal to that of the sleep—wakefulness cycle (*see* Chapter 2 and *Fig.* 5.6).
This is probably an example of the 'masking' effect of sleep, but, as
discussed in Chapter 5, can be considered during spontaneous internal
desynchronization to be due to an endogenous oscillator rather than due
to an exogenous factor.

2.4. Are there more than two internal oscillators?

In the sense that cellular rhythmicity can be demonstrated, a vast number
of internal oscillators exist potentially; however, under normal circumstances, the system acts in accord with the concept of a hierarchy such
that all rhythms are controlled by a single 'master oscillator'. Under
special conditions, most notably spontaneous internal desynchronization,
evidence for two 'master oscillators' exists. However, there seems no need
to postulate the presence of more than two 'master oscillators' since no

case of internal desynchronization has been reported in which the simultaneous effects of more than two independent oscillators have been observed. Similarly, in the entrainment studies upon squirrel monkeys by Moore-Ede and his colleagues (1977) (discussed later) no evidence in favour of more than two oscillators was found.

3. Properties of the Endogenous Oscillator(s)

3.1. Reliability

One of the most important properties of an effective internal clock must be its reliability. In lower animals in which homeothermy (endothermy) is not developed, the observation that the free-running period of the circadian clock is temperature-compensated (that is, the period varies little with changes in temperature) is remarkable (Aschoff, 1965; Bunning, 1967; Palmer, 1976; Saunders, 1977). In humans and other homeotherms, such an independence is less surprising but no less important.

The properties of the internal clock are best investigated under free-running conditions by studying subjects in temporal isolation. Nearly all the work upon humans has been performed by Aschoff and Wever and their colleagues and the results have recently been assembled by Wever (1979a). A summary of these findings follows:

1. The free-running period of $25 \cdot 0 \pm 0 \cdot 5$ hours was not influenced by the age or sex of the subject.

2. No consistent effect upon the period was found when the intensity or wavelength of the (continuous) lighting was changed or when the amount of physical or mental work performed by the subjects was changed (*see also* Wever, 1979b).

3. The period could be lengthened by allowing subjects to control the lighting and heating of the isolation unit (for example, by turning off the lights when retiring and turning them on when rising), by performing the experiment upon pairs of subjects rather than upon a single subject and by shielding the subject from electromagnetic fields (possible in one of the two isolation units).

4. The period could be shortened by exposing the subject in the shielded room to a 10 Hz AC field.

5. The phenomenon of spontaneous internal desynchronization was found equally in both sexes but more commonly in aged or neurotic subjects (*see also* Lund, 1974).

6. Conditions increasing the free-running period—(2) and (3) *above*—increased the frequency of internal desynchronization. In addition, internal desynchronization was more frequently seen in the shielded room and applying a 10 Hz AC field—(4) *above*—could promote internal synchronization in cases where this had spontaneously broken down.

The explanation of many of these data is still awaited, and any implications that might arise when the site of the internal clock and the mechanisms by which it might operate are considered have yet to be explored. For example, in those cases in which internal synchronization was present—(1) to (4) *above*—it cannot be decided whether the observed properties apply equally to both oscillators or whether the effect was directly upon one oscillator only, the other following by virtue of being coupled to it. However, preliminary results from some recent experiments suggest that a 10 Hz AC field can exert at least some effects upon the 'temperature oscillator' (Wever, 1979a). Thus, subjects were placed on a 28-hour day and forced internal desynchronization took place since the zeitgeber period was beyond the range of entrainment of the temperature oscillator. Accordingly, the temperature rhythm free-ran with a period that was about 25 hours and independent of the imposed sleep–wakefulness rhythm. There was some evidence that the free-running period of the temperature rhythm was slightly shorter in the presence of the electric field than in its absence—(4) *above*; the effect of the electric field upon the sleep–wakefulness oscillator cannot be estimated from such a protocol, of course.

The observation that spontaneous internal desynchronization was more likely under certain circumstances—(5) and (6)—implies either that the coupling between the two had been decreased or that the period of (presumably) the sleep–wakefulness oscillator had been changed to a value that no longer enabled mutual entrainment or coupling to take place.

3.2. Where are the clocks?

One 'classic' way to investigate the function of a particular organ is to ablate it. When such a protocol is applied to the search for the internal clock, it is fraught with a number of interpretative problems. Thus: (*a*) if the ablation results in a loss of a particular rhythm, it must then be established that it is the clock itself rather than the transmission of its effects that has been removed; (*b*) if no change is produced by the ablation, then this might indicate that the timing mechanism has not been removed or that its function has been taken over by another (normally subsidiary) clock, especially since there is evidence that more than one oscillator is present; and (*c*) if the rhythm persists but is changed in some way, then one cannot decide whether there has been partial removal of the clock (or its transmission processes) or whether the compensatory mechanisms that have taken over are imperfect.

If, instead, recordings are made of either electrical or chemical activity, again a decision has to be taken as to whether any rhythmicity is due to the clock itself or to transmission processes. Clearly, such techniques are inappropriate to a detailed study of man but observations have been made of altered rhythms after organic or physical damage to areas of the brain (*see*, for example, Reinberg et al., 1973; Zurbrügg, 1976; Ron et al.,

1980; Chapter 11). However, it is rarely possible to decide how directly any changes are related to the 'clock' itself.

Experiments upon species other than man have recently been reviewed by Block and Page (1978), Menaker et al. (1978), Handler and Konopka (1979) and Rusak and Zucker (1979). There is general agreement, reached on the basis of results from transplantation studies, that the 'brain' of cockroaches, fruit-flies and silkmoths can transfer period and phase information to an arrhythmic host. Work on rodents and birds has indicated more circumscribed areas of the brain to have clock-like functions. For example, the isolated sparrow pineal gland has been shown to manifest rhythmicity *in vitro* (*see* Chapter 7) and this can be transferred to a host by transplantation (Takahashi and Menaker, 1979).

However, in other birds as in rodents, interest has centred rather upon part of the hypothalamus, the suprachiasmatic nucleus (SCN). The role proposed for this area has changed as experimental protocol and analysis of the results have been refined. Initial claims, based upon experiments with *groups* of animals, were made that SCN ablation produced arrhythmicity in locomotor activity. When *individuals* were studied it was observed that each animal did not become arrhythmic but instead showed either a free-running period or an erratic synchronization to a light—dark cycle; in this latter case, when placed in constant light or dark, the animal showed rhythmicity indicative of a mixture of ultradian components (*see* Fig. 12.3). The partial 'synchronization' to a light—dark cycle has been attributed to a direct 'masking' effect of the external environment (Rusak, 1979), and the mixture of ultradian components is generally interpreted to indicate that the SCN, even if not the site of the clock itself, is the region of the brain where oscillatory activity is co-ordinated. (The nature of this oscillatory activity will be considered later.) It is important to realize that a *group* of animals, each individually showing the kind of activity pattern shown in *Fig.* 12.3, would together produce a mean rhythm of reduced amplitude, this problem—the relationship between the average rhythm of a group and the rhythms of individuals when these are not synchronized—has been raised on a number of previous occasions.

Some data relevant to the distinction between the SCN acting as the clock itself or rather as a co-ordinator have recently been obtained in rats by Inouye and Kawamura (1979). They have measured the electrophysiological activity of hypothalamic 'islands' (including the SCN). These are groups of cells which have been surgically isolated from neural connections with the rest of the brain. The authors have concluded that circadian rhythmicity continues to be shown by such 'islands', though not by adjacent, separated regions of the brain. This result, which needs independent confirmation, strongly implies that the SCN is something more than a co-ordinator or relay station but is rather acting as the clock itself.

In humans, such data are lacking of course, but the observation that

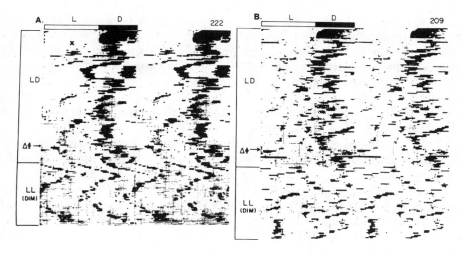

Fig. 12.3. The activity data from two hamsters in which SCN lesions were performed. The record of the animals' running-wheel activity was cut into 24-hour segments and each segment placed under the previous day's. The data were double-plotted such that the first line shows days 1 and 2. The second line days 2 and 3 and so on. The black marks indicate the times the animals were active. The SCN lesions were performed at the time indicated by x. The animal was studied after this in a light–dark cycle (LD), then after a phase-shift of the light–dark cycle (Δ ϕ) and finally in constant light (LL). The breakdown of the circadian rhythm into ultradian components can clearly be seen especially during LL (*From* Rusak, 1977, Fig. 5.)

circadian rhythms with high endogenous components are found in variables closely associated with the hypothalamus (*see* s. 1) is taken to imply a similarity between man and rodents. A recent histological study upon preserved human brains has indicated the presence of diffuse groups of neurones situated lateral to the third ventricle and above the optic chiasma (Lydic et al., 1980).

3.3. The relationship between ultradian and circadian rhythms

As Chapter 1 has indicated, rhythms with a period of less than 21 hours are defined as ultradian, and, for example, the 'post-prandial dip' has been speculated to be some manifestation of an ultradian rhythm (*see* Chapter 6 and Colquhoun, 1971). The presence of ultradian as well as circadian components of rhythms has been recorded in a wide variety of variables (*see* Kripke, 1974), including cardiovascular (Wilson et al., 1977), renal (Lavie and Kripke, 1977), thermoregulatory (Lovett-Doust, 1979) and psychometric (Lovett-Doust et al., 1978).

   In assessing the possible relationship between such ultradian components and circadian rhythms, two main problems arise. One of these stems from the definition of the word 'ultradian' which includes rhythms with a wide range of periodicities. Thus, periods of 90 minutes are common, especially in sleep studies (*see* Chapter 5) but values much shorter than this (7 minutes in reaction time, Lovett-Doust et al., 1978) or considerably longer (12 hours or other simple fractions of 24 hours) can exist. It is not clear whether all these rhythms with such different periods are related or rather independent of each other.

This problem becomes acute when Kleitman's 'rest–activity cycle' is considered (Kleitman, 1963; Othmer and Hayden, 1974). Thus, Kleitman has proposed that the REM/non-REM cycle (which has been described in Chapter 5) is one manifestation of a rest–activity cycle that continues throughout the nychthemeron. In spite of the elegance of being able to link diurnal ultradian rhythms with the nocturnal REM/non-REM cycle as proposed by this hypothesis, some difficulties require consideration. Thus:

1. As has been described above, there is a range of period lengths associated with diurnal ultradian rhythms.

2. The length of the REM/non-REM cycle varies considerably both within and between individuals (*see* Chapter 5).

3. Recent evidence (Lavie, 1979) has indicated that, within REM sleep episodes, there is a further rhythm of eye movement with a period of between 10 and 20 minutes.

In brief, therefore, it is not certain whether such a multiplicity and variability of ultradian oscillations can be accommodated within Kleitman's proposal (*see* Kripke, 1974).

Nevertheless, there is some evidence to indicate that brain activity associated with the REM/non-REM cycle during sleep continues in waking subjects. Othmer and Hayden (1974) measured the EEG of three subjects who stayed awake overnight but with their eyes shut. EEG activity could be divided into 'waking-REM' and 'non-REM' states; the former was associated with rapid eye movement, a fall in muscle tone, a preponderance of alpha-rhythm and dramatic, visual experience, while the 'non-REM' state showed opposite features and abstract, non-visual sensation. The appearance of these two states in the subjects was not random, but episodic (compare with the REM/non-REM cycle).

This concept of 'waking-REM' and 'non-REM' states has been developed by Klein and Armitage (1979). They base their theory on two sets of data. The first is the limited amount of data that suggests that the two cerebral hemispheres behave differently in REM and non-REM states and the second is a far larger body of evidence that the cognitive functions differ between the cerebral hemispheres (Sperry, 1968; Milner, 1971). Klein and Armitage (1979) then postulated that there was an alternation between left- and right-hemisphere activity during both sleep and wakefulness and that this could be measured polygraphically (compare with the results of the study of Othmer and Hayden, 1974). Some evidence in favour of a cyclic alternation between left- and right-hemisphere-dominated performance tasks comes from the simultaneous performance of a verbal task (left hemisphere) and a spatial task (right hemisphere) at 15-minute intervals for 8 hours (Klein and Armitage, 1979). Both tasks showed ultradian rhythms (with a period of 96 minutes) with the two rhythms 180° out of phase (*Fig.* 12.4). The relationship between these rhythms and the 'waking-REM/non-REM' cycle in these subjects was not assessed.

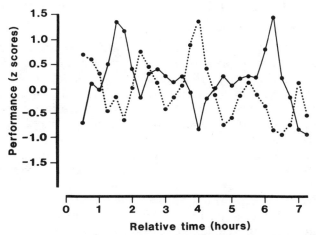

Fig. 12.4. Ultradian oscillations in verbal (solid line) and spatial (dotted line) performance. Data represent means from 8 subjects. Note that although both variables show an about-90-minute oscillation, the two are 180° out of phase (*From* Klein and Armitage, 1979, Fig. 1. Copyright 1979 by the American Association for the Advancement of Science.)

Even if it is accepted that some link between the REM/non-REM cycle of sleep and (at least some of the) ultradian rhythms during wakefulness does exist, then there remains a second problem in relating ultradian to circadian rhythms. This is whether or not the two rhythms are synchronized. Oscillator theory can account for the simultaneous presence of a fundamental frequency and its harmonics (Aschoff, 1965), but these are normally in phase with one another. Some discussion of this problem in humans was given in Chapter 5 when the question of whether the REM/non-REM cycle was synchronized to sleep was considered. The answer is uncertain at the moment but, clearly, if ultradian and circadian rhythms are not synchronized, then a close link between the two becomes more difficult to accept. This problem has recently been considered by Kass et al. (1980) with reference to urinary rhythms of potassium excretion in squirrel monkeys.

Ultradian components in rhythms were mentioned also when prolonged sojourns in caves (Chapter 5), the development of circadian rhythms in infancy (Chapter 8) and the effects of SCN ablation were considered (s. 3.2). Indeed, there seems to be a certain similarity (which may of course be fortuitous only) between *Figs*. 5.5, 8.2 and 12.3. It is an attractive, but wholly speculative, hypothesis that ultradian components are co-ordinated by the SCN, that such an area matures during infancy and that destruction of it leads to a desynchronization of ultradian rhythms, as does prolonged isolation from a rhythmic environment.

3.4. Mechanisms by which the oscillator activity might originate

Homeostasis, oscillation and rhythmicity

In any homeostatic system, the controlled variable tends to oscillate about its set-point. It is conceivable therefore that circadian rhythmicity could originate from the activity of negative feedback loops. Some evidence against such a mechanism accounting for circadian rhythms is as follows:

1. The rhythmic secretion of ACTH by the pituitary continues after adrenalectomy even though mean levels are raised; that is, the rhythmic secretion of the hormone does not require feedback control from cortisol. Indeed, Weitzman's view that the episodic secretion of hormones and their circadian rhythms are not under homeostatic control has been discussed in Chapter 7.

2. Nocturnal urine is normally more acid than that excreted during the daytime. This oscillation can be 'interrupted' by ingestion of bicarbonate thus rendering the urine alkaline throughout the nychthemeron. It is observed that when the acid load has been excreted the circadian rhythm continues with normal nychthemeral phasing; this suggests that the circadian oscillator had continued unchanged throughout, only the manifestation of its effect being masked by the homeostatic requirement to excrete an alkaline urine. The 'homeostatic' theory of circadian rhythms would have predicted that, after the excretion of the alkali load, the rhythmicity would have started from the 'alkaline phase' of its oscillation whatever the time of day.

3. Negative feedback mechanisms tend to result in a constant amplitude of oscillation about a constant mean, the period being variable and dependent upon the (variable) time taken to pass from the peak to the trough of the oscillation; circadian rhythms are very different, the period being constant and the mean and amplitude varying (*see* Chapter 4).

Cellular mechanisms

These approaches have led to the belief that there is no direct connection between homeostatic mechanisms and circadian rhythms. Further, the observations that isolated tissues show circadian rhythmicity—which argues against a mechanism involving the interaction between different organs (*see* (1) *above*)—and that circadian rhythmicity can be demonstrated in unicellular plants and animals have led to attempts to understand the origin of circadian rhythms in terms of mechanisms at the cellular level.

One way in which circadian rhythms have been postulated to be generated is by periodic activity of some components of the DNA–RNA–Protein Synthesis machinery (Palmer, 1976). Evidence indirectly in favour of this has been outlined in Chapter 8 in which experiments upon the inheritance of mutant clocks in *Drosophila* spp. were considered (Konopka and Benzer, 1971).

More direct evidence has come from studies involving blockers of DNA-dependent RNA synthesis or of protein synthesis. In interpreting the results of such experiments, the problems associated with the use of blockers in general should be borne in mind, especially when processes as fundamental as DNA transcription and RNA translation are being investigated. For example, a loss of rhythmicity is an ambiguous result since it might indicate no more than that the manifestation of clock-like processes had been blocked. (Compare with the earlier arguments relating to the function of the SCN as a clock.) However, recent studies with blockers, mainly upon the eye of *Aplysia* (Jacklet, 1977, 1980; Rothman and Strumwasser, 1977) have produced changes in phase of the circadian clock in imitation of effects produced by light, a natural zeitbeger for this system (*see* s. 4.2). There is some evidence that such treatments did not exert their effect through any electrophysiological damage.

The way in which the machinery for converting genetic information into protein might produce a clock-like effect is not known, but a model that has often been invoked is the 'Chronon concept' (Ehret and Trucco, 1967). In brief, this postulates that the protein product from the expression of one gene initiates transcription of the next and so on until a protein is produced that initiates transcription of the first gene again. Evidence that there is a circadian change of messenger RNA or proteins within cells is sparse (for example, Ehret, 1974) but the technical and interpretative problems are enormous. Thus, the implications of the recent demonstration of a diurnal fluctuation in the rate of incorporation of L-methionine into rat brain (Burnet, 1980) are not decisive since it cannot be deduced whether such a result is a direct or indirect expression of clock acitvity or (as the Chronon concept would require) part of the clock itself (compare with the interpretative problems mentioned in s. 3.2).

Alternatively, the importance of the cell membranes has been stressed by others. This view arises because many substances which have an effect upon the clock (for example, ionophores, ethanol and lithium) are believed to change the flux of materials across membranes (Njus et al., 1976; Palmer, 1976) or the composition of the membrane (Brody and Martins, 1976). In addition, claims have been made for circadian changes in the distribution of ions across the cell membrane (for example, Kiyosawa, 1979) and for phase shifts produced by treatments that were believed to change the potential difference across excitable membranes (Eskin, 1979). As with the 'Chronon concept', it cannot easily be established whether the changes reflect the activity of the clock mechanism itself or instead are manifestations of the clock.

Recently, there has been a tendency to see both the 'protein synthesis' and 'membrane' models as not being mutually exclusive (Jacklet, 1978; Surowiak, 1978); thus the 'membrane' model might require the synthesis of particular proteins just as changes in membrane function might affect the expression of part of the genome via (say) changes in intracellular composition.

Adjusting the oscillator period

A difficulty in establishing the cellular basis for the circadian oscillator, certainly where oscillating chemical systems are concerned, is that oscillating systems are known in which the period of oscillation is very much shorter than this. Accordingly means of lengthening the output frequency are necessary (Palmer, 1976). One possible mechanism consists of having a large number of stages in the oscillating pathway (e.g. the 'Chronon concept'). An alternative model requires the interaction of a number of oscillators, the integrated output from which shows a period that is greater than that of the original inputs (Brown et al., 1975). Such a mechanism might apply in rodents in which the 'co-ordinating' role of the SCN, together with the simultaneous presence of ultradian components, has already been described.

Clearly this is all highly speculative but some indirect evidence has been obtained from the eye of *Aplysia* and is summarized by Block and Page (1978), Menaker et al. (1978), Eskin (1979) and Rusak and Zucker (1979). Thus it has been found that the free-running period of the frequency of the action potentials recorded in continuous darkness decreases as the number of elements in the eye decreases. This result would not be expected if the rhythm was determined by cells that individually discharged with a circadian rhythmicity; instead it might indicate that some form of interaction takes place between elements that individually possess ultradian periodicities. However, the free-running period of the individual components of the *Aplysia* eye is not known nor is the form that the proposed interaction might take.

The relevance to man of this work upon cellular mechanisms is unknown of course; present data do not enable a detailed model of the clock mechanism in any system to be described. Even if such a model were available, its application to man would be conjectural; thus we do not know whether a model like that suggested for the eye of *Aplysia* or one applicable to a unicellular organism is more appropriate (or even if a radically different model is required). This, combined with the difficulties of deciding how many clocks are present in man and where they are sited, leads inexorably to the conclusion that the situation is very far from even a partial explanation.

4. Entrainment and Zeitgeber

4.1. Entrainment

In Chapter 1 it was explained that the role of zeitgeber or synchronizers was to entrain or adjust the free-running circadian rhythm to the external environment. As the importance of entrainment is to ensure that the peaks of the rhythms are appropriately phased with respect to the environment, interest has centred on the relationship that exists between the phase of the zeitgeber and that of the circadian rhythms (Wever, 1972, 1973, 1979a). *Fig.* 4.6 shows the phase relationship that exists between a number of circadian rhythms when zeitgeber periods of 24 and 26·67 hours were used. The figure indicates that, as the zeitgeber period lengthens, so the phase of the circadian rhythm advances relative to it; with shorter zeitgeber periods the acrophases become delayed. These results accord with theoretical predictions when oscillator properties are considered (Aschoff, 1965; Sollberger, 1965).

However, as this figure indicates, the position is rather more complex since not all rhythms show the same changes of phase when the zeitgeber period is modified. Thus the rhythms of rectal temperature and urinary potassium show more marked changes than do the rhythms of activity (sleep—wakefulness) and urinary flow (water). The explanation that is generally offered is that the result gives some idea of the relative strengths of the zeitgeber and internal oscillator that are being considered (Hoffmann, 1969; Aschoff, 1979; Wever, 1979a). For example, if the zeitgeber is relatively strong in comparison to the oscillator, then the phase angle of the oscillator will not change much with respect to the zeitgeber; if the zeitgeber is weaker, then the phase angle will change considerably. Some of the evidence that the oscillator controlling sleep and wakefulness is weaker than that controlling the rhythm of deep body temperature has already been considered. With respect to a light—dark cycle as a zeitgeber, therefore, *Fig.* 4.6 confirms this view. The further observation that the potassium rhythm follows that of temperature accords with the view that it has a strong endogenous component and the finding that the urinary water rhythm follows that of the sleep—wakefulness oscillator accords with the view that water is largely exogenous and influenced much by the sleep—wakefulness cycle (*see* Chapter 4). Presumably, a variable that was influenced equally by both oscillators would show intermediate phase changes; in addition, if the experiment were repeated with a (say) weaker zeitgeber, then the gradient of all lines would change, greater phase shifts being produced by changes in zeitgeber period.

A further implication of *Fig.* 4.6 is that it enables one to gain some idea of the range of entrainment to a light—dark zeitgeber for any variable.

Since the activity oscillator (by reason of its weakness) shifts little in relation to the zeitgeber, a considerable range of zeitgeber periods will be able to entrain this oscillator; in the case of deep body temperature, the large phase shifts (due to the strength of the oscillator involved) will allow only a small range of entrainment.

Recently, Wever (1979a) has attempted to compare the range of entrainment of the temperature and sleep–wakefulness rhythms by a light–dark cycle by progressively increasing the length of the cycle beyond 24 hours. The temperature oscillator was entrained by a period of up to 26 hours in length, and the sleep–wakefulness cycle by one of up to 35 hours. Such a result provides further evidence that the temperature oscillator is stronger than that controlling the sleep–wakefulness cycle. Incidentally, in all of these experiments, the subjects were unaware of any manipulation of their 'day' length suggesting that the sense of passage of time was equally labile (*see* Chapter 5).

Since the free-running period of most subjects is more than 24 hours, the nychthemeral condition will entail entraining the oscillators to a period that is shorter than the free-running value. This will produce a greater phase-advance of the (weaker) sleep–wakefulness rhythm than that of deep body temperature. Therefore the temperature maximum will occur after mid-activity and the minimum about mid-sleep (*see* Chapters 2 and 5). Under free-running conditions, in the absence of these effects of entrainment, the (stronger) temperature oscillator will delay less than will that affecting the sleep–wakefulness cycle so that the temperature maximum will now precede mid-activity and the minimum will coincide with sleep onset (*see also* the work of Kriebel, 1974).

4.2. The phase-response curve

In lower animals and plants (Palmer, 1976; Saunders, 1977) the adjustment of internal timing produced by the zeitgeber has been found to depend upon the stage of the circadian cycle when the external stimulus is applied; at some stages a phase-advance, at others a phase-delay, and at others no change in timing of the rhythm is produced. This relationship is called a 'phase-response curve'. Such a curve has not yet been obtained in man; this is because the measurement of the phase-response curve is very time-consuming. Thus the organism is placed in constant conditions in the absence of the zeitgeber (say, in the dark) and an external stimulus (light pulse) is given at different stages of the circadian cycle. The effect of this stimulus upon some easily measured rhythm is assessed (for example, locomotor activity or leaf movement). The experiment must then be repeated a number of times giving the stimulus at different stages of the circadian cycle. Recently a phase-response curve for the rat has been described (Honma et al., 1978); it is clear that performance of the experiment required many weeks.

It is generally accepted that such curves exist for man if only because they enable one to speculate upon the process by which a non-24-hour clock can be entrained to a 24-hour day by a zeitgeber. Thus, the presence of a zeitgeber early in the waking span will tend to advance the internal clock and its presence late in the waking span will tend to delay it. As a result, if the clock is running slow, sleep will be curtailed by the new day (man is a diurnal animal) and if the clock is running fast, the passage of rhythms through their declining phase will be delayed by the zeitgeber.

4.3. Means of identifying zeitgeber

There are two principle ways in which zeitgeber can be identified. The first is illustrated by some elegant experiments of Moore-Ede and his colleagues (Sulzman et al., 1977). They used squirrel monkeys that were placed in constant conditions so that their circadian rhythms free-ran. Then each putative zeitgeber was tested singly by imposing a regular 'present—absent' routine (for example, noise—quiet, eating—fasting, light—dark etc.). *Fig.* 12.5 shows a record of locomotor activity during such an experiment; the effectiveness of eating—fasting and light—dark can be seen clearly, other external rhythms having no obvious effect. A second approach by which a zeitgeber can be identified is illustrated by some work upon rats by Krieger and Hauser (1978). Circadian rhythms were monitored in animals which were subjected to all the normal zeitgeber, but one or more of these was shifted or reversed in the different stages of the experiment. Since a reversal of the light—dark regimen (with food freely available) reversed the corticosteroid and temperature rhythms, it was concluded that the light—dark cycle could act as a zeitgeber for these two rhythms. However, if feeding was restricted (21^{30} to 23^{30}), shifts in rhythms produced by reversal of the light—dark cycle were much less effective, the feeding routine seeming to be a more powerful zeitgeber.

A number of interpretative problems arise with both approaches and with the study of zeitgeber in general. Thus:

1. Clearly a zeitgeber acts only by virtue of exerting an effect upon the internal clock, but the process of entrainment (which influences the phase and period of the oscillator) differs from that of 'masking' (which is an interfering factor that exerts no effect upon it). Entrainment will alter the phase of the internal clock so that when the zeitgeber is removed the internal clock will free-run again but starting from a position determined by the zeitgeber; this can be seen in *Fig.* 12.5 when the locomotor activity before, during and after exposure to 'eating—fasting' cycles is considered. 'Masking' will not affect the internal clock so that it will continue to free-run throughout; when this external interference is removed, the internal clock will be uncovered (unmasked), its phase being independent of the disturbing rhythm. A phenomenon similar to this has been found in rats after SCN ablation. Although they do not manifest circadian rhythms

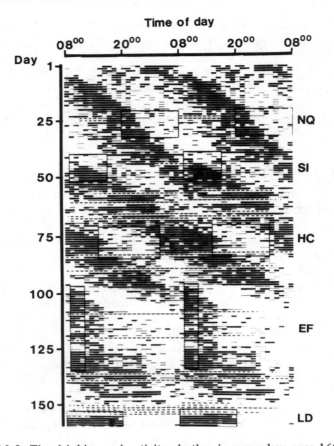

Fig. 12.5. The drinking and activity rhythm in a monkey over 160 days. The small vertical marks indicate the times at which the monkey was active or drinking with the frequency of marks being proportional to the amount of activity or drinking in hourly periods. Continuous black bar indicates active or drinking throughout the hour. Different zeitgeber were changed during the experiment with all other zeitgeber held constant and the animal continually lit at 600 lux: days 21–32 (NQ), cycles of 12 hours of noise (08⁰⁰–20⁰⁰) and 12 hours quiet (20⁰⁰–08⁰⁰); days 40–51 (SI), cycles of 8 hours of social interaction (09⁰⁰–17⁰⁰) and 16 hours of isolation (17⁰⁰–09⁰⁰); days 71–84 (HC), hot and cold temperature cycles with 28 °C (03⁰⁰–15⁰⁰) and 20 °C (15⁰⁰–03⁰⁰); days 97–134 (EF), eating and fasting cycles with food available at 09⁰⁰–12⁰⁰ only; days 153–160 (LD), light–dark cycles with light 08⁰⁰–20⁰⁰ and dark from 20⁰⁰–08⁰⁰. (*From* Sulzman et al., 1977, Fig. 1.)

in constant conditions (*see* p. 282), an 'induced diurnal rhythm' can be produced by the restriction of feeding times in rats; this rhythm disappears with removal of the periodic stimulus suggesting that, in this case, the feeding times acted as a 'masking' influence (Krieger, 1979; Moore, 1980).

A distinction between a 'masking' influence and a zeitgeber can be made also by considering the phenomenon of 'range of entrainment' (Wever, 1979a). As already described, such a range is larger (but presumably finite) with a stronger zeitgeber whereas a 'masking' influence will exert its effect immediately (that is, with no 'transients') and at any frequency.

2. It is likely that under normal circumstances, a number of zeitgeber of different strengths entrains the internal clock. The approach indicated in *Fig.* 12.5 indicates that the 'light—dark' cycle *by itself* is sufficient to entrain locomotor activity but that the 'noise—quiet' cycle is not. Nevertheless, in normal circumstances, it is possible that the 'noise—quiet' cycle, with other zeitgeber, contributes to the process of entrainment as a whole. The alternative approach, by which many potential zeitgeber are held constant and another is changed (Krieger and Hauser, 1978), suffers from the disadvantage that no change in rhythms might indicate that the combined force of the unchanged zeitgeber outweighs that of the changed one. This is a problem which has already been considered when adaptation to shift work was considered (see Chapter 10) and will be discussed again in s. 5.6.

3. The evidence that there is more than one oscillator, and that they are of different strengths, has already been commented upon. It is quite possible that any zeitgeber will have a different efficacy upon each oscillator and, since any circadian rhythm is probably influenced by both oscillators, each rhythm will be affected by a group of zeitgeber in a complex manner. This is a problem that has been given considerable attention in the squirrel monkey by Moore-Ede and his associates (Moore-Ede et al., 1977; Sulzman et al., 1978a,b,c). They have measured a number of circadian rhythms (including those of drinking, urinary water and potassium loss and deep body temperature) under conditions in which the effectiveness of rhythmic changes in 'light—dark' and 'eating—fasting' cycles could be compared. Both phase-shifts and changes in period of each zeitgeber were investigated, the outcome of the experiments being that the temperature rhythm was influenced more by the 'light—dark' cycle and those of drinking and urinary excretion more by the cycle of eating and fasting.

4.4. Means by which zeitgeber might affect rhythms

Since the zeitgeber is required to entrain the internal clock, some link between the two is necessary. Although the relative importance of different zeitgeber depends upon the species under consideration, 'light—dark' and 'feeding—fasting' rhythms seem to be important generally.

The pathway by which the 'light—dark' cycle might act as a zeitgeber has been investigated by retrograde axonal transport. This has been used to show that there is a direct link between the retina and the suprachiasmatic nucleus in a number of animals, though this link has not yet been demonstrated in man (Moore, 1979). Cutting this tract has led to animals' losing the ability to be entrained by environmental light–dark cycles and showing instead a free-running rhythm. Although it is generally accepted that this is the pathway by which the light–dark cycle can act as a zeitgeber, in some birds the position is more complex, there being evidence for an extra-retinal photoreceptor at least closely connected with the pineal, if not the pineal itself. (It will be recalled that there is evidence that the pineal gland acts as an oscillator in some avian species, s. 3.2.)

Another zeitgeber is the rhythm of feeding and fasting; it has been established in rats (Fernstrom, 1979) that the concentration of the transmitters 5-hydroxytryptamine (5-HT) and catecholamines depends upon brain levels of tryptophan and tyrosine respectively. These levels depend in turn upon the ratio of tryptophan (or tyrosine) to neutral amino acids (valine, leucine, iso-leucine and phenylalanine). In rats, insulin lowers the concentrations in plasma of neutral amino acids more than those of tryptophan or tyrosine. As a consequence a low-protein meal will modify the ratio of plasma amino acids in favour of tryptophan and tyrosine, thereby promoting their uptake and synthesis into 5-HT and catecholamines, respectively, by the brain. (Which of these two transmitters will be synthesized in greater amounts will depend in turn upon their relative plasma concentrations.) By contrast, ingestion of protein will decrease the synthesis of these transmitters since the concentrations of neutral amino acids will have been raised.

There is limited evidence for such a scheme in humans (*Fig.* 12.6) when the effect of diet upon the ratio of concentrations of different amino acids in the plasma is considered; but the effect of this ratio upon brain uptake and transmitter synthesis is not known. Furthermore, and this reservation applies to all species, it is not known how these changes in transmitter synthesis will affect the activity of the internal clock. Nevertheless, it is intellectually satisfying to be able to postulate such a link between an external zeitgeber and the internal clock. Similarly, one could speculate that the rhythmic release of transmitters could form a link between the clock and hormone release from the hypothalamus, as has been implied in Chapter 7 (*see also* Perry et al., 1977; Carlsson et al., 1980.)

As will be seen in the next section, important zeitgeber in man are social awareness and the effect that this might exert upon the pattern of sleep and wakefulness. Pathways by which these factors might influence the internal clock can be only speculative at the moment, especially in the case of an influence as (neurophysiologically at least) nebulous as 'social awareness'.

Finally, the cellular mechanisms by which the internal clock might be affected by the zeitgeber pathway are, of course, quite unknown, but the observation that phase-response curves can be mimicked by the use of ionophores or protein synthesis inhibitors has been interpreted to indicate that the cellular effect of a zeitgeber is to shift some cyclic phenomenon

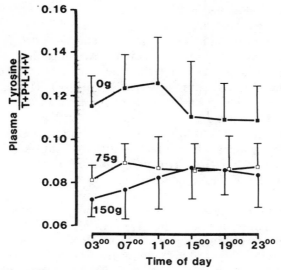

Fig. 12.6. Relationship between dietary protein content and plasma ratio of tyrosine to the sum of its neutral amino acid competitors in normal human subjects. Subject consumed three meals per day at 08^{00}, 12^{00} and 17^{00}, each containing 0, 25 or 50 g of protein. Each diet was ingested for 4 consecutive days. Plasma samples were collected on the third and fourth days and analysed for the various amino acids. Data indicate means ± standard deviations. T, tryptophan; P, phenylalanine; L, leucine; I, isoleucine; V, valine. (*From* Fernstrom, 1979, Fig. 15. Reproduced from Krieger D. T. (ed.) *Endocrine Rhythms,* 1979, by kind permission of Raven Press, New York.)

(connected with membrane activity and/or protein synthesis) from one point on the cycle to another (Aschoff, 1965; Njus et al., 1976).

5. Zeitgeber in Man

At the outset it must be stated that studies upon zeitgeber in man have rarely distinguished the process of entrainment from that of 'masking' and experiments in which a single zeitgeber has been introduced into an environment that was otherwise constant have not been performed. In practice the experimental protocol has generally consisted of a comparison of circadian rhythms after a putative zeitgeber has been shifted or removed, that is, the second approach described in s. 4.3. Reviews of zeitgeber in man are to be found in Aschoff et al. (1975), Aschoff (1979), Wever (1979a) and, in relation to hormones, in Daly and Evans (1974).

5.1. Light as a zeitgeber

Evidence that can be construed as support for the view that light acts as a zeitgeber has come from a number of sources.

1. The effect of changes in the time of sunrise was studied in a subject whose rhythms were measured during the course of a year which he spent isolated in the countryside and with no social constraints (Mills and Waterhouse, 1973). He showed a higher waking plasma 17-hydroxy-corticosteroid level in the winter (when he rose at about sunrise) than in the summer (when he rose some hours after sunrise). This suggests that his steroid rhythm was phased earlier in summer than in the winter and had fallen for some time before waking in the summer.

2. In the continuous light of an Arctic summer, waking at a normal time but remaining awake in a blacked-out tent mimicked the effect of changing the time of 'sunrise'. This delayed the morning fall in circulating eosinophils and the rises in urine flow and sodium excretion (Sharp, 1960a, b).

A more elaborate series of experiments was performed by Orth et al. (1969). Three experiments were involved; in all cases there was a control phase with subjects sleeping from 22^{00} to 06^{00} in the dark. Then: (a) subjects slept from 22^{00} to 06^{00}, laboratory in darkness from 10^{00} to 18^{00}; or (b) subjects slept from 22^{00} to 06^{00}, laboratory in darkness from 22^{00} to 10^{00}; or (c) subjects slept from 22^{00} to 06^{00}, laboratory in darkness except from 18^{00} to 19^{00}. These schedules were maintained for at least 10 days, and then blood samples were taken for cortisol analysis, some samples being taken when the subjects were asleep. In the control phase, normal hormone rhythms were obtained with minimum levels at sleep onset and peak values about the time of lights-on and waking (see Chapter 7). On the first protocol, the rhythm was little changed, an additional peak being measured when lights were turned on; on the second, the peak of steroid concentration was delayed, occurring when the lights were turned on; on the third, a subsidiary peak was observed when the lights were turned on but the period of maximum secretion was associated with waking.

3. Lobban (1967, 1977) described the rhythms of urinary excretion in Indians and Eskimoes from polar regions where the normal alternation of light and dark is greatly reduced. The rhythms were reduced in amplitude and their timing from day to day was more erratic than the excretory rhythms of control subjects living in non-polar regions or of British visitors sojourning in the Arctic. Simpson and Bohlen (1973) have reviewed the evidence relating to the effect of latitude in general upon the amplitude of circadian rhythms. The evidence in favour of the view that latitude and rhythm amplitude are inversely related is contradictory. By way of example, a group of Wainwright Eskimoes showed high-amplitude rhythms whereas a longitudinal study of two Britons near the equator and pole in successive summers showed a fall in amplitude in the Arctic. The extent to which these results are influenced by other factors (for example genetic differences, climate and activity) is unknown.

4. Studies of urinary excretory rhythms (Lobban and Tredre, 1964), of

oral temperature rhythms (Tokura and Takagi, 1974) and of plasma 11-hydroxycorticosteroid rhythms (Krieger and Rizzo, 1971) in blind and partially sighted subjects have indicated rhythms that were of lower amplitude and more irregularly phased than those in sighted controls.

However, in his review of the role of the light–dark cycle as a zeitgeber in humans, Wever (1979a) concluded that the role was a small one. In essence he based these conclusions upon three arguments.

First, he accepted that blind subjects *as a group* had rhythms which varied more than control subjects but maintained that, when considered individually, each blind subject had a rhythm of similar amplitude and day-to-day reliability as sighted subjects. He attributed the difference between blind individuals to their greater reliance on 'social factors' which he assumed were not uniform for all subjects. Without doubt, if there is a greater inter-individual variation between blind subjects (whatever its cause) then pooling data will result in a mean rhythm of low amplitude; however, not all studies have found the intra-individual reliability among the blind suggested by Wever (*see also* Pauly et al., 1977) and it is not known exactly what are the 'social factors' that control the (individually different) phases of rhythms in the blind.

Secondly, a number of circadian rhythms (urinary excretion, rectal temperature, psychometric tests) were measured in subjects on a strict schedule of sleep and wakefulness, meal times and psychometric testing times, first in 16 hours of light (waking) and 8 hours of darkness (sleeping) and then in constant darkness (Aschoff et al., 1974; Giedke et al., 1974). No differences in the rhythms in the two conditions were found, from which it was concluded that the light–dark cycle was unimportant.

Thirdly, some subjects isolated from time-cues free-ran even in the presence of a 24-hour light–dark cycle (*Fig.* 12.7). At first this would seem strong evidence against the light–dark cycle acting as a zeitgeber until it is realized that the subjects could use auxiliary lighting at 'night'. Therefore, provided that they were prepared to sleep in the light, subjects could become independent of the light–dark cycle and no marked disadvantages would arise as a result of their ignoring it. In any case, a small effect of the light–dark cycle can be seen in the figure, the phenomenon of 'relative co-ordination' being manifest; thus the temperature and sleep–wakefulness rhythms showed a period closer to 24 hours when they were in phase with the light–dark cycle (days 1 to 8 and 22 to 26) and a period most in excess of 24 hours when they were completely out of phase (days 12 to 18.

These data do not allow an unambiguous assessment of the importance in man of the 'light–dark' cycle as a zeitgeber to be made. In part the difficulty arises because the protocols that have been used, as has already been stated, give rise to interpretative difficulties. By way of examples: the lack of effect of changes in the light–dark cycle in the study of Aschoff et al. and Giedke et al. (*above*) might have been because the strict

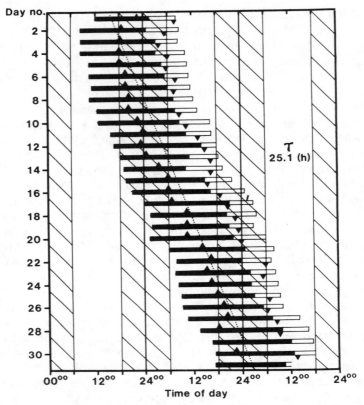

Fig. 12.7. The sleep–wakefulness and rectal temperature rhythm in a subject living without environmental time cues but under the influence of an artificial light–dark cycle. The activity rhythm is presented as bars (black = activity; white = sleep). The temperature rhythm is represented by triangles, indicating the position of maximum (▲) and minimum values (▼). Successive days are represented from above downwards. Shaded area indicates the dark period. Dotted line indicates the average period of the rhythms, 25·1 hours. (*From* Wever, 1970, Fig. 1.)

schedule was sufficient to entrain the rhythms (*see below*); also, the changes produced in the study of Orth et al. (1969) might have been 'masking' effects.

5.2. Social factors' as zeitgebers

The concept that 'social factors' are the most important zeigeber in man has already been mentioned when Wever's account of rhythms in the blind were considered; such a view has been enthusiastically championed by Aschoff and Wever (Aschoff, 1979; Wever, 1979a). The evidence in favour of this view can be summarized as follows:

When pairs of subjects were studied together in an isolation unit (Aschoff, 1969; Wever, 1975b) the subjects' rhythms of sleep–wakefulness and temperature free-ran in synchrony even in an experiment in which both subjects showed internal desynchronization with identical periods for the temperature rhythm (25·1 hours) and activity rhythm (31·5 hours). Because it is extremely improbable that the free-running periods of subjects studied singly would be identical, it seems that the subjects in these experiments became mutually entrained due to 'social interaction', especially since the pairs of subjects were not instructed to remain synchronized. The similar synchrony in the experiments of Mills et al. (1974, Groups H and J) cannot be interpreted the same way; in these experiments the subjects were instructed to agree upon their schedule.

Secondly, when the protocol of *Fig.* 12.7 was repeated but with the addition of regular gong strokes (to indicate times at which psychometric tests were to be performed and urine samples given) it was observed that subjects' rhythms became entrained to the 24-hour cycle of 'light–dark plus gong strokes' (Wever, 1970). Furthermore, by appropriate modification of the light–dark cycle and gong intervals, it was possible to entrain the activity and temperature rhythms of subjects to days of lengths from 23 to 27 hours (Wever, 1979a). Questioning the subjects elicited that they 'perceived the gong signals as personal calls by the experimenter (i.e. social contacts)' (Wever, 1979a, p. 151). Wever considers that they used such signals to structure their day though it has not been established if the gong alone can entrain the rhythms. Wever describes the cycle of the mixture of 'light–dark plus gong strokes' as a 'socially enriched' zeitgeber. As was mentioned in s. 4.3, the combined effects of the two factors might act as a synchronizer even though neither alone is sufficient.

5.3. The importance of the sleep–wakefulness cycle

When Wever describes the effect of 'socially enriched' zeitgeber as enabling the subjects to structure their day, the distinction between the effects of 'social influences' and of 'activity' becomes blurred since most people arrange their hours of sleep and wakefulness to coincide with other people or with what they consider to be (or have been told are) the requirements of the experimenters. The impossibility of doing this when on shift work and some of its consequences in terms of disturbances of circadian rhythms have been discussed in Chapter 10.

Throughout this book the advantages in terms of comparability of expressing rhythms in terms of mid-sleep rather than block time have been stressed. In the cases of cardiovascular and respiratory rhythms (Chapter 3) and of growth hormone (Chapter 7) this was especially important since they had a large exogenous component. For cortisol too (which had a large endogenous component) the reproducibility of the circadian rhythm was increased when expressed relative to times of waking or onset of sleep (*see* Czeisler, 1978; Aschoff, 1979). Additionally, in Lobban's Arctic studies (Lobban, 1967) the urinary excretory rhythms showed amplitudes in the winter that were greater than in summer and this could be attributed to the more regular habits of the subjects in winter than in summer. Such a conclusion agrees with that from a recent study in the Arctic by Buck (1980). Performance in a pursuit tracking task was assessed on 4 separate days during the course of the year in subjects who lived throughout within a rigid, imposed time structure. Circadian rhythmicity was retained with a similar amplitude whether the natural light–dark cycle during the previous months had been continuous light, continuous dark or an alternation between light and dark. The observed stability was attributed to the 'nychthemeral' existence that had been led by the participants.

The importance of the sleep rather than the activity component of the sleep–wakefulness cycle in determining the phase of the rhythms has been stressed by Reinberg (Reinberg, Vieux et al., 1979), and by Knauth et al. (1978). Thus they found that the circadian rhythms of shift workers adjusted so that the acrophase bore a normal relation to mid-sleep even if the relation to mid-work was 'inappropriate' (*see* Chapter 10). Minors and Waterhouse (1980a) also have stressed the role of sleep in this respect.

Since it is known that sleep exerts a direct effect upon circadian rhythms (Aschoff, 1978; Mills et al., 1978) the effects that have just been described might be more accurately described as 'masking' rather than 'entrainment'. Indeed, in some studies by Orth et al. (1967), subjects' plasma corticosteroids were measured when they were placed on days of abnormal length (12, 19 or 33 hours). The marked influence on such schedules upon the corticosteroid rhythm (especially since the 'day' lengths used would be expected to be well beyond the range of entrainment of the 'strong' steroid oscillator) would accord with such a view.

However there is some evidence that sleep can act as a zeitgeber as well as a 'masking' influence. The role of 'anchor sleeps' in stabilizing circadian rhythms of subjects on irregular routines has already been described (Chapter 10). Further experiments were performed by Minors and Waterhouse (1980b) in which the endogenous component of rhythms was investigated after some days on an 'anchor sleep' by placing the subjects on a 'constant routine' (*see* Chapter 1). Evidence was obtained that the endogenous component had retained a 24-hour day and that its timing was similar to that before anchor sleeps were taken (*see* Chapter 2, s. 4). As has been mentioned before, it is not known which component of sleep (changes in neurophysiology, food and water intake, posture etc.) was acting as the zeitgeber, in these or other circumstances.

5.4. Knowledge of 'real' time

One aspect of 'social influences' that might seem to be important is a knowledge of the passage of real time. However the volunteer Workman, who lived in isolation in a cave (Mills, 1964), took a watch with him which he intended to use to structure his days. After a while he found he was lying in bed unable to sleep when the watch indicated it was past bedtime and was staying in bed progressively later each morning. Finally, he gave up trying to follow the dictates of the watch and free-ran with a period of about 24·7 hours (*see Fig.* 5.2). Nevertheless, the watch did have some influence, since his period of sleep and wakefulness when he was using it was less, only just in excess of 24 hours (this part of his sleep–wakefulness record is not included in *Fig.* 5.2). Similar results have been reported in another study (Migraine et al., 1974). Therefore, it seems likely that awareness of real time by itself is insufficient to act as a powerful zeitgeber; what is needed is some relationship with the outside world that makes it *worth while* and *relevant* to adhere (entrain) to real time and to override the natural tendency for the internal clock to run slow. Possibly the 'socially enriched' environment of a cycle of 'light–dark plus gong strokes' fulfilled this requirement. In connection with this, Wever (1979a, Fig. 20) described the case of one 'isolated' subject who nevertheless showed a 24-hour period of sleep–wakefulness and body temperature because of the possibility of being able to make contact with an assistant from outside the isolation unit at a regular time each day.

Certainly for many isolated subjects the external world becomes irrelevant; 'real time' becomes the time as dictated by their subjective assessment of sleep and wakefulness. Thus:

1. When isolated subjects manifest 'anomalous' sleep–wakefulness cycles in free-running conditions or are entrained to days of abnormal length, the sense of passage of time accords with their sleep–wakefulness pattern. This even extends to calling sleeps 'naps' when they appear during the 'day time', even if objective estimates indicate that these 'naps' have a regular circadian frequency or are as long as subjectively assessed 'sleep' (*see* Chapter 5).

2. When putative zeitgeber do not entrain circadian rhythms, the rhythmicity of these external factors is perceived as meaningless and irregular. Thus, the subjects from whom the data of *Fig.* 12.7 were derived, considered the (24-hour) light–dark cycle to be quite erratic (Wever, 1979a).

5.5. Meal times as a zeitgeber

As described in s. 4.3, there is evidence in squirrels and rats, as well as in other mammals, that feeding times act as an important zeitgeber; the evidence in man is far weaker. There is no doubt that the timing of meals affects the activities of the gut, of its associated organs of digestion and of the secretion of some hormones associated with the metabolism of ingested foodstuffs, particularly glucagon and insulin (*see* Chapter 3). In addition, there are marked changes in the plasma concentrations of substances originating from diet, such as iron, amino acids, glucose and fatty acids. However these are direct 'masking' effects and the point at issue is whether, directly or indirectly, these affect other circadian rhythms in man. The evidence here is less convincing.

1. In the case of plasma cortisol, whereas Quigley and Yen (1979) found that a secondary peak of secretion at about noon was accentuated by lunch, Haus (1976) found no change in the rhythm of this hormone when food was taken as a single meal at different times of the day.

2. Apfelbaum et al. (1976) found little change in circadian rhythms when normal meals were replaced by identical meals taken at 08^{00}, 12^{00}, 16^{00} and 20^{00} or when only one meal was taken at breakfast or suppertime.

3. Reinberg, Migraine et al. (1979), in their studies of shift workers on a number of shift systems, found comparatively little change in the pattern of food intake though many other circadian rhythms had adjusted considerably to changes in the sleep—wakefulness schedule (see Chapter 10).

4. On the irregular pattern of sleep and wakefulness already described in Chapter 10 (Minors and Waterhouse, 1980a)—in which circadian rhythms seemed to 'free-run' with a period significantly greater than 24 hours—the composition and timing of meals were maintained throughout on a schedule appropriate to a 24-hour day.

5. Schaefer et al. (1979) investigated submariners on an 18-hour work—rest cycle while retaining a 24-hour pattern for meal times; circadian rhythms appeared to develop periods of 25—28 hours at the expense of a 24-hour component (see Chapter 10).

All these results argue against meal times being an effective zeitgeber; however, strong proponents of the view that meal times are effective zeitgeber in man are Graeber and his colleagues (1978). In a large study upon military personnel, the effect of different feeding regimens (ad libitum, dinner only and breakfast only) upon a wide variety of circadian rhythms was tested. Some rhythms associated with food intake (for examples: plasma insulin, plasma iron and blood urea nitrogen) changed by large amounts; other rhythms (a variety of simple psychometric tests, blood cortisol and lymphocyte count) were unaffected; and other rhythms (temperature, diastolic blood pressure, mood and leucocyte count) showed small shifts that were far less than the amount by which meal times were changed. Graeber et al. (1978) concluded from this that meal times could act as an effective zeitgeber. However, it is not known the extent to which these changes would be accounted for if subjects' activity had declined during times of fasting; some evidence exists for such a decline in so far as the acrophase of the rhythm of pulse rate was 9 hours earlier when breakfast only was eaten than when the only meal was evening dinner.

5.6. The relative importance of different zeitgeber in man

Clearly it is not possible to assess the relative importance to man of the different putative zeitgeber; the discrepancies in the data and difficulties of interpretation prevent this. Nevertheless, some comment can be made relating to the manner in which zeitgeber as a group act upon the process of synchronization.

When all zeitgeber change simultaneously, as after a time-zone shift (Chapter 9), then the process of adaptation is complete and fairly rapid. The importance of 'social influences'—that is contact with the new society—in speeding up the process of adaptation has already been described. It seems reasonable to suppose that other cases of strengthening zeitgeber (feeding and lighting patterns) would also promote adaptation.

When the putative zeitgeber do not all change simultaneously, as on shift work (Chapter 10), then adaptation is slower and in some respects incomplete. Such an incomplete adaptation can be understood more easily if it is accepted that a number of zeitgeber affect each internal oscillator. The observation that the acrophase of a rhythm eventually adjusts to a value 'appropriate' for mid-sleep argues for the importance of some component of sleep in phasing rhythms in the human. Of course such a view assumes that the adaptation is 'real' rather than 'apparent' due to the 'masking' effect of sleep; the very limited amount of evidence indicating that adaptation was 'real' was considered in Chapter 10, s. 5.3. The seemingly less important role of the light—dark cycle might be spurious since an artificial cycle of light (electricity) and dark (drawing curtains and shutting eyes) is produced fairly easily. The exact effect of social influences is particularly difficult to assess since, although the social alienation of the shift worker has been stressed, some of the effects of this dissociation from the majority of society can be ameliorated if the worker's shopping etc. is performed by other members of his family or if his leisure-time pursuits are of a more introvert nature.

Another occasion on which not all possible zeitgeber change simultaneously is in the adjustment of the clocks in autumn and spring. As Monk and Aplin (1980) point out, these changes, in which the external 'light—dark' cycle does not change but all other zeitgeber (social, sleep—wakefulness, feeding) do, have not received much study even though, throughout the world, they influence very many people twice yearly. Results from these studies (*see also* Monk, 1980; Monk and Folkard, 1976; Nicholson and Stone, 1978) indicate that a surprisingly long period (approaching a week) is required for adjustment to occur. Before a strong effect of the light—dark rhythm is inferred (in spite of the use of artificial light—dark cycles) two questions need consideration. First, would not one expect the shift of 1 hour to be within the range of entrainment of the oscillator and so adaptation to be immediate? Secondly, to what extent might the changes (in temperature and psychometric performance rhythms and amount of stage 4 sleep) result from a changed amount of sleep that the clock shift imposed upon the subjects rather than from a changed phasing of several zeitgeber?

A few subjects that live a diurnal existence do not seem to be entrained to it (Weitzman, 1981). Thus Miles et al. (1977) described in detail a male, blind from birth, in whom all rhythms (cortisol, pulse, urine, temperature and alertness) showed a period of 24·9 hours. Attempts to entrain this subject to a 24-hour day by placing him on a strict routine met with only limited success, since there was evidence for

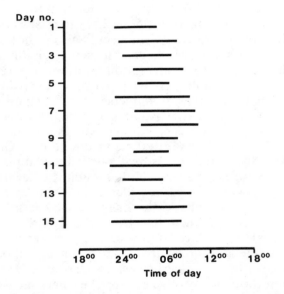

Fig. 12.8. The sleep—wakefulness rhythm in a subject who, despite living in a customary nychthemeral environment, was unable to conform to conventional sleeping times. Hence this rhythm free-ran for several days until his sleep times were quite different from the remainder of society when he tried to re-adjust them to conventional times. Black bars indicate the times in bed. (Data of Elliott et al., 1970.)

the persistence of 25-hour rhythms in sleep—wakefulness and plasma cortisol. Orth et al. (1979) described a single female subject who had been blind for 18 years; a very extensive series of cortisol measurements indicated that she too had a cortisol rhythm with a period in excess of 24 hours (other rhythms were not measured) at the same time as a 24·0-hour rhythm of sleep and wakefulness. Furthermore, the cortisol showed a 24·0-hour component (due to the exogenous or 'masking' effect of the sleep—wakefulness cycle) as a result of which a 'beat' phenomenon was observed. These results argue in favour of a role of the light—dark cycle as a zeitgeber. The observation that the subject studied by Miles and his colleagues (1977) had been blind from birth implies a role of the light–dark cycle that cannot be substituted in an individual even if the zeitgeber was never experienced by him. These authors maintain that desynchronization is not uncommon among blind subjects. This view and the demonstration of a similar desynchronization in the subject studied by Orth et al. (1979) point to the ineffectiveness of the sleep–wakefulness rhythm as a zeitgeber in at least some blind subjects, a position which is in disagreement with the views of Wever (*see* s. 5.1).

Finally, the importance to a normal individual of his interaction with society acting as a zeitgeber can be inferred from a report by Elliott et al. (1970). They studied a single man who was not blind but normally went to bed and woke later each day. When the asynchrony between his habits and those of his colleagues and the rest of society was great enough to be inconvenient, he then suddenly adjusted his routine generally by having an abnormally short waking span (*Fig.* 12.8). When questioned as to the reason for this sudden adjustment in his rhythm of sleep and wakefulness, he acknowledged this to be due to the need to remain in phase with the rest of society. Very similar inferences can be drawn from a more recent study upon two supposed healthy students, one of whom recorded his times of sleep for 4 years (Weber et al., 1980). However, there is limited evidence that in schizophrenic subjects (who might be surmised as being less influenced than normal subjects by social factors), the phases of their temperature rhythms are influenced by the time of sunset (Morgan et al., 1980).

The conclusion from all of this must be that more than one zeitgeber is made use of and that the relative importance of these zeitgeber will differ between individuals and circumstances; it implies that much research in the field of human circadian rhythms remains to be done.

References

Apfelbaum M., Reinberg A. and Lacatis D. (1976) Effects of meal timing on circadian rhythms in 9 physiologic variables of young healthy but obese women on a calorie restricted diet. *Int. J. Chronobiol.* **4**, 29–37.

Aschoff J. (ed.) (1965) *Circadian Clocks.* Amsterdam, North-Holland.

Aschoff J. (1969) Desynchronization and resynchronization of human circadian rhythms. *Aerospace Med.* **40**, 844–9.

Aschoff J. (1976) Complexity and order of the human circadian system. *Bull. Soc. Ital. Biol. Sper.* **52**, Suppl., 1–11.

Aschoff J. (1978) Features of circadian rhythms relevant for the design of shift schedules. *Ergonomics* **21**, 739–54.

Aschoff J. (1979) Circadian rhythms: general features and endocrinological aspects. In: Krieger D. T. (ed.) *Endocrine Rhythms.* New York, Raven Press, pp. 1–61.

Aschoff J., Fatranska M., Gerecke U. et al. (1974) Twenty-four-hour rhythms of rectal temperature in humans: effects of sleep-interruptions and of test sessions. *Pflügers Arch.* **346**, 215–22.

Aschoff J., Hoffmann K., Pohl H. et al. (1975) Re-entrainment of circadian rhythms after phase-shifts of the zeitgeber. *Chronobiologia* **2**, 23–78.

Aschoff J. and Wever R. (1976) Human circadian rhythms: a multi-oscillatory system. *Fed. Proc.* **35**, 2326–32.

Block G. D. and Page T. L. (1978) Circadian pacemakers in the nervous system. *Ann. Rev. Neurosci.* **1**, 19–34.

Brody S. and Martins S. A. (1976) Circadian rhythms in Neurospora: the role of unsaturated fatty acids (abstr.). In: Hastings J. W. and Schweiger H.-G. (ed.) *The Molecular Basis of Circadian Rhythms.* Report of the Dahlem Workshop, Berlin, 1975. Berlin, Abakon Verlagsgesellschaft, pp. 245–6.

Brown B. H., Duthie H. L., Horn A. R. et al. (1975) A linked oscillator model of electrical activity of human small intestine. *Am. J. Physiol.* **229**, 384–8.

Buck L. (1980) Circadian rhythms of performance among workers in the Arctic. *Aviat. Space Environ. Med.* **51**, 805–8.

Bünning E. (1967) *The Physiological Clock,* 2nd ed. New York, Longmans.

Burnet F. R. (1980) Diurnal fluctuation in the rate of synthesis of a specific protein fraction in the rat brain. *Experientia* **36**, 34–6.

Carlsson A., Svennerholm L. and Winblad B. (1980) Seasonal and circadian monoamine variations in human brains examined *post mortem. Acta Psychiat. Scand.* **61**, Suppl. 280, 75–83.

Colquhoun W. P. (1971) Circadian variations in mental efficiency. In: Colquhoun W. P. (ed.) *Biological Rhythms and Human Performance.* London, Academic Press, pp. 39–107.

Conroy R. T. W. L. and Mills J. N. (1970) *Human Circadian Rhythms.* London, Churchill.

Czeisler C. A. (1978) Human circadian physiology: internal organisation of temperature, sleep–wake and neuroendocrine rhythms monitored in an environment free of time cues. PhD Thesis, Stanford University.

Daly J. R. and Evans J. I. (1974) Daily rhythms of steroid and associated pituitary hormones in man and their relationship to sleep. *Adv. Steroid Biochem. Pharmacol.* **4**, 61–110.

Ehret C. F. (1974) The sense of time: evidence for its molecular basis in the eukarotic gene-action system. In: Lawrence J. H. and Gofman J. W. (ed.) *Advances in Biological and Medical Physics.* New York, Academic Press, vol. 15, pp. 47–77.

Ehret C. F. and Trucco E. (1967) Molecular models for the circadian clock. I. The chronon concept. *J. Theoret. Biol.* **15**, 240–62.

Elliott A. L., Mills J. N. and Waterhouse J. M. (1970) A man with too long a day. *J. Physiol.* **212**, 30P–31P.

Eskin A. (1979) Circadian system of the *Aplysia* eye: properties of the pacemaker and mechanisms of its entrainment. *Fed. Proc.* **38**, 2573–9.

Fernstrom J. D. (1979) The influence of circadian variations in plasma amino acid concentration on monoamine synthesis in the brain. In: Krieger D. T. (ed.) *Endocrine Rhythms.* New York, Raven Press, pp. 89–122.

Giedke H., Fatranska M., Doerr P. et al. (1974) Tagesperiodik der Rectaltemperatur sowie der Ausscheidung von Elektrolyten, Katecholammin-metaboliten und 17-Hydroxycorticosteroiden mit dem Harn beim Menschen mit und ohne Lichtzeitgeber. *Int. Arch. Arbeitsmed.* **32**, 43–6.

Graeber R. C., Gatty R., Halberg F. et al. (1978) *Human Eating Behavior: Preferences, Consumption Patterns, and Biorhythms.* Food Sciences Laboratory, Technical Report Natick/TR-78/022, United States Army Natick Research and Development Command, Natick, Massachusetts.

Handler A. M. and Konopka R. J. (1979) Transplantation of a circadian pacemaker in *Drosophila. Nature* **279**, 236–8.

Haus E. (1976) Pharmacological and toxicological correlates of circadian synchronization and desynchronization. In: Rentos P. G. and Shephard R. D. (ed.) *Shift Work and Health.* Washington, DC, US Department of Health, Education and Welfare, pp. 87–117.

Hoffman K. (1969) Die relative Wirksamkeit von Zeitgebern. *Oecologia (Berl.)* **3**, 184–206.

Honma K., Katabami F. and Hiroshige T. (1978) A phase response curve for the locomotor activity rhythm of the rat. *Experientia* **34**, 1602–3.

Inouye S.-I. T. and Kawamura H. (1979) Persistence of circadian rhythmicity in a mammalian hypothalamic 'island' containing the suprachiasmatic nucleus. *Proc. Natl Acad. Sci. USA* **76**, 5962–6.

Jacklet J. W. (1977) Neuronal circadian rhythm: phase shifting by a protein synthesis inhibitor. *Science* **198**, 69–71.

Jacklet J. W. (1978) Cellular mechanisms for circadian clocks. *Trends Neurosci.* **1**, 117–19.

Jacklet J. W. (1980) Protein synthesis requirement of the *Aplysia* circadian clock. Tested by active and inactive derivatives of the inhibitor anisomycin. *J. Exp. Biol.* **85**, 33–42.

Kass D. A., Sulzman F. M., Fuller C. A. et al. (1980) Are ultradian and circadian rhythms in renal potassium excretion related? *Chronobiologia* **7**, 343–55.

Kiyosawa K. (1979) Unequal distribution of potassium and anions within the *Phaseolus* pulvinus during circadian leaf movement. *Plant Cell Physiol.* **20**, 1621–34.

Klein R. and Armitage R. (1979) Rhythms in human performance: 1½-hour oscillations in cognitive style. *Science* **204**, 1326–7.

Kleitman N. (1963) *Sleep and Wakefulness,* 2nd ed. Chicago, University of Chicago Press.

Knauth P., Rutenfranz J., Herrmann G. et al. (1978) Re-entrainment of body temperature in experimental shift-work studies. *Ergonomics* **21**, 775–83.

Konopka R. J. and Benzer S. (1971) Clock mutants of *D. melanogaster. Proc. Natl. Acad. Sci. USA* **68**, 2112–16.

Kriebel J. (1974) Changes in internal phase relationships during isolation. In: Scheving L. E., Halberg F. and Pauly J. E. (ed.) *Chronobiology.* Tokyo, Igaku Shoin, pp. 451–9.

Krieger D. T. (1979) Restoration of corticosteroid periodicity in obese rats by limited a.m. food access. *Brain Res.* **171**, 67–75.

Krieger D. T. and Hauser H. (1978) Comparison of synchronization of circadian corticosteroid rhythms by photoperiod and food. *Proc. Natl Acad. Sci. USA* **75**, 1577–81.

Krieger D. T. and Rizzo F. (1971) Circadian periodicity of plasma 11-hydroxycorticosteroid levels in subjects with partial and absent light perception. *Neuroendocrinology* **8**, 165–79.

Kripke D. F. (1974) Ultradian rhythms and sleep: introduction. In: Scheving L. E., Halberg F. and Pauly J. E. (ed.) *Chronobiology.* Tokyo, Igaku Shoin, pp. 475–7.

Lavie P. (1979) Rapid eye movements in REM sleep—more evidence for a periodic organisation. *Electroencephal. Clin. Neurophysiol.* **46**, 683–8.

Lavie P. and Kripke D. F. (1977) Ultradian rhythms in urine flow in waking humans. *Nature* **269**, 142–3.

Lobban M. C. (1967) Daily rhythms of renal excretion in Arctic-dwelling Indians and Eskimos. *Q. J. Exp. Physiol.* **52**, 401–10.

Lobban M. C. (1977) Seasonal changes in daily rhythms of renal excretion and activity patterns in an Arctic Eskimo community. *J. Interdiscipl. Cycle Res.* **8**, 259–63.

Lobban M. C. and Tredre B. (1964) Renal diurnal rhythms in blind subjects. *J. Physiol.* **170**, 29P–30P.

Lovett-Doust J. W. (1979) An ultradian periodic servo-system of thermoregulation in man. *J. Interdiscipl. Cycle Res.* **10**, 95–103.

Lovett-Doust J. W., Payne W. D. and Podnieks I. (1978) An ultradian rhythm of reaction time measurements in man. *Neuropsychobiology* **4**, 93–8.

Lund R. (1974) Personality factors and desynchronization of circadian rhythms. *Psychosom. Med.* **36**, 224–8.

Lydic R., Schoene W. C., Czeisler C. A. et al. (1980) Suprachiasmatic region of the human hypothalamus: homolog to the primate circadian pacemaker? *Sleep* **2**, 355–61.

Menaker M., Takahashi J. S. and Eskin A. (1978) The physiology of circadian pacemakers. *Ann. Rev. Physiol.* **40**, 501–26.

Migraine C., Reinberg A. and Migraine C. (1974) Persistance des rythmes circadiens de l'alternance veille-sommeil et du comportement alimentaire d'un homme de 20 ans pendant son isolement souterrain, avec et sans montre. *CR Acad. Sci. (Paris)* **279**, 331–4.

Miles L. E. M., Raynal D. M. and Wilson M. A. (1977) Blind man living in normal society has circadian rhythms of 24·9 hours. *Science* **198**, 421–3.

Mills J. N. (1964) Circadian rhythms during and after three months in solitude underground. *J. Physiol.* **174**, 217–31.

Mills J. N. (1966) Human circadian rhythms. *Physiol. Rev.* **46**, 128–71.

Mills J. N. (1973) Transmission processes between clock and manifestations. In: Mills J. N. (ed.) *Biological Aspects of Circadian Rhythms*. London, Plenum Press, pp. 27–84.

Mills J. N., Minors D. S. and Waterhouse J. M. (1974) The circadian rhythms of human subjects without timepieces or indication of the alternation of day and night. *J. Physiol.* **240**, 567–94.

Mills J. N., Minors D. S. and Waterhouse J. M. (1978) The effect of sleep upon human circadian rhythms. *Chronobiologia* **5**, 14–27.

Mills J. N. and Waterhouse J. M. (1973) Circadian rhythms over the course of a year in a man living alone. *Int. J. Chronobiol.* **1**, 73–9.

Milner B. (1971) Interhemispheric differences in the localization of psychological processes in man. *Br. Med. Bull.* **27**, 272–7.

Minors D. S. and Waterhouse J. M. (1980a) Anchor sleep as a synchronizer of rhythms on abnormal routines. *Int. J. Chronobiol.* **7** (in the press).

Minors D. S. and Waterhouse J. M. (1980b) Does 'anchor sleep' entrain the internal oscillator that controls circadian rhythms? *J. Physiol.* **308**, 92P–93P.

Minors D. S. and Waterhouse J. M. (1981) Endogenous and exogenous components of circadian rhythms when living on a 21-h day. *Int. J. Chronobiol.* **8** (in the press).

Monk T. H. (1980) Traffic accident increases as a possible indicant of desynchronosis. *Chronobiologia* **7**, 527–9.

Monk T. H. and Aplin L. C. (1980) Spring and autumn daylight-saving time changes: studies of adjustment in sleep timings, mood and efficiency. *Ergonomics* **23**, 167–78.

Monk T. H. and Folkard S. (1976) Adjusting to the changes to and from Daylight-Saving Time. *Nature* **261**, 688–9.

Moore R. Y. (1979) The anatomy of central neural mechanisms regulating endocrine rhythms. In: Krieger D. T. (ed.) *Endocrine Rhythms*. New York, Raven Press, pp. 63–87.

Moore R. Y. (1980) Suprachiasmatic nucleus, secondary synchronizing stimuli and the central neural control of circadian rhythms. *Brain Res.* **183**, 13–28.

Moore-Ede M. C., Kass D. A. and Herd J. A. (1977) Transient circadian internal desynchronization after light–dark phase shift in monkeys. *Am. J. Physiol.* **232**, R31–R37.

Moore-Ede M. C., Schmelzer W. S., Kass D. A. et al. (1976) Internal organization of the circadian timing system in multicellular animals. *Fed. Proc.* **35**, 2333–8.

Moore-Ede M. C. and Sulzman F. M. (1977) The physiological basis of circadian timekeeping in primates. *Physiologist* **20**, 17–24.

Morgan R., Minors D. S. and Waterhouse J. M. (1980) Does light rather than social factors synchronize the temperature rhythm of psychiatric patients? *Chronobiologia* **7**, 331–5.

Nicholson A. N. and Stone B. M. (1978) Adaptation of sleep to British Summer Time. *J. Physiol.* 275, 22P–23P.

Njus D., Gooch Van D., Mergenhagen D. et al. (1976) Membranes and molecules in circadian systems. *Fed. Proc.* 35, 2353–7.

Orth D. N., Besser G. M., King P. H. et al. (1979) Free-running circadian plasma cortisol rhythm in a blind human subject. *Clin. Endocrinol.* 10, 603–17.

Orth D. N., Island D. P. and Liddle G. W. (1967) Experimental alteration of the circadian rhythm in plasma cortisol concentration in man. *J. Clin. Endocrinol. Metab.* 27, 549–55.

Orth D. N., Island D. P. and Liddle G. W. (1969) Light synchronization of the circadian rhythm in plasma cortisol concentration in man. *J. Clin. Endocrinol. Metab.* 29, 479–86.

Othmer E. and Hayden M. (1974) Rapid eye movements during sleep and wakefulness. In: Scheving L. E., Halberg F. and Pauly J. E. (ed.) *Chronobiology.* Tokyo, Igaku Shoin, pp. 491–4.

Palmer J. D. (1976) *An Introduction to Biological Rhythms.* New York, Academic Press.

Pauly J. E., Scheving L. E., Burns E. R. et al. (1977) Studies of the circadian system in blind human beings. In: *Proceedings XII International Conference of the International Society for Chronobiology,* Washington, 1975. Milan, Il Ponte, pp. 19–28.

Perry E. K., Perry R. H., Taylor M. J. et al. (1977) Circadian variation in human brain enzymes. *Lancet* 1, 753–4.

Quigley M. E. and Yen S. S. C. (1979) A mid-day surge in cortisol levels. *J. Clin. Endocrinol. Metab.* 49, 945–7.

Reinberg A., Gervais P., Pollack E. et al. (1973) Circadian rhythms during drug-induced coma (transverse study of rectal temperature, heart rate, systolic blood pressure, urinary water and potassium). *Int. J. Chronobiol.* 1, 157–62.

Reinberg A., Migraine C., Apfelbaum M. et al. (1979) Circadian and ultradian rhythms in the eating behavior and nutrient intake of oil refinery operators (Study 2). *Chronobiologia* 6, Suppl. 1, 89–102.

Reinberg A., Vieux N., Andlauer P. et al. (1979) Concluding remarks. Shift work tolerance: perspectives based upon findings derived from chronobiologic field studies on oil refinery workers. *Chronobiologia* 6, Suppl. 1, 105–10.

Ron S., Algom D., Hary D. et al. (1980) Time-related changes in the distribution of sleep stages in brain injured patients. *Electroencephal. Clin. Neurophysiol.* 48, 432–41.

Rothman B. S. and Strumwasser F. (1977) Manipulation of a neuronal circadian oscillator with inhibitors of macromolecular synthesis. *Fed. Proc.* 36, 2050–5.

Rusak B. (1977) The role of the suprachiasmatic nuclei in the generation of circadian rhythms in the golden hamster, *Mesocricetus auratus. J. Comp. Physiol.* 118, 145–64.

Rusak B. (1979) Neural mechanisms for entrainment and generation of mammalian circadian rhythms. *Fed. Proc.* 38, 2589–95.

Rusak B. and Zucker I. (1979) Neural regulation of circadian rhythms. *Physiol. Rev.* 59, 449–526.

Saunders D. S. (1977) *An Introduction to Biological Rhythms.* Glasgow, Blackie.

Schaefer K. E., Kerr C. M., Buss D. et al. (1979) Effect of 18-h watch schedules on circadian cycles of physiological functions during submarine patrols. *Undersea Biomed. Res.* Submarine Suppl., S81–S90.

Sharp G. W. G. (1960a) Reversal of diurnal leucocyte variations in man. *J. Endocrinol.* 21, 107–14.

Sharp G. W. G. (1960b) The effect of light on the morning increase in urine flow. *J. Endocrinol* 21, 219–23.

Simpson H. W. and Bohlen J. G. (1973) Latitude and human circadian system. In: Mills J. N. (ed.) *Biological Aspects of Circadian Rhythms.* London, Plenum Press, pp. 85–120.

Sollberger A. (1965) *Biological Rhythm Research.* Amsterdam, Elsevier.

Sperry R. W. (1968) Mental unity following surgical disconnection of the cerebral hemispheres. *Harvey Lectures* Ser. 62, 293–323.

Sulzman F. M., Fuller C. A. and Moore-Ede M. (1977) Environmental synchronizers of squirrel monkey circadian rhythms. *J. Appl. Physiol.* **43**, 795–800.

Sulzman F. M., Fuller C. A. and Moore-Ede M. C. (1978a) Extent of circadian synchronization by cortisol in the squirrel monkey. *Comp. Biochem. Physiol.* **59A**, 279–83.

Sulzman F. M., Fuller C. A., Hiles L. G. et al. (1978b) Circadian rhythm dissociation in an environment with conflicting temporal information. *Am. J. Physiol.* **235**, R175–R180.

Sulzman F. M., Fuller C. A. and Moore-Ede M. C. (1978c) Comparison of synchronization of primate circadian rhythms by light and food. *Am. J. Physiol.* **234**, R130–R135.

Surowiak J. F. (1978) Circadian rhythms. *Med. Biol.* **56**, 117–27.

Takahashi J. S. and Menaker M. (1979) Physiology of avian circadian pacemakers. *Fed. Proc.* **38**, 2583–8.

Tokura H. and Takagi K. (1974) Comparison of circadian oral temperature rhythms between blind and normal subjects. *J. Physiol. Soc. Japan* **36**, 255–6.

Weber A. L., Cary M. S., Connor N. et al. (1980) Human non-24-hour sleep–wake cycles in an everyday environment. *Sleep* **2**, 347–54.

Weitzman E. D. (1981) Sleep and its disorders. *Ann. Rev. Neurosci.* **4**, 381–417.

Wever R. (1970) Zur Zeitgeber-Stärke eines Licht-Dunkel-Wechsels für die circadiane Periodik des Menschen. *Pflügers Arch.* **321**, 133–42.

Wever R. (1972) Mutual relations between different physiological functions in circadian rhythms in man. *J. Interdiscipl. Cycle Res.* **3**, 253–65.

Wever R. (1973) Internal phase-angle differences in human circadian rhythms: causes for changes and problems of determinations. *Int. J. Chronobiol.* **1**, 371–90.

Wever R. (1975a) The circadian multi-oscillator system of man. *Int. J. Chronobiol.* **3**, 19–55.

Wever R. (1975b) Autonomous circadian rhythms in man. Singly versus collectively isolated subjects. *Naturwissenschaften* **62**, 443–4.

Wever R. (1979a) *The Circadian System of Man. Results of Experiments under Temporal Isolation.* Berlin, Springer-Verlag.

Wever R. (1979b) Influence of physical workload on free-running circadian rhythms of man. *Pflügers Arch.* **381**, 119–26.

Wilson D. M., Kripke D. F., McClure D. K. et al. (1977) Ultradian cardiac rhythms in surgical intensive care unit patients. *Psychosom. Med.* **39**, 432–435.

Zurbrügg R. P. (1976) Hypothalamic-pituitary-adrenocortical regulation: a contribution to its assessment, development and disorders in infancy and childhood with special reference to plasma circadian rhythm. In: *Monographs in Paediatrics* 7. Basle, Karger.

appendix   *Statistical Analysis of Rhythms*

The statistical analysis of time series is extremely complex as evidenced by the numerous textbooks devoted to this subject (for example, Box and Jenkins, 1970; Anderson, 1971); and new texts are still appearing (Priestley, 1981). This complex situation is exacerbated for the chronobiologist because classical methods of time series analysis have been developed for physical or economic systems where the time series are very long, usually encompassing very many cycles, whereas most chronobiological time series, particularly those in the human, are short. The problems associated with the analysis of short time series using classical methods have been reviewed by Sollberger (1970, 1973). It is beyond the scope of this book to describe the many statistical procedures which may be used to analyse time series. Rather, we will limit our discussion to those methods which are most popularly used and referred to in our text.

Probably the most frequently used technique is the cosinor analysis developed by Halberg, which may be used to define the rhythm parameters in a population—the population cosinor (Halberg et al., 1967); a group—the group mean cosinor (Tong et al., 1973); or an individual—the single cosinor (Halberg et al., 1972). The method is based upon the least-squares regression of a cosine function of the form:

$$f(t) = M + A \cdot \cos(\omega t + \phi),$$

where $f(t)$ is the value at time t of the function regression coefficients M (the mean level, termed the *mesor*), A (the amplitude, half the range of oscillation), ω (the angular frequency $= 360/\tau$, where τ is the period, so that $360°$ represents one complete cycle) and ϕ (the time of maximum, termed the *acrophase*). In the cosinor analysis a linear multiple regression analysis is performed so that the period (τ) must be known or assumed *a priori* and the remaining parameters (M, A and ϕ) are estimated from the regression analysis.

Using this model, a biological time series may be represented by function plus random noise. Thus the value of Y_i of the variable at time t_i may be represented as:

$$Y_i = M + A \cdot \cos(\omega t_i + \phi) + e_i; \quad i = 1, 2, 3, \ldots n,$$

311

where e_i, the error terms, are assumed to be independent normal deviates with a normal distribution of zero mean and unknown variance.

This equation may be expanded to the form:

$$Y_i = M + A \cdot \cos \phi \cdot \cos \omega t_i - A \cdot \sin \phi \cdot \sin \omega t_i + e_i$$

and transformed by substituting

$$x_{1i} = \cos \omega t_i; \quad x_{2i} = \sin \omega t_i$$

$$\beta = A \cdot \cos \phi; \quad \gamma = -A \cdot \sin \phi$$

to yield:

$$Y_i = M + \beta x_{1i} + \gamma x_{2i} + e_i,$$

which is linear in coefficients β and γ and thus becomes amenable to a multiple linear least squares regression analysis (i.e. minimizing $\sum_{i=1}^{i=n} = e_i^2$).

Most standard statistics textbooks describe methods by which the equation can be solved for M, β and γ. The estimates of the amplitude (A) and acrophase (ϕ) are derived from:

$$A = (\beta^2 + \gamma^2)^{\frac{1}{2}} \text{ and } \phi = \tan^{-1}(-\gamma/\beta) + K,$$

where K is a multiple of $90°$ to designate the correct trigonometric quadrant and depends on the signs of β and γ.

If the single cosinor is being performed the rhythm may be represented on rectangular co-ordinates as the co-ordinate (β, γ) or on polar co-ordinates as a vector with length proportional to the amplitude (A) and direction to the acrophase (ϕ). A single quadrant of a polar diagram superimposed on rectangular co-ordinates is shown in *Fig. A.1*.

The significance of the fitted cosine curve can be assessed by testing the null hypothesis $\beta = \gamma = 0$ (equivalent to $A = 0$) that is, the data are better represented by a cosine curve than a straight line, using a variance ratio test where:

$$F = \frac{\text{variance due to regression}}{\text{residual variance}} = \frac{\Sigma (y_i - \overline{Y})^2/2}{\Sigma (y_i - Y_i)^2/(n-3)} .$$

This has $(2, n - 3)$ degrees of freedom and where Y_i is the observed value and y_i the predicted value at time t_i.

Furthermore, a joint elliptical confidence region for the parameters β and γ can be calculated (*Fig. A.1*) and hence 95 per cent confidence limits for the acrophase and amplitude derived. The way in which this confidence region and 95 per cent confidence limits are calculated, as well as the use of the cosinor analysis for derivation of rhythm parameters in a group or population, has recently been reviewed (Nelson et al., 1979).

The advantage of using the cosinor analysis is that it can be used to

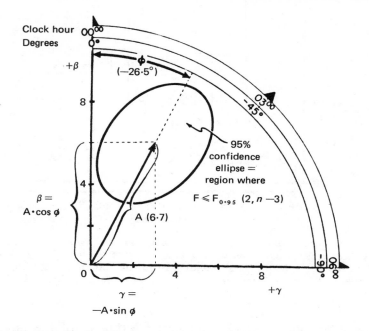

Fig. A.1. Rectangular (β, γ) and polar (A, ϕ) representation of circadian rhythm parameter estimates with joint confidence region, based on an abstract example with a 24-hour cosine function, $Y(t) = \text{Mesor} + A \cdot \cos{(\omega t + \phi)}$ fitted by least squares. The confidence ellipse encompasses all points $(\beta, \gamma$ or $A, \phi)$ where the F-ratio calculated from the ratio of (Variance due to regression)/(Residual variance) is less than or equal to 95th percentile of an F-distribution with $(2, n - 3)$ degrees of freedom. (Halberg et al., 1972, Fig. 5.)

analyse time series where the data are non-equidistantly sampled. Furthermore, it can be used not only on data discretely sampled but also on integrative values such as rates of urinary excretion, where mean rates over short time intervals only are available. In this latter case, however, an integrated cosine model should be used:

$$f(t) = \int_{t_{i-1}}^{t_i} M + A \cdot \cos{(\omega t + \phi)} \cdot dt$$

and the variance ratio to test for zero amplitude derived as described by Fort and Mills (1970).

The use of the cosinor analysis, however, has often been criticised when the data are non-sinusoidal. Thus, fitting a sine wave to the waveform shown in *Fig.* 1.1C would give an erroneous estimate of the time of peak value. An example of cosinor analysis of urinary data containing two peaks is shown in *Fig.* A.2; it is obvious that the fitted cosine curve although accurately representing the time of minimum excretion does not define the peaks of the rhythm accurately. In such cases, it has been

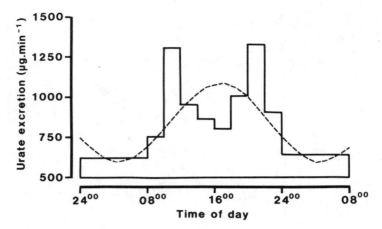

Fig. A.2. Urinary urate excretion in a single subject living a conventional nychthemeral routine. The dashed line indicates the best-fitting 24-hour cosine curve fitted by least squares. Note that the acrophase does not reliably indicate the times of maximal urate excretion. (Data of Minors, 1975.)

suggested (Tong et al., 1977) that better estimates of the phase of the rhythm may be derived by adding harmonics of the fundamental period. One must proceed with caution here, however, since Fourier's theorem (Bloomfield, 1976) states that any waveform (rhythm) can be reproduced if a sufficient number of sine terms with different periods are added. Since any observed rhythm results from true rhythmicity superimposed upon which is random noise, one would thus detect periods where they do not really exist. Furthermore, by accurately representing data by a series of sine terms, one does not explain anything about the mechanisms controlling the rhythms.

However, there are circumstances where the data may be suspected or known to have multiple frequencies; for example with the simultaneous presence of ultradian and circadian components (*see* Chapter 12) or the secretion of several hormones (*see* Chapter 7) which is subject to high frequency (episodic) variation as well as circadian variation. The inappropriateness of cosinor analysis under such circumstances has been discussed elsewhere (Van Cauter and Huyberechts, 1973; Van Cauter, 1974).

One further disadvantage of the use of the cosinor analysis is that it is based on *a priori* knowledge of the period of the rhythm. When a subject is living a conventional nychthemeral existence, a 24-hour period may safely be assumed, but under other circumstances (*see* Chapter 1, s. 6.2) this assumption may not be warranted. Under circumstances when the period is unknown, then the cosinor analysis can be performed repeatedly with a set of trial periods and a least squares spectrum (Halberg et al., 1965) constructed in which the residual errors are plotted against the test periods (*see also Fig.* 1.4). The period of the rhythm is then assessed as the period of the cosine function which yields the minimum residual error. The disadvantage of this method, however, is that there is no statistical test to determine whether the 'best-fitting' period is significantly different from any other near-period. To test whether a period represents a true periodicity, more classic techniques of spectral analysis can be used, as described in the next paragraph. For the moment, however, it may be mentioned that the least squares spectrum may be used in circumstances in which more than one periodicity is present in the time series. Thus, it may be used to analyse data from a subject living on a non-24-hour day when it is suspected that an exogenous as well as an endogenous component (*see* Chapter 1 s. 6.2) is present. *Fig.* 1.4 is a least squares spectrum of such data from a subject who was living on a 21-hour day. It can be seen from *Fig.* 1.4 that minima in residual error are found when cosine curves are fitted with periods of 21 and 24·5 hours, indicating that both these periodicities may be present in the data. Using this method, however, it has been assumed that the two periodic components are independent of one another, which may not be the case. Therefore, a better analysis of this data would be to perform a multiple linear regression analysis of a function containing two cosine terms each with a different period. In the case of the above data, where the 21-hour component is suspected *a priori*, the period of the first cosine term may be fixed at 21 hours and a set of trial periods used for the second to determine that which minimizes the residual error (*see* Chapter 4). An alternative approach is to perform a combined linear–nonlinear least squares regression analysis as described by Rummel et al. (1974).

When the period of the rhythm is unknown and/or the time series contains multiple frequencies then classical methods of spectral analysis may be used. That which is most frequently used is based upon a search technique introduced by Schuster (1898) to detect 'hidden periodicities' and is termed the periodogram. Classically, this is performed by a harmonic analysis of the form:

$$Y(t) = \sum_{i=0}^{i=N/2} A_i \cdot \cos(\omega_i t + \phi_i),$$

where N is the total number of (equidistant) data points; ω_i, the angular frequencies, $= 2\pi i/N$, that is all frequencies are submultiples of the total series length; $A_i = $ the amplitude of the ith component; $\phi_i = $ the phase of the ith component. The amplitude, A_i, of each harmonic component is estimated from:

$$A_i = \left[(\frac{2}{N}\sum_{t=1}^{t=N} Y_t \cdot \cos \omega_i t)^2 + (\frac{2}{N}\sum_{t=1}^{t=N} Y_t \cdot \sin \omega_i t)^2 \right]^{1/2}.$$

The periodogram is then a plot of the normalized amplitudes, $(N/2)A_i^2$ (termed the sample intensity) against the frequency (i/N) or period (N/i).

Periodicities present in the data are detected by peaks in this plot. This method of analysis is well suited to long time series but suffers several disadvantages:

1. The data must be equidistantly sampled, though a method where data are periodically missing, as over periods of sleep, is described by Haggan (1977).

2. The spectral resolution is limited by the length of the time series. To distinguish between periods of x and y hours would require $(x \cdot y)/(x \sim y)$ hours of data. Thus, for example, to distinguish between a period of 24 and 25 hours would require 25 days of data.

3. Periods which are integral multiples of the total duration of the data only are detected. However, as described by van Cauter (van Cauter and Huyberechts, 1973; van Cauter, 1974), methods are available to test whether peaks in the periodograms represent true periodicities and non-integral periods may be tested by modifying the length of the time series analysed.

A conceptually much simpler method of period detection has been described by Enright (1965a, b). This method is based upon the super-imposition of corresponding points from different cycles in a longitudinal study and maximizing the standard deviation of the mean values obtained at each time-point around the overall mean of all data points. This can best be visualized by construction of a 'Buys–Ballot' table. Thus for N data points (Y_i) obtained at hourly intervals the average waveform may be educed for the period, τ, from:

Cycle 1	Y_1	Y_2	$\cdots$	Y_τ
Cycle 2	$Y_{\tau + 1}$	$Y_{\tau + 2}$	$\cdots$	$Y_{2\tau}$
$\cdot$	$\cdot$	$\cdot$	$\cdot$	$\cdot$
Last cycle	$Y_{(N - \tau + 1)}$	$Y_{(N - \tau + 2)}$	$\cdots$	Y_N
Average	X_1	X_2		X_τ

The standard deviation of the means (X_s) around the overall mean (termed the root mean square amplitude by Enright) is calculated as:

$$X_s = \left[\frac{1}{n} \cdot \sum_{i=1}^{i=\tau} (X_i - \overline{X})^2 \right]^{1/2}$$

where $\overline{X}$ is the overall mean ($= \Sigma_{i=1}^{i=N} Y_i$) and n is the number of rows in the Buys–Ballot table. The calculation is then repeated for different values of τ and the root mean square amplitude plotted against period. A periodicity in the original data is shown by a peak in this plot and the significance of a peak can be derived (Dörrescheidt and Beck, 1975). Similar methods to the Enright technique have been used by Citta (1977), Czeisler (1978) and Williams and Naylor (1978) each maximizing a different statistic. It is evident, however, that with this method, in all cases, equidistant data are again required.

In all the methods described thus far, it has been assumed that the data are derived from a stationary time series, that is, the rhythm parameters (period, phase, amplitude) are not changing. This may not always be the case, however. Thus in transition from an entrained to free-running state (*see* Chapter 1, s. 6.2) the period of the rhythm may change, or, following a time-zone transition (*see* Chapter 9), the phase of the rhythm will change. In these cases the data must be fractionated into sections so as to determine whether the rhythm parameters vary as a function of time. Such division of the data into sections very much limits the analytical methods available, however, since the classical spectral analyses (for example, periodogram) require long time series. The division of the data into smaller sections, however, can still be used with the cosinor analysis when it has been termed a serial section analysis (Halberg et al., 1965). Thus one can fit a cosine curve to the data day by day. However, when the number of measurements per day is small, significant fits cannot often be obtained. This problem may be overcome by dividing the data into longer sections and overlapping successive sections. For example, a cosine curve may be fitted to the first three days of data, then to days 2–4, 3–5 and so on. When the data are thus overlapping, the analysis is referred to as a *pergressive* serial section analysis. It should be noted, however, that using this method the rhythm parameters derived for consecutive sections are not independent (since they share some data) and a degree of smoothing results. None the less, this technique does allow us to confirm that a change in rhythm parameter has occurred.

A final technique worthy of mention is that of cross-correlation introducing various phase lags (Horton and West, 1977; Mills et al., 1978). This technique allows a comparison of the rhythm before and after some experimental intervention so as to detect any changes in the timing of the rhythm without making any assumption as to its shape (*see* Chapter 9, s. 4.2 for an example of its use). Data suitable for this type of analysis, however, must be collected at an integral number of equal intervals throughout each cycle.

It is obvious from the foregoing discussion that any statistical analysis of rhythms is often limited by the type of data and the mathematical expertise available. For example, many methods are unavailable if the data are not equidistantly sampled or from long time series. It must always be borne in mind, therefore, that considerable information about a rhythm can be deduced with minimal reduction of the raw data. Thus simply plotting the time at which the variable is at a maximum or a minimum can yield information on the period or changes in phase of the rhythm (for examples, *see Figs.* 2.3 and 12.7). The relative merits of using this technique as opposed to more objective statistical analyses are discussed further in Chapter 9, s. 2.2.

References

Anderson T. W. (1971) *The Statistical Analysis of Time Series.* New York, Wiley.

Bloomfield P. (1976) *Fourier Analysis of Time Series: An Introduction.* New York, Wiley.

Box G. and Jenkins G. (1970) *Time Series Analysis.* San Francisco, Holden-Day.

Citta M. (1977) Time series analysis: the profilogram method. *Int. J. Chronobiol.* **4**, 171–83.

Czeisler C. A. (1978) Human circadian physiology: internal organization of temperature, sleep–wake and neuroendocrine rhythms monitored in an environment free of time cues. PhD Thesis, Stanford University.

Dörrescheidt G. J. and Beck L. (1975) Advanced methods for evaluating characteristic parameters ($\tau\ \alpha\ \rho$) of circadian rhythms. *J. Math. Biol.* **27**, 107–21.

Enright J. T. (1965a) Accurate geophysical rhythms and frequency analysis. In: Aschoff J. (ed.) *Circadian Clocks.* Amsterdam, North Holland, pp. 31–42.

Enright J. T. (1965b) The search for rhythmicity in biological time-series. *J. Theoret. Biol.* **8**, 426–68.

Fort A. and Mills J. N. (1970) Fitting sine curves to 24 h urinary data. *Nature* **226**, 657–8.

Haggan V. (1977) The detection of circadian rhythms in physiological data with periodically missing observation points. *J. Interdiscipl. Cycle Res.* **8**, 161–74.

Halberg F., Engeli M., Hamburger C. et al. (1965) Spectral resolution of low-frequency, small-amplitude rhythms in excreted 17-ketosteroids; probable androgen-induced circasepten desynchronization. *Acta Endocrinol.* Suppl. 105, 5–54.

Halberg F., Johnson E. A., Nelson W. et al. (1972) Autorhythmometry—procedures for physiologic self-measurements and their analysis. *Physiology Teacher* **1**, 1–11.

Halberg F., Tong Y. L. and Johnson E. A. (1967) Circadian system phase—an aspect of temporal morphology; procedures and illustrative examples. In: von Mayersbach H. (ed.) *The Cellular Aspects of Biorhythms.* Berlin, Springer-Verlag. pp. 20–43.

Horton B. J. and West C. E. (1977) Shifting phase analysis correlation in the determination of reproducibility of biological rhythms. Its use in the study of patterns of rabbit and rat feeding. *J. Interdiscipl. Cycle Res.* **8**, 65–76.

Mills J. N., Minors D. S. and Waterhouse J. M. (1978) Adaptation to abrupt time shifts of the oscillator(s) controlling human circadian rhythms. *J. Physiol.* **285**, 455–70.

Minors D. S. (1975) PhD Thesis, University of Manchester.

Nelson W., Tong Y. L., Lee J-K. et al. (1979) Methods for cosinor-rhythmometry. *Chronobiologia* **6**, 305–23.

Priestley M. B. (1981) *Spectral Analysis.* London, Academic Press.

Rummel J., Lee J-K., Halberg F. (1974) Combined linear-nonlinear chronobiologic windows by least squares resolve neighbouring components in a physiologic rhythm spectrum. In: Ferin M., Halberg F., Richart R. M. et al. (ed.) *Biorhythms and Human Reproduction.* New York, Wiley, pp. 53–82.

Schuster A. (1898) On the investigation of hidden periodicities with application to a supposed 26 day period of meteorological phenomena. *Terrestr. Magn.* **3**, 13–45.

Sollberger A. (1970) Problems in the statistical analysis of short periodic time series. *J. Interdiscipl. Cycle Res.* **1**, 49–88.

Sollberger A. (1973) Statistical aspects of autorhythmometry. *J. Interdiscipl. Cycle Res.* **4**, 179–83.

Tong Y. L., Lee J. and Halberg F. (1973) Number-weighted mean cosinor technique resolves phase- and frequency- synchronized rhythms with differing mesors and amplitudes. *Int. J. Chronobiol.* **1**, 365–6.

Tong Y. L., Nelson W. L., Sothern R. B. et al. (1977) Estimation of the orthophase (timing of high values) on a non-sinusoidal rhythm—illustrated by the best timing for experimental cancer chemotherapy. In: *Proceedings of the XII International Conference of the International Society for Chronobiology*. Milan, Il Ponte, pp. 765–9.

Van Cauter E. (1974) Methods for the analysis of multifrequential biological time series. *J. Interdiscipl. Cycle Res.* 5, 131–48.

Van Cauter E. and Huyberechts S. (1973) Problems in the statistical analysis of biologic time series: the cosinor test and the periodogram. *J. Interdiscipl. Cycle Res.* 4, 41–57.

Williams J. A. and Naylor E. (1978) A procedure for the assessment of significance of rhythmicity in time-series data. *Int. J. Chronobiol.* 5, 435–44.

glossary

Acrophase: Time of maximum of a fitted cosine curve given as a delay from a given phase reference. Often expressed in clock time (phase reference, midnight) or degrees (with $360° \equiv$ period (q.v.) of rhythm).

The acrophase may also be specified with the phase reference being the point of an environmental cycle (for example, mid-light time) or the acrophase of another rhythm in the same individual (for example, mid-sleep time).

Amplitude: The measurement of the peak of the rhythm above the mean level estimated by a mathematical function (for example, the difference between the maximum and mesor (q.v.) of a best-fitting cosine curve). When a mathematical function has not been fitted to the data, the amplitude usually refers to the range of oscillation—the distance between crest and trough values.

Antiphase: Consistent difference of $180°$ in the timing of two rhythms exhibiting the same frequency.

Circadian: Relating to biological variations or rhythms (q.v.) with a period (q.v.) of 24 ± 4 hours.

Cosinor analysis: The fitting of a cosine curve to a rhythm (q.v.) by the method of least squares regression. The rhythm amplitude (q.v.) and acrophase (q.v.) are displayed on polar co-ordinates by the length and the angle of a directed line, respectively, shown with a bivariate statistical confidence region. (*See Fig.* A.1.)

Cycle: Synonymous with rhythm (q.v.).

Desynchronization: A steady state in which different rhythms run with different periods (q.v.). When the desynchronization is between a biological rhythm and an environmental cycle, *External desynchronization* is shown. When the desynchronization is between two biological rhythms in the same individual, *Internal desynchronization* is shown.

Dissociation, internal: Temporary change in phase relationship between different biological rhythms in the same individual.

Diurnal: Relating to events occurring during the hours of daylight. The opposite of nocturnal (q.v.).

Endogenous rhythm: Biological rhythm driven by an internal timing mechanism. Such a rhythm will continue self-sustained oscillations in the absence of external rhythms.

Entrainment: The coupling of a biological self-sustained (endogenously generated) rhythm with an external rhythm (zeitgeber, q.v.), with the result that both rhythms run synchronously with the same period.

Exogenous rhythm: Biological rhythm driven by an external oscillation. Such a rhythm will not continue oscillating in the absence of the external oscillation.

Free-running rhythm: A biological rhythm which is continuing self-sustained oscillations with an inherent frequency at least slightly different from that of known environmental frequencies.

Frequency: The number of cycles per unit time; the reciprocal of period (q.v.).

Infradian rhythm: Rhythms with a period longer than that of a circadian rhythm.

Mesor: The average value of rhythmic variable over a single cycle determined as the mean of a fitted cosine curve.

Nocturnal: Relating to events occurring during the hours of darkness. The opposite of diurnal (q.v.).

Nychthemeral: Relating to the alternation of day and night. A nychthemeral rhythm is a biological rhythm observed when the organism is exposed to all the fluctuations associated with the solar day.

Oscillation: Synonymous with cycle and rhythm (q.v.).

Period: The time to complete one cycle of a rhythm. The reciprocal of frequency.

Phase: Instantaneous value of a rhythm at a fixed time. Also used to describe the temporal relationship between two rhythms where a particular phase of one rhythm is consistently temporally related to that of another rhythm.

Phase-shift: A displacement of a rhythm along the time axis. A phase shift may be qualified as a *Phase-advance* when all aspects of the rhythm occur earlier in time, or a *Phase-delay* when all aspects of the rhythm occur later in time.

Rhythm: A sequence of events which in a steady state repeat themselves in time in the same order and same interval. Synonymous with oscillation and cycle.

Synchronizer: Synonymous with zeitgeber (q.v.).

Ultradian rhythm: Rhythm with a period shorter than that of a circadian rhythm.

Zeitgeber: An external oscillation which is capable of entraining an endogenously generated biological rhythm.

Index